Studies in Classification, Data Analysis, and Knowledge Organization

For further volumes:
http://www.springer.com/series/1564

Paolo Giudici • Salvatore Ingrassia
Maurizio Vichi
Editors

Statistical Models for Data Analysis

Springer

Editors
Paolo Giudici
Department of Economics and Management
University of Pavia
Pavia
Italy

Salvatore Ingrassia
Department of Economics and Business
University of Catania
Catania
Italy

Maurizio Vichi
Department of Statistics
University of Rome "La Sapienza"
Rome
Italy

ISSN 1431-8814
ISBN 978-3-319-00031-2 ISBN 978-3-319-00032-9 (eBook)
DOI 10.1007/978-3-319-00032-9
Springer Cham Heidelberg New York Dordrecht London

Library of Congress Control Number: 2013941993

Springer is part of Springer Science+Business Media (www.springer.com)

Preface

This volume contains revised versions of the selected papers presented at the 8th biannual meeting of the Classification and Data Analysis Group (CLADAG) of the Italian Statistical Society, organized by the Department of Economics and Management of the University of Pavia, in September 2011.

The conference has encompassed 170 presentations, organized in 3 plenary talks and 46 sessions. With 230 attendees from 10 different countries, the conference provided an attractive interdisciplinary international forum for discussion and mutual exchange of knowledge. The topics of all plenary and specialized sessions were chosen, in a peer-review process, to fit the mission of CLADAG which is to promote methodological, computational and applied research, within the fields of classification, data analysis and multivariate statistics.

The contributions in this volume were selected in a second peer-review process, after the conference. In addition to the fundamental areas of clustering and discrimination, multidimensional data analysis and data mining, the volume contains manuscripts concerning data analysis and statistical modelling in application areas like economics and finance, education and social sciences and environmental and biomedical sciences.

We would like to express our gratitude to all members of the scientific program committee, for their ability in attracting interesting contributions. We also thank the session organizers, the invited speakers, the chairpersons and the discussants of all sessions for a very stimulating scientific atmosphere. We are very grateful to the referees, for their careful reviews of the submitted papers and for the time spent in this professional activity.

We gratefully acknowledge financial support from the Italian Ministry of Research (PRIN programme), the University of Pavia, the Credito Valtellinese banking group and the IT company ISED. We also thank PRAGMA Congressi for the precious support in the organization of the conference.

A special thanks is due to the local organizing committee and, in particular, to its coordinator, Dr. Paola Cerchiello, for a very well organized conference, with the related scientific proceedings. Finally we would like to thank Ruth Milewski and

Dr. Martina Bihn of Springer-Verlag, Heidelberg, for the support and dedication to the production of this volume.

Pavia, Italy Paolo Giudici
Catania, Italy Salvatore Ingrassia
Roma, Italy Maurizio Vichi
7 December 2012

Contents

Contributors

G. Adelfio Dipartimento di Scienze Statistiche e Matematiche "S. Vianelli", Universitá degli Studi di Palermo, Palermo, Italy

Claudio Agostinelli Department of Environmental Science, Informatics and Statistics, Ca' Foscari University of Venice, Venice, Italy

Massimo Attanasio Dipartimento di Scienze Statistiche e Matematiche "Silvio Vianelli", Universit degli Studi di Palermo, Palermo, Italy

Antonio Balzanella Second University of Naples, Naples, Italy

Tonio Di Battista University G. D'Annunzio, Rome, Italy

Annamaria Bianchi DMSIA, University of Bergamo, Bergamo, Italy

Silvia Biffignandi DMSIA, University of Bergamo, Bergamo, Italy

Giuseppe Boari Dipartimento di Scienze statistiche, Universitá Cattolica del Sacro Cuore, Milano, Italy

Andrea Bonanomi Dipartimento di Scienze statistiche, Universitá Cattolica del Sacro Cuore, Milano, Italy

Stefano Bonnini Department of Economics, University of Ferrara, Ferrara, Italy

Giovanni Boscaino Dipartimento di Scienze Statistiche e Matematiche "Silvio Vianelli", Universit degli Studi di Palermo, Palermo, Italy

Anne-Laure Boulesteix Biometry and Epidemiology of the Faculty of Medicine, Department of Medical Informatics, University of Munich, Munich, Germany

Giuseppe Bove Dipartimento di Scienze dell'Educazione, Rome, Italy

Riccardo Bramante Department of Statistical Sciences, Catholic University of Milan, Milan, Italy

Gabriele Cantaluppi Dipartimento di Scienze statistiche, Universitá Cattolica del Sacro Cuore, Milano, Italy

Andrea Capotorti Dip. Matematica e Informatica, Universitá di Perugia, Perugia, Italy

Carlo Capuano University of Naples Federico II, Napoli, Italy

Vincenza Capursi Dipartimento di Scienze Statistiche e Matematiche "Silvio Vianelli", Universit degli Studi di Palermo, Palermo, Italy

Paola Cerchiello Department of Economics and Management, University of Pavia, Lombardy, Italy

M. Chiodi Dipartimento di Scienze Statistiche e Matematiche "S. Vianelli", Universitá degli Studi di Palermo, Palermo, Italy

Rosario D'Agata University of Catania, Catania, Italy

A. D'Alessandro Istituto Nazionale di Geofisica e Vulcanologia, Centro Nazionale Terremoti, Terremoti, Italy

Livia Dancelli Department of Quantitative Methods, University of Brescia, Brescia, Italy

Maria Rosaria D'Esposito University of Salerno, Fisciano (SA), Italy

Carlo Drago University of Naples "Federico II" Complesso Universitario Monte Sant'Angelo via Cinthia, Naples, Italy

Lara Fontanella University G. d'Annunzio, Chieti-Pescara, Italy

S. Fontanella University G. d'Annunzio, Chieti-Pescara, Italy

C. Fusilli University G. d'Annunzio, Chieti-Pescara, Italy

L.A. García-Escudero Facultad de Ciencias, Universidad de Valladolid, Valladolid, Spain

A. Gordaliza Facultad de Ciencias, Universidad de Valladolid, Valladolid, Spain

Alessandra Guglielmi Politecnico di Milano, Milano, Italy

Francesca Ieva Politecnico di Milano, Milano, Italy

Domenica Fioredistella Iezzi Tor Vergata University, Rome, Italy

L. Ippoliti University G. d'Annunzio, Chieti-Pescara, Italy

Antonio Irpino Dipartimento di Studi Europei e Mediterrranei, Seconda Universit degli Studi di Napoli, Caserta, Italy

Valen E. Johnson University of Texas MD Anderson Cancer Center, Houston, TX, USA

Frank Klawonn Ostfalia, University of Applied Sciences, Wolfenbüttel, Germany

Rudolf Kruse Otto-von-Guericke University Magdeburg, Magdeburg, Germany

Francesco Lagona University Roma Tre, Rome, Italy

Carlo Lauro University of Naples “Federico II” Complesso Universitario Monte Sant’Angelo via Cinthia, Naples, Italy

Vincent Lemaire Orange Labs, Lannion, France

Caterina Liberati Economics Department, University of Milano-Bicocca, Milan, Italy

Eleonora Lorenzini Department of Economics and Management, University of Pavia, Lombardy, Italy

Pietro Giorgio Lovaglio Department of Quantitative Methods, University of Bicocca-Milan, Milan, Italy

D. Luzio Dipartimento di Scienza della Terra e del Mare, Universitá degli Studi di Palermo, Palermo, Italy

Mario Mastrangelo Sapienza University, Rome, Italy

C. Matrán Facultad de Ciencias, Universidad de Valladolid, Valladolid, Spain

Marica Manisera Department of Quantitative Methods, University of Brescia, Brescia, Italy

Paolo Mariani Statistics Department, University of Milano-Bicocca, Milan, Italy

Federico Marsich Trieste, Italy

A. Mayo-Iscar Facultad de Ciencias, Universidad de Valladolid, Valladolid, Spain

Angelo Mazza Dipartimento di Economia e Impresa, Universit di Catania, Catania, Italy

Paul D. McNicholas University of Guelph, Guelph, ON, Canada

Alfredo Del Monte University of Naples Federico II, Napoli, Italy

Isabella Morlini Department of Economics, University of Modena and Reggio Emilia, Modena, Italy

Riccardo Di Nisio University G. D’Annunzio, Rome, Italy

Eugenia Nissi Dipartimento di Economia, Università “G. d’Annunzio” di Chieti-Pescara, Pescara, Italy

Margret-Ruth Oelker Department of Statistics, University of Munich, Munich, Germany

Biometry and Epidemiology of the Faculty of Medicine, Department of Medical Informatics, University of Munich, Munich, Germany

Akinori Okada Graduate School of Management and Information Sciences, Tama University, Tama, Japan

Anna Maria Paganoni Politecnico di Milano, Milano, Italy

Marco Picone University Roma Tre, Rome, Italy

Marine Service, Ispra, Italy

Antonella Plaia Dipartimento di Scienze Statistiche e Matematiche "Silvio Vianelli", Universit degli Studi di Palermo, Palermo, Italy

Mariano Porcu Dipartimento di Scienze Sociali e delle Istituzioni, Universitá di Cagliari, Cagliari, Italy

Antonio Punzo Dipartimento di Economia e Impresa, Universit di Catania, Catania, Italy

Emanuela Raffinetti Department of Economics, Management and Quantitative Methods, Università degli Studi di Milano, Milano, Italy

Agnese Rapposelli Dipartimento di Economia, Università "G. d'Annunzio" di Chieti-Pescara, Pescara, Italy

Lidia Rivoli University of Naples Federico II, Naples, Italy

Luca La Rocca Dipartimento di Comunicazione e Economia, University of Modena and Reggio Emilia, Reggio Emilia, Italy

Mario Romanazzi Department of Environmental Science, Informatics and Statistics, Ca' Foscari University of Venice, Venice, Italy

Elvira Romano Second University of Naples, Caserta, Italy

David Rossell Institute for Research in Biomedicine of Barcelona, Barcelona, Spain

Fabrizio Ruggeri CNR IMATI Milano, Milano, Italy

Luigi Salmaso Department of Management and Engineering, University of Padova, Vicenza, Italy

Christophe Salperwyck Orange Labs, Lannion, France

LIFL, Université de Lille 3, Villeneuve d'Ascq, France

Scipione Sarlo Sapienza University, Rome, Italy

A. Sarra University G. D'Annunzio, Rome, Italy

Monica Scannapieco Istat, Rome, Italy

Germana Scepi University of Naples "Federico II" Complesso Universitario Monte Sant'Angelo via Cinthia, Naples, Italy

Klaus B. Schebesch Faculty of Economics, Vasile Goldiş, Western University Arad, Arad, Romania

Gabriella Schoier Dipartimento di Scienze Economiche Aziendali Matematiche e Statistiche, Universitá di Trieste, Trieste, Italy

Ricardo Silva University College London, London, UK

Francesca Solmi Department of Statistical Sciences, University of Padova, Padova, Italy

Ralf Stecking Department of Economics, Carl von Ossietzky University Oldenburg, Oldenburg, Germany

Domenico De Stefano University of Trieste, Trieste, Italy

Isabella Sulis Dipartimento di Scienze Sociali e delle Istituzioni, Universitá di Cagliari, Cagliari, Italy

Donatello Telesca University of California, Los Angeles, CA, USA

Venera Tomaselli University of Catania, Catania, Italy

Hiroyuki Tsurumi College of Business Administration, Yokohama National University, Yokohama, Japan

Barbara Vantaggi Dip. Scienze di Base e Applicate per l'Ingegneria, Universit´a La Sapienza, Roma, Italy

Rosanna Verde Second University of Naples, Naples, Italy

Dipartimento di Studi Europei e Mediterrranei, Seconda Universit degli Studi di Napoli, Caserta, Italy

Maria Prosperina Vitale University of Salerno, Fisciano (SA), Italy

Giorgio Vittadini Department of Quantitative Methods, University of Bicocca-Milan, Milan, Italy

Marika Vezzoli Department of Quantitative Methods, University of Brescia, Brescia, Italy

Roland Winkler German Aerospace Center, Braunschweig, Germany

Angelo Zanella Dipartimento di Scienze statistiche, Universitá Cattolica del Sacro Cuore, Milano, Italy

Diego Zardetto Istat, Rome, Italy

Bruno D. Zumbo University of British Columbia, Vancouver, BC, Canada

Ordering Curves by Data Depth

Claudio Agostinelli and Mario Romanazzi

Abstract Application of depth methods to functional data provides new tools of analysis, in particular an ordering of curves from the center outwards. Two specific depth definitions are band depth and half-region depth (López-Pintado & Romo (2009). *Journal of the American Statistical Association, 104*, 718–734; López-Pintado & Romo (2011). *Computational Statistics & Data Analysis, 55*, 1679–1695). Another research area is local depth (Agostinelli and Romanazzi (2011). *Journal of Statistical Planning and Inference, 141*, 817–830.) aimed to identify multiple centers and dense subsets of the space. In this work we suggest local versions for both band and half-region depth and illustrate an application with real data.

1 Introduction

The data considered here are a set of n curves $y_i(t) \equiv y_i, t \in T \subseteq \mathbb{R}, i = 1, \ldots, n$ to be interpreted as IID observations of a functional variable $y(t)$ defined in an infinite dimensional space. Examples arise in chemometrics (spectrometrics curves), environmetrics (time trajectories of concentrations of pollutants), finance (stock prices), medicine (ECG and EEG curves) and many other fields. Two comprehensive references on the statistical methods are Ramsay and Silverman (2005) and Ferraty and Vieu (2006). A general problem is to define an ordering for functional data able to rank the curves according to centrality. Useful notions of mean curve and median curve are already available (e.g., Ferraty and Vieu (2006, p. 127)) but a more general solution is in terms of functional depth, an extension of the order statistic.

C. Agostinelli (✉) · M. Romanazzi
Department of Environmental Science, Informatics and Statistics, Ca' Foscari University of Venice, Venice, Italy
e-mail: claudio@unive.it; romanaz@unive.it

P. Giudici et al. (eds.), *Statistical Models for Data Analysis*, Studies in Classification, Data Analysis, and Knowledge Organization, DOI 10.1007/978-3-319-00032-9_1,

Data depth was established at the end of the last century as a general nonparametric methodology for multivariate numerical data (Liu 1990; Zuo and Serfling 2000). The extension to functional data is a more recent research area (López-Pintado and Romo 2009) with some useful achievements including a center-outwards ordering of curves and the notion of depth regions to be interpreted similarly to interquartile intervals. An important application is supervised classification of a curve into one of several classes (López-Pintado and Romo 2005).

In this work we concentrate on local versions of functional depth. Unlike the usual definition, local depth functions (Agostinelli and Romanazzi 2011) allow recognition of multiple centers and dense subsets of the reference space, a topic of interest also in the context of functional data. An obvious application is curve clustering. The content of the paper is the following. The basic definitions of functional depth are reviewed in Sect. 2 and their local versions are described in Sect. 3. Some real data applications are discussed in Sect. 4.

2 Functional Depth

The basic tool in data depth is the depth function which maps each object of the reference space to a non negative real number—the depth value—to be interpreted as a centrality rank. Depth ranks are different from the standard ranks generated by the univariate order statistic. The maximum depth identifies the center whereas the minimum depth corresponds to the outskirts of the distribution. In the case of the euclidean space $\mathbb{R}^p$ some popular definitions are Mahalanobis', halfspace, projection and simplicial depth (Zuo and Serfling 2000). Half-region and band depth can be interpreted as extensions of halfspace and simplicial depth, respectively, to the functional setting. Roughly speaking, half-region depth is the minimum probability that a randomly sampled curve lies below or above the curve under consideration whereas band depth is the probability that the curve is covered by the band determined by a suitable number of random copies of the underlying stochastic process. Alternative, less restrictive, definitions are obtained by replacing probabilities with the expected *time* the curve satisfies either constraint and are called *modified* depths. Precise notation and definitions for the sample situation are given below. The population versions can be found in the original papers.

The observed data are a collection of functions $\mathbf{y}_n = \{y_i \in \mathscr{C}(T) : i = 1, \cdots, n\}$, with $\mathscr{C}(T)$ denoting some functional space, e.g., the space of all continuous functions on some compact interval $T \subset \mathbb{R}$. The graph of a function y is the subset of the space $G(y) = \{(t, y(t)) : t \in T\}$.

We start with the definition of half-region depth. The hypograph (epigraph) of a curve y is the set of points lying on or below (on or above) its graph. Let $R_{hypo}(y; \mathbf{y}_n)$ ($R_{epi}(y; \mathbf{y}_n)$) be the sample proportion of graphs belonging to the hypograph (epigraph) of y.

Definition 1 (Half-region depth, López-Pintado and Romo 2011). For a functional dataset $\mathbf{y}_n$, the half-region depth of a curve y is

$$d_{HR}(y;\mathbf{y}_n) = \min(R_{hypo}(y;\mathbf{y}_n), R_{epi}(y;\mathbf{y}_n)) \,. \tag{1}$$

Let $\lambda(\cdot)$ denote Lebesgue measure. The modified version of (1) is

$$d_{MHR}(y;\mathbf{y}_n) = \min(EL(y;\mathbf{y}_n), HL(y;\mathbf{y}_n)), \tag{2}$$

where

$$EL(y;\mathbf{y}_n) = \tfrac{1}{n\lambda(T)} \textstyle\sum_{i=1}^n \lambda\left(t \in T : y(t) \le y_i(t)\right), \tag{3}$$

$$HL(y;\mathbf{y}_n) = \tfrac{1}{n\lambda(T)} \textstyle\sum_{i=1}^n \lambda\left(t \in T : y(t) \ge y_i(t)\right). \tag{4}$$

We proceed with the definition of band depth. For any choice of $k \le n$ functions $y_{i_1}, y_{i_2}, \cdots, y_{i_k}$ out of the n functions $y_1, y_2, \cdots, y_n$, a band $B(y_{i_1}, y_{i_2}, \cdots, y_{i_k})$ is the smallest region of the plane including all of them, that is

$$\begin{aligned} B(y_{i_1}, y_{i_2}, \cdots, y_{i_k}) &= \left(\cup_{l=1}^k \text{hypo}(y_{i_l})\right) \cap \left(\cup_{m=1}^k \text{epi}(y_{i_m})\right) \\ &= \cup_{l=1}^k \cup_{m=1}^k \left(\text{hypo}(y_{i_l}) \cap \text{epi}(y_{i_m})\right) \,. \end{aligned} \tag{5}$$

The quantity

$$d_B^{(k)}(y;\mathbf{y}_n) = \binom{n}{k}^{-1} \sum_{1 \le i_1 < i_2 < \cdots < i_k \le n} \mathbf{1}(G(y) \subseteq B(y_{i_1}, y_{i_2}, \cdots, y_{i_k})) \tag{6}$$

is the proportion of bands $B(y_{i_1}, y_{i_2}, \cdots, y_{i_k})$ containing the whole graph of y. Here, $\mathbf{1}(\cdot)$ is the indicator function of its argument.

Definition 2 (Band depth, López-Pintado and Romo 2009). Let $2 \le K \le n$ be a fixed value. The band depth of a curve y is

$$d_{B,K}(y;\mathbf{y}_n) = \sum_{k=2}^{K} d_B^{(k)}(y;\mathbf{y}_n) \,. \tag{7}$$

The modified version of (7) is

$$d_{MB,K}(y;\mathbf{y}_n) = \sum_{k=2}^{K} d_{MB}^{(k)}(y;\mathbf{y}_n) \,, \tag{8}$$

where

$$d_{MB}^{(k)}(y;\mathbf{y}_n) = \binom{n}{k}^{-1} \frac{1}{\lambda(T)} \sum_{1 \le i_1 < i_2 < \cdots < i_k \le n} \lambda(t \in T : G(y) \subseteq B(y_{i_1}, y_{i_2}, \cdots, y_{i_k})) \,. \tag{9}$$

An often used value is $K = 2$.

Together with the depth values, another basic statistic is the deepest—or median—curve. A useful generalization is the $100p\%$ depth region $\hat{D}_n(p)$, defined as the subset of the $100p\%$ deepest curves, $0 < p < 1$. Frequently, $p = 0.5$ and $\hat{D}_n(0.5)$ is a possible generalization of the interquartile interval. We end this section recalling that band and half-region depth, and their maximizers, are consistent estimators of their population counterparts (López-Pintado and Romo 2009, Th. 4, López-Pintado and Romo 2011, Th. 3 and 4).

3 Local Functional Depth

The rationale of local depth is to measure depth conditional on a bounded neighbourhood of the objects under consideration, so as to capture features of the nearby portion of the space, only. It depends on a tuning parameter τ, constant across the space, measuring the size of the neighbourhoods. When τ grows higher, local depth becomes more similar to ordinary depth. For simplicial depth, the tuning parameter is just simplex size measured, e.g., by diameter or volume. For halfspace depth, it is the width of minimum probability slabs. The reader is referred to Agostinelli and Romanazzi (2011) for more details. With functional data sets, local versions of both half-region and band depth are obtained by constraining the widths of hypo/epigraphs and bands to a finite value. Setting τ equal to the p-th order percentile of pairwise distances of curves works well in practice, with p in the range 10%–30%. Accordingly, local half-region depth of a curve $y(t)$ is the minimum proportion of curves lying at a distance not greater than τ below or above $y(t)$ and local band depth is the sample proportion of bands covering $y(t)$ formed by K curves with pairwise distances not greater than τ. The computing formulae are given in Eqs. (10) and (11) below.

$$ld_{HR}(y; \mathbf{y}_n, \tau) = \min\left(\frac{1}{n}\sum_{i=1}^{n}\mathbf{1}(G(y_i) \subset \text{hypo}(y;\tau)), \frac{1}{n}\sum_{i=1}^{n}\mathbf{1}(G(y_i) \subset \text{epi}(y;\tau))\right), \tag{10}$$

where

$$\text{hypo}(y;\tau) = \text{hypo}(y) \cap \text{epi}(y-\tau) = \{(t,z) : t \in T, y(t) - \tau \le z \le y(t)\},$$
$$\text{epi}(y;\tau) = \text{hypo}(y+\tau) \cap \text{epi}(y) = \{(t,z) : t \in T, y(t) \le z \le y(t) + \tau\}.$$

$$\begin{aligned} ld_{B,K}(y; \mathbf{y}_n, \tau) &= \sum_{k=2}^{K} ld_B^{(k)}(y; \mathbf{y}_n, \tau) \qquad (11)\\ &= \sum_{k=2}^{K}\binom{n}{k}^{-1}\mathbf{1}(G(y) \subseteq B(y_{i_1}, \cdots, y_{i_k}) \cap s(B(y_{i_1}, \cdots, y_{i_k})) \le \tau). \end{aligned}$$

In the previous equation, $s(B(y_1, \cdots, y_k))$ is a scalar measure of band size, like the diameter

$$\text{diam}(B(y_1, \cdots, y_k)) = \max_{t \in T} \mid \max(y_1(t), \cdots, y_k(t)) - \min(y_1(t), \cdots, y_k(t)) \mid .$$

The *modified* versions are defined in a similar way as the global counterparts and are omitted. For $x, y \in \mathscr{C}(T)$, two often-used distances are

$$\delta_1(x, y) = \sup_{t \in T} \mid x(t) - y(t) \mid ,$$
$$\delta_2(x, y) = (\int_T (x(t) - y(t))^2 dt)^{1/2} .$$

The consistency of the local depth functions defined in this section can be proved along the same lines as Agostinelli and Romanazzi (2011).

4 Applications

We use two data sets to illustrate curve ranking. The first one includes absorbance curves of 240 samples of meat recorded in the wavelength range 850–1,050 nm. See Ferraty and Vieu (2006) for more details. From Fig. 1, the curves appear very regular and, apart from a vertical shift, they exhibit a similar behaviour. The second data set is the gross domestic product (GDP) dataset. Observed data are time trajectories over the period 1970–2009 of log GDP pro capite of 83 countries (yearly data in dollars; source: World Bank). Graphical inspection of the curves (see Fig. 2) shows different locations but a largely common shape. Rich countries form a fairly separated cluster, notably stable over time.

Our analysis is based on modified half-region depth, the tuning parameter of the local version being the 10-th percentile of pairwise δ_1 distances of trajectories.

For the meat data, the ranks provided by global and local depth largely agree and there is no evidence of partial centers or clusters. Both functions reach the maximum value on the dense subset of curves in the lower half of the plot and they steadily decrease when proceeding towards the lower or the upper extreme of the absorbance range. The shift of local central region towards the lower absorbance values reflects the typical behaviour of local depth with *asymmetric* distributions (Agostinelli and Romanazzi 2011). A referee suggested that the depth ordering of *derivatives* of the curves could provide information on curve shape. In the present instance this proves to be true because local depth of the second derivatives recognizes three classes of curves corresponding to those found by Ferraty and Vieu (2006), p. 136–137.

A different picture arises from GDP data (see Fig. 2). The deepest curves according to global depth are Botswana and Colombia and the extreme curves are Burundi, Malawi and Norway. In most cases local depth is very different from global depth, with variations in both directions, which supports the hypothesis that GDP is a compound process. The (normalized) local and global depth values of

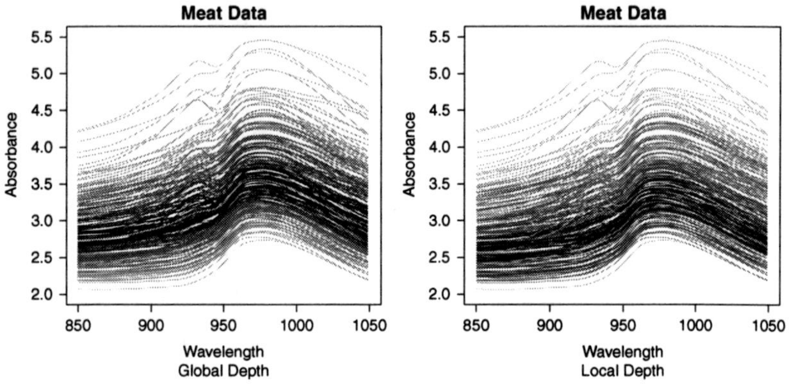

Fig. 1 Absorbance curves with gray level proportional to depth (darker means deeper; *dashed line*: deepest curve). *Left*: modified half-region depth, *right*: local version

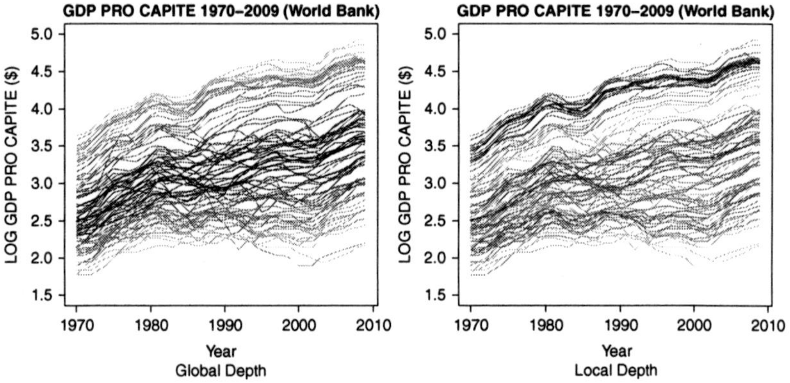

Fig. 2 GDP curves with gray level proportional to depth (darker means deeper). *Left*: modified half-region depth (*dashed line*: Botswana, *dotted line*: Colombia); *right*: local version (*dashed line*: Austria, *dotted line*: Guatemala)

Botswana are 0.078 and 1, respectively, whereas the corresponding values of Austria are 1 and 0.194. Indeed, Austria can be considered the centroid of the group of richest countries. The second highest value of local depth is Guatemala, also shown in Fig. 2. The different behaviour of global and local depth is best evaluated from Fig. 2 where the gray level of GDP curves is proportional to (normalized) depth (black corresponds to maximum depth). It is confirmed that the richest countries form a densely packed group. Another two dense regions are suggested, but they are more dispersed. However, the gap between rich and non rich countries is filled with several curves with low depth values that can mask the group structure.

Previous results suggest that meat curves are determinations of a homogeneous process whereas GDP curves can come from a mixture of (at least) two processes describing GDP dynamics of rich and non rich countries, respectively. An important

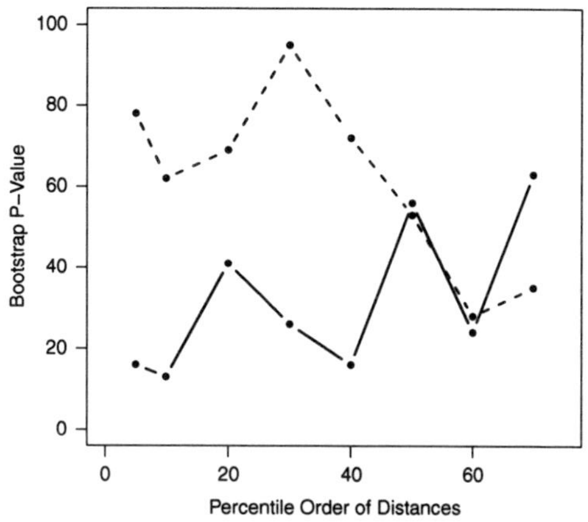

Fig. 3 Bootstrap p-value of homogeneity test with $T_{3,n}(\tau)$ test function (*solid*: GDP data, *dashed*: meat data)

problem is to test the (null) hypothesis that the underlying process is homogeneous. The *rank transformation* of the local and global depth values are obvious candidates to build a test function because, under the null hypothesis, local and global depth produce the same ordering. Two natural test functions are the sample mean $T_{1,n}(\tau)$ of the absolute differences of local and global ranks,

$$T_{1,n}(\tau) = n^{-1} \sum_{i=1}^{n} \mid d_i^{(R)} - ld_i^{(R)}(\tau) \mid ,$$

and the sample correlation $T_{2,n}(\tau)$ of local and global ranks,

$$T_{2,n}(\tau) = \mathrm{corr}(d^{(R)}, ld^{(R)}(\tau)) .$$

Here, for $i = 1, \cdots, n$, $d_i^{(R)}$ ($ld_i^{(R)}(\tau)$) stands for the rank of the i-th global (local) depth value and corr denotes Pearson's correlation. A more selective version of $T_{1,n}(\tau)$ arises from consideration of rank differences corresponding to maximizers of local and global depth functions, only. This test function is denoted with $T_{3,n}(\tau)$.

The homogeneity test was performed on both data sets, using the three test functions with the percentile order of distances ranging from 5% to 70%. The p-values are estimated through bootstrap resampling with 500 replicates. The p-values corresponding to $T_{3,n}(\tau)$ are shown in Fig. 3. For meat data, according to the initial guess, the test is never significant, whatever test function or τ value are used. For GDP data, the p-values corresponding to lower values of τ confirm the observed differences between local and global ranks to be important but they always remain above the standard critical values. Accordingly, the differences between rich and non rich countries could just be explained in terms of chance error. Another possibility, supported by simulations not discussed here, is that the power of the three test functions is poor when mixture components are not well separated.

References

Agostinelli, C., & Romanazzi, M. (2011). Local depth. *Journal of Statistical Planning and Inference, 141*, 817–830.

Ferraty, F., & Vieu, P. (2006). *Nonparametric functional data analysis*. New York: Springer.

Liu, R. (1990). On a notion of data depth based on random simplices. *The Annals of Statistics, 18*, 405–414.

López-Pintado, S., & Romo, J. (2005). *Depth-based classification for functional data*. Working Paper 2005.56, Departamento de Estadistica, Universidad Carlos III de Madrid

López-Pintado, S., & Romo, J. (2009). On the concept of depth for functional data. *Journal of the American Statistical Association, 104*, 718–734.

López-Pintado, S., & Romo, J. (2011). A half-region depth for functional data. *Computational Statistics & Data Analysis, 55*, 1679–1695.

Ramsay, J. O., & Silverman, B. W. (2005). *Functional data analysis*. New York: Springer

Zuo, Y., & Serfling, R. (2000). General notions of statistical depth function. *Annals of Statistics, 28*, 461–482.

Can the Students' Career be Helpful in Predicting an Increase in Universities Income?

Massimo Attanasio, Giovanni Boscaino, Vincenza Capursi, and Antonella Plaia

Abstract The students' academic failure and the delay in obtaining their final degree are a significant issue for the Italian universities and their stakeholders. Based on indicators proposed by the Italian Ministry of University, the Italian universities are awarded a financial incentive if they reduce the students' attrition and failure. In this paper we analyze the students' careers performance using: (1) aggregate data; (2) individual data. The first compares the performances of the Italian universities using the measures and the indicators proposed by the Ministry. The second analyzes the students' careers through an indicator based on credit earned by each student in seven academic years. The primary goal of this paper is to highlight elements that can be used by the policy makers to improve the careers of the university students.

1 Introduction

The Article 5 of the Law 537 of 1993 (Legge 2003) issued by the Italian Government is the first step towards the evaluation of the Italian University System (IUS). It marks the beginning of the financial autonomy of the universities with the hope to contain public expenditures. The law identifies the evaluation criteria of the various aspects of the IUS with the aim to obtain a fair distribution of the financial resources. The same Article has created an Evaluation Committee (*"Nuclei di Valutazione"*)

M. Attanasio (✉) · G. Boscaino · V. Capursi · A. Plaia
Dipartimento di Scienze Economiche, Aziendali e Statistiche e Matematiche "Silvio Vianelli", Università degli Studi di Palermo, Palermo, Italy
e-mail: massimo.attanasio@unipa.it; giovanni.boscaino@unipa.it; vincenza.capursi@unipa.it; antonella.plaia@unipa.it

P. Giudici et al. (eds.), *Statistical Models for Data Analysis*, Studies in Classification, Data Analysis, and Knowledge Organization, DOI 10.1007/978-3-319-00032-9_2,

within each university and a National Assessment Committee, named "*Osservatorio Permanente*", whose primary scope is the management of the resources allocation.

Since late 1993 the law has experienced many modifications, such as the Ministerial Decree 509 in 1999. These reforms were created in order to improve the efficiency and the effectiveness of the universities due to the high number of drop outs and of long survivor students, defined as those students who stayed much more over the legal duration of the degree course (Lambert and Butler 2006). To cope with these aspects, the M.D. 509 introduced a new structure of university curricula, introducing (1) University Educational Credit (*Credito Formativo Universitario*—CFU), as a measurement student academic workload; (2) two qualifications (for both public and private institutions): first degree (L) or 3-year Bachelor, second degree (LS) corresponding to two cycles of courses or 2-year Master degree. Since 2004 a part of the National University Funding System (*Fondo di Finanziamento Ordinario*—FFO) have been distributed to the universities following new criteria on *university performance*, essentially based on teaching and research. Inspite of the fact that the 509 reform was implemented in the academic year 2001/2002 and was revised in 2004, the problems of dropouts and of long survivor students is still unchanged.

Monitoring students' careers allows us to monitor our system in order to improve efficiency for the benefit of the students, and it allows real time monitoring of the performance indicators provided by the Ministry of Education of the University to better distribute the competitive funds (which was around 10% of the Total Funds—FFO—in 2010). Indeed, monitoring of the students' CFUs rates is both a commitment and a source of essential knowledge for those responsible for the creation of university degree programs. It also allows to collect information on the strengths and weaknesses of specific education programs; to acquire important data for programming and prevention and to establish checks and confirmations in order to create a process that is constantly under control. Furthermore, in order to improve educational services and identify appropriate ways to improve weak students' performance it is important to analyze the determining factors of university courses successes and failures (Boscaino et al. 2007).

In this paper the student career performance (SCP) is analyzed using both aggregate data, taking into account the FFO criteria and the outcome of the 2010 universities fund allocation (Sect. 2), and individual longitudinal data, taking into account the heterogeneity of students and the monitoring of their careers (Sect. 3).

The SCP data analysis raises several questions. For instance, is it possible to figure out simple and straightforward actions to accelerate the improvement FFO's indicators? Is it possible to figure out a policy to improve FFO's indicators in the long term? Are there simple numbers or indicators, useful to address university policies, which can be *extracted* from individual SCP analysis? All these questions lead to the ultimate purpose of this paper: which policy *my* university will/could adopt to improve the SCPs?

2 Aggregate Data: FFO Structure and Criteria

Italian FFO has been divided in three parts since 2009: the *QB* (*Quota Base*) corresponding to the general budget, the *MP* (*Modello Premiale*) corresponding to the competitive funds (introduced to reward teaching and research quality), and a residual part mostly related to salary increase (*RE*) (D. M. 2009, 2010). Every year the Ministry sets the FFO amount, the *MP* percentage amount and how to compute the *QB*, while *RE* is obtained as residual.

In 2010 the Ministry established the amount of the funds to be distributed to the universities and set the MP to be 10% of the yearly FFO while the QB was the 80% of the 2009 FFO. The 2010 FFO is composed by 10% *MP*, 83% *QB*, and 7% *RE*. Moreover, The Ministry has announced that the MP will increase to 12% of the 2011 FFO .

The *MP* award is proportional to 6 performance indicators, 2 for teaching and 4 for research. The IUS 54 universities have received the *MP* according to (1):

$$MP = 0.17 \times A1 + 0.17 \times A2 + 0.23 \times B1 + 0.10 \times B2 + \times 0.20 \times B3 + 0.13 \times B4 \tag{1}$$

The two indicators relevant to the "quality of teaching", $A1$ and $A2$, will be analyzed in detail in this paper, while we do not analyze the others because $B1 - B4$ represent the quality of the research outcome, which is not relevant to the paper aims. $A1$ and $A2$ consider respectively the demand of each University Education and the number of CFUs earned by the students:

$$A1 = (4 \times STUD_A + 3 \times STUD_B + 2 \times STUD_C + 1 \times STUD_D) \times (K_T + K_A) \tag{2}$$

$$A2 = CFU\_P_A + CFU\_P_B + CFU\_P_C + CFU\_P_D \tag{3}$$

with:

$$CFU\_P_i = \frac{\frac{CFU\_E_i}{CFU\_T_i}}{Median_i} \times CFU\_E_i \tag{4}$$

where:

1. i = A, B, C, D (ministerial course classification based on the financial aid allocated to each student).
2. $STUD_i$ is the number of "active students" attending type i course.
3. An "active student" is defined as a student with the following features: they have been in the university system for a number of years less or equal to the legal duration of the course; and they have earned at least 5 CFUs per year.

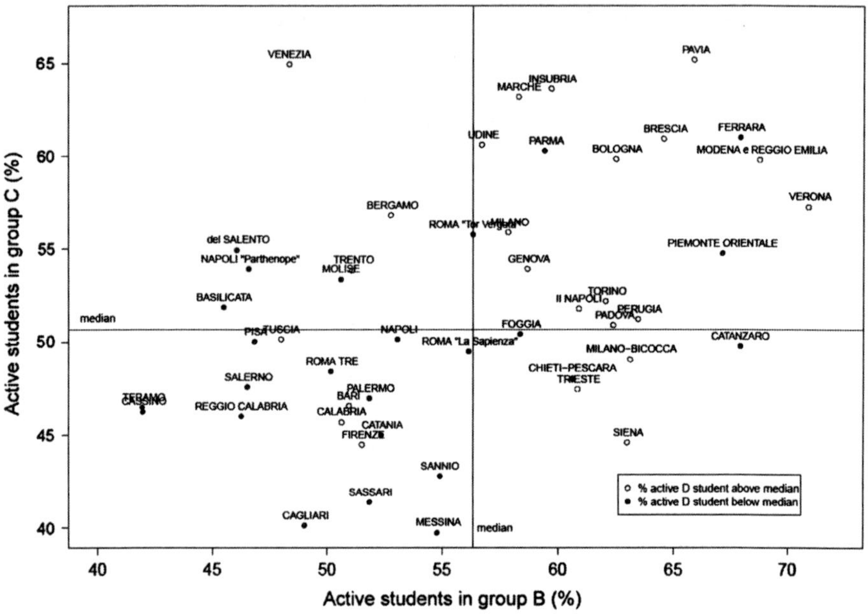

Fig. 1 Active students in groups B, C, and D (%)

4. K_T and K_A are correction factors related to the local context (based on the net income distribution of the university region) and to the sustainability of the courses (the ratio between the number of teachers and the number of Degree Courses), respectively.
5. CFU_E_i are the CFU earned in a year by the students of the i-th group.
6. CFU_T_i are the CFU correlated to the workload of the full-time students per year in the i-th group.
7. $Median_i$ is the median over the rates CFU_E_i/CFU_T_i computed for the 54 universities, in each group.

The *MP* could cause high losses/gains which will influence the FFO in the subsequent years, since the *QB* is computed as a percentage (80%) of the previous year FFO. In this way, successive losses are summed up each year since the *QB* is progressive, so it is crucial to undertake actions to improve $A1$ and $A2$ in order to avoid extra losses in the successive years. But how universities can improve their own performance? According to $A1$, universities are funded proportionally to the number of "active students", but nothing is said about the "inactive students": how do these students affect the total number of students? The graphical representation of the "location" of the single universities (Fig. 1) allows to summarize and compare quickly the students' performance (in terms of active students percentage) with respect to the groups B, C, and D. Group A is not included because it covers just 4% of the students.

The worst universities performance correspond to the black dots in the III quadrant in the scatterplot (Fig. 1). These are the universities that need to improve performances for the benefit of the groups (B to D) by increasing the number of students earning at least 5 CFUs per year and/or by reducing the number of long survivor students. In practice, a university should increase the number of "active students" or, eventually, decrease the number of "inactive students": this target can be achieved by avoiding students to stay in the university system for a period not longer than the legal duration of the course, and by letting them accumulate at least 5 CFUs in a year. This second task, which allows to stimulate earning 5 CFUs per year, will improve A2 too. A great advantage is to be able to identify early those students who could become "inactive" or could earn few CFUs each year.

To cope with these problems, a cohort study, as described in the next section, can give useful information.

3 Individual Data: Student Career Performance

In this section we investigate the student career performance applying individual longitudinal data. We will use the CFUs accumulation over 7 years.

The data concern the cohort of the Palermo University freshmen, enrolled in 2002/2003 and followed up till the 31st May 2010. For simplicity's sake, we analyze only those students who never change faculty during 7 years, who have payed university fees in the legal terms every year and never dropped out (the core students). Moreover, we examine 3 faculties (Economics, Engineering, and Arts) belonging to three different cultural areas. The performance is defined as the number of CFUs earned by the i-th students at the end of the first j years ($j = 1, 2, \ldots, J$) ($CFU_i(j)$). Table 1 reports the distribution of freshmen enrolled in 2002/2003 classified according to the student career status in 2009. Our classification of the students is the following:

Core	The student who never change faculty.
Mover	The student who changed faculty, course, and/or university.
Withdrawal	The student who quits.
Lost	The student who never quitted, but whose follow up ended before 7th year.

The latter is largely represented in Economics and Arts Faculties, but for the purpose of this paper we will not investigate this group issue because it deserves in-depth examination.

Table 1 shows that the most efficient faculty at Palermo University is the Faculty of Engineering. In this Faculty the graduation rate for the bachelor degree is greater than the one of two other faculties, reaching 46% at the end of the sixth year (may be it will be higher if a part of the lost is considered as withdrawal). The low success rate aetiology is difficult to be proved due to its complexity. To simplify we focus our attention to few aspects, with the aim of finding out simple and useful information for the universities.

Based on these findings we investigate the number of CFUs earned at the end of the first year (CFU(1)) versus the years needed to obtain the degree.

Table 1 Distribution of Palermo University freshmen enrolled in 2002/2003 by Status and Faculty

	Faculty					
Status in 2009	Economics	(%)	Arts	(%)	Engineering	(%)
Core	536 (341[a])	49.1	1,339 (960[a])	54.2	733 (526[a])	63.7
Mover to other Faculties	53 (7[a])	8.8	67 (12[a])	5.1	88 (18[a])	9.7
Mover to other Universities	43		60		24	
Withdrawal	141	42.1	304	40.6	95	26.6
Lost	318		700		211	
Total	1,091	100	2,470	10	1,151	100

[a]Students who took a degree (bachelors)

This investigation is needed to figure out if CFU(1) is a good predictor. We restrict our analysis to those students who never changed Degree Course (core) in 7 years. Following Cozzucoli and Ingrassia (2005), we define:

B The bachelors, or those students that graduated within 7 years.

O The others (with respect to the B's), those students not yet graduated in 2009.

$CFU_i(j)$ The number of CFUs earned by the i-th students at the end of the first j years ($j = 1, 2, \ldots, J; J = 7$).

X_i The number of years expected to obtain a degree by the i-th student, considering the number of CFUs earned by the i-th student at the end of the first year ($CFU_i(1)$), that is $180/CFU_i(1)$, where 180 is the amount of CFUs needed to get a degree. We group X into 4 classes: $\leq$4; 4-|5; 5-|6; and >6; obviously, the students whose CFU(1) $= 0$ (12% of core students) have been excluded.

Y_i The number of years observed to obtain a degree for i-th bachelor (B).

EY_i The number of years expected to obtain a degree for the others (O). EY_i is computed extrapolating the student's annual earning speed (υ_i) based on $CFU_i(j)$, and followed to the attainment of 180 CFU:

$$\upsilon_i = \frac{CFU_i\,(j)}{j} \tag{5}$$

$$EY_i = \frac{180 - CFU_i\,(j)}{\upsilon_i} + j \tag{6}$$

EY_i first term is the number of years beyond the j-th to "get" the degree. For instance, if a student earned 105 CFUs at the end of the first 7 years ($j = 7$), we expect that they will get their degree by the 12th year of attendance:

$$\upsilon_i = \frac{CFU_i\,(7)}{7} = \frac{105}{7} = 15$$

$$EY_i = \frac{180 - CFU_i\,(7)}{\upsilon_i} + 7 = \frac{180 - 105}{15} + 7 = 12$$

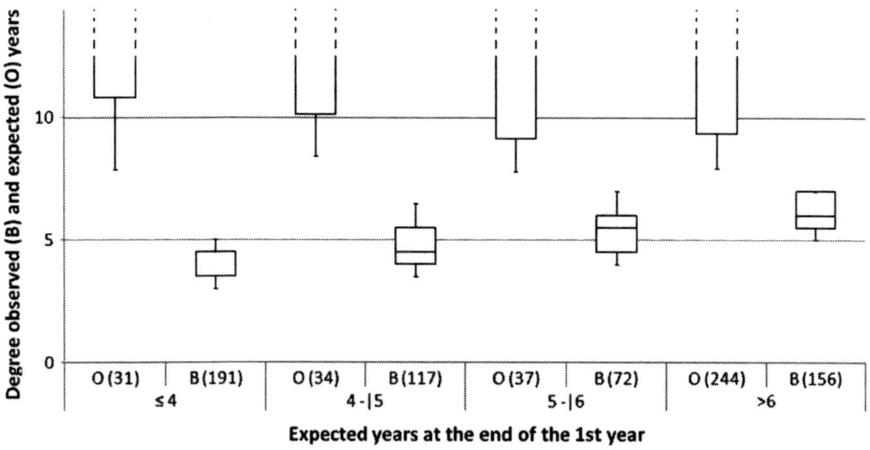

Fig. 2 Box-plots of degree observed or expected years, by expected years at 1st year and student's status (O and B)—Faculty of Engineering, Palermo University. Upper O's boxes are truncated for reason of space

Figure 2 summarizes the results for the Faculty of Engineering because the other faculties' results are quite similar. The X axis is the number of the years to obtain the degree, as expected at the end of the first academic year (computed through X), for bachelors and others. The Y axis reports: for B's the observed number of years to obtain the degree; for O's the expected number of years to obtain the degree (computed through EY). Figure 2 clearly shows in most cases that a good start is a good predictor for success. For instance, 117 B students obtained the degree in 4-|5, as expected by their performance in the 1st year, and only 34 O students did not "keep the promise" held in the 1st year. Interestingly, the ratio between B's and O's decreases towards the class 5-|6, showing the "best" promise for class ≤ 4. These results suggest the importance of the first year CFUs as predictor of success. The box-plots relative to the expected years to obtain the degree (O's) are very large and their upper bounds are always over 20. This unusual number is due to the large number of students who have earned very few CFUs in 7 years. This dysfunction obviously affects the denominator of EY_i.

4 Concluding Remarks and Future Developments

In conclusion, Palermo University aggregate career students data show that the number of earned CFUs are below the national median while the individual SCPs show the dramatic slowness to obtain a degree.

These analysis are the first step to investigate the current IUS Evaluation. This Evaluation system presents several incongruities, whose analysis deserves specific attention. For instance, there is a need for different policies within the single university and, more importantly, competitive policies within the different Italian universities. A good example is the FFO implementation by the Ministry for the four types of degree courses A, B, C and D. Nevertheless this is not enough. It is also very important to underline that FFO distributes the competitive fund on the basis of crude indicators. In fact, the process of comparison is conducted using the same set of indicators for all universities, without taking into account the different characteristics of universities and students. Moreover this process is somehow unfair, because good universities can attract more easily good students, and a vicious circle could be boosted by this system of awards (Lambert and Butler 2006).

Further statistical investigations with individual career data will provide detailed information on other covariates that may influence the success/failure of the students. For instance, Fasola (2011) applied a discrete-time competing risks models to the cohort of freshmen of 2002/2003. This model gets in a discrete-time setting simultaneous estimates of the degree and failure risks, including several covariates, such as the high school grades; the high school qualification; the age and the gender (in some faculties). These results may be useful as a basis to create a policy able to address specific actions for specific types of students with the aim to improve the quality of teaching outcomes and to provide recommendations to improve FFO indicators.

Acknowledgments The article is the result of the productive collaboration among the authors. In particular, paragraph 1 can be ascribed to Vincenza Capursi, paragraph 2 can be ascribed to Antonella Plaia, paragraph 3 can be ascribed to Giovanni Boscaino, and paragraphs 4 can be ascribed to Massimo Attanasio.

This paper has been supported from Italian Ministerial grant PRIN 2008 "Measures, statistical models and indicators for the assessment of the University System", n. 2008BWXSLH

References

Boscaino, G., Capursi, V., & Giambona, F. (2007). La performance delle carriere di una coorte di studenti universitari, DSSM Working Papers - n. 2007.1.

Cozzucoli, P. C., & Ingrassia, S. (2005). Indicatori Dinamici di Efficienza Didattica dei Corsi di Laurea Universitari. *Statistica e Applicazioni, III*, special issue 1, 61–68.

Fasola, S. (2011). *Laurea o abbandono: il buongiorno si vede dal mattino*? Graduation Thesis, University of Palermo.

Lambert, R., & Butler, N. (2006). *The future of European universities: Renaissance or decay?* London: Centre for European Reform.

Legge. (2003). 24-12-3, n. 537, Interventi correttivi di finanza pubblica. Italian Government.

D. M. (2009). 23-9-2009, n. 45, Decreto criteri di Ripartizione del Fondo di Finanziamento Ordinario (FFO) delle Università per l'anno 2009. Italian Government.

D. M. (2010). 21-12-10, n. 655, Decreto criteri di Ripartizione del Fondo di Finanziamento Ordinario (FFO) delle Università per l'anno 2010. Italian Government.

Model-Based Classification Via Patterned Covariance Analysis

Luca Bagnato

Abstract This work deals with the classification problem in the case that groups are known and both labeled and unlabeled data are available. The classification rule is derived using Gaussian mixtures where covariance matrices are given according to a multiple testing procedure which asesses a pattern among heteroscedasticity, homometroscedasticity, homotroposcedasticity, and homoscedasticity. The mixture models are then fitted using all available data (labeled and unlabeled) and adopting the EM and the CEM algorithms. The performance of the proposed procedure is evaluated by a simulation study.

1 Introduction

The main purpose of discriminant analysis (DA) is to assign an object to one of the K groups $G_1, \ldots, G_K$, according to a rule which is based on a vector $\mathbf{x} = \left(x_1, \ldots, x_p\right)$ of observations of p variables. Following the well-known Bayes rule, the object with measurement $\mathbf{x}$ is assigned to G_k when

$$k = \underset{1 \leq j \leq K}{\operatorname{argmax}} \ \pi_j f_j(\mathbf{x}), \tag{1}$$

where π_j is the unconditional prior probability of observing a class j member and $f_j(\mathbf{x})$ denotes the j-th group conditional density of $\mathbf{x}$. Then, when π_j and $f_j(\mathbf{x})$ for all $j = 1, \ldots, K$ are known (or estimated), an object can be classified using the rule (1). The most common DA methods assume that the data within group j are

L. Bagnato (✉)
Dipartimento di Discipline Matematiche, Finanza Matematica e Econometria, Università Cattolica del Sacro Cuore, Milano, Italy
e-mail: luca.bagnato@unicatt.it

P. Giudici et al. (eds.), *Statistical Models for Data Analysis*, Studies in Classification, Data Analysis, and Knowledge Organization, DOI 10.1007/978-3-319-00032-9_3,

generated from a p-variate normal with mean $\boldsymbol{\mu}_j$ and covariance matrix $\boldsymbol{\Sigma}_j$. Then, the global density can be written as a mixture of normal components:

$$f(\mathbf{x}) = \sum_{j=1}^{K} \pi_j f_j(\mathbf{x}), \tag{2}$$

with $f_j(\mathbf{x}) = (2\pi)^{-p/2} \left|\boldsymbol{\Sigma}_j\right|^{-1/2} \exp[-1/2(\mathbf{x}-\boldsymbol{\mu}_j)'\boldsymbol{\Sigma}_j^{-1}(\mathbf{x}-\boldsymbol{\mu}_j)]$. These methods differ depending on how $\boldsymbol{\Sigma}_j$, $j = 1, \ldots, K$, are defined. While linear discriminant analysis (LDA) assumes homoscedasticity, quadratic discriminant analysis (QDA) assumes heteroscedasticity. Operationally, in order to opt for LDA, assumptions of normality and homoscedasticity must be checked; with this aim, a possible solution consists in adopting the test proposed by Hawkins (1981). Unfortunately, this is a rigid practice, since intermediate configurations between heteroscedasticity and homoscedasticity can be observed in the data. Methods such as regularized DA (RDA) and eigenvalue decomposition DA (EDDA) (see Friedman 1989 and Bensmail and Celeux 1996, respectively) are discriminant methodologies offering different ways to modelize such situations. Another approach proposed in literature arises from a test on the between-group covariance structure: some sub-models of EDDA, such as proportional covariance matrices (PCMs) (see Flury 1986) and common principal components (CPCs) (see Flury 1984), have been singularly applied in DA. Also equal correlation matrices (Manly and Rayner 1987) have been used to this purpose in literature (see McLachlan 1992, Sect. 5.4 and the references therein for applications of these and other related testing-based approaches of DA).

The traditional discriminant analysis involves samples of known origin (labeled data) and provides classification rules for samples of unknown origin (unlabeled data). These methodologies suffer whenever only a few known observations are available. Furthermore, unlabeled data could contain important information in order to define the classification rule. Our proposal is to adopt the idea presented in Dean et al. (2006) where both labeled and unlabeled data are involved in the estimation procedure. In particular, we use the family of models (as structures for DA) and the multiple testing procedure (as model selection criteria) recently introduced in Greselin et al. (2011). The combination of these two approaches (Dean et al. 2006; Greselin et al. 2011) results in a novel procedure composed by two steps: firstly the common covariance structure is chosen by means of a test of hypothesis and then the model parameters are obtained by constrained estimation. Conversely, in Dean et al. (2006) the choice of the underlying model is performed *a posteriori* according to the highest BIC- value after having estimated all the models. It is worth noticing that the new procedure accommodates also for the case of equal orientation covariance matrices (the homotroposcedastic model), which has never been used in the context of mixture models, despite it has been frequently observed in real data. Then, the obtained classification rule permits to achieve parameter reduction and to increase the interpretability of the results. In Sect. 2, after a brief introduction to the adopted family of models and the estimation procedure, two real examples are presented. Furthermore, some considerations about the relation with the CPC model are provided. Finally, in Sect. 3, the validity of the proposed procedure is illustrated by a simulation study.

2 Patterned Covariance Analysis and Model Estimation

Suppose that the density of the data is given by (2) and let $\boldsymbol{\Sigma}_j = \boldsymbol{\Gamma}_j \boldsymbol{\Lambda}_j \boldsymbol{\Gamma}'_j$ be the spectral decomposition of the matrix $\boldsymbol{\Sigma}_j$, $j = 1, \ldots, K$, where $\boldsymbol{\Lambda}_j$ is the diagonal matrix of the eigenvalues of $\boldsymbol{\Sigma}_j$ sorted in non-increasing order, and $\boldsymbol{\Gamma}_j$ is the $p \times p$ orthogonal matrix whose columns are the normalized eigenvectors of $\boldsymbol{\Sigma}_j$ ordered according to their eigenvalues. Each component of the spectral decomposition has a different morphologic interpretation in terms of the group scatters: $\boldsymbol{\Gamma}_j$ governs the *orientation*, while $\boldsymbol{\Lambda}_j$ controls the *size-shape*. By allowing $\boldsymbol{\Gamma}_j$ and $\boldsymbol{\Lambda}_j$ to be equal (E) or variable (V) between groups, we obtain four parsimonious and easily interpreted models which are appropriate to describe various practical situations (see Table 1). Thus, for example, by imposing $\boldsymbol{\Lambda}_1 = \cdots = \boldsymbol{\Lambda}_K = \boldsymbol{\Lambda}$ we obtain homometroscedasticity (model EV in our notation), while by constraining $\boldsymbol{\Gamma}_1 = \cdots = \boldsymbol{\Gamma}_K = \boldsymbol{\Gamma}$ we have homotroposcedasticity (model VE). We propose to use this family of models which allows to exploit the augmentation multiple testing procedure in Greselin et al. (2011) as model selection criteria (instead of the common use of AIC or BIC as in Dean et al., 2006). Without going into details, the main idea of this procedure is based on giving two separate tests for EV and VE which are combined by using the relation $EV \cap VE = EE$. In particular, EV is tested exploiting the relationship between the eigenvalues in $\boldsymbol{\Lambda}_j$ and the variances of the principal components in the j-th group, $j = 1, \ldots, K$. The test for VE is obtained by adding some modification to the CPC test proposed by Flury and Gautschi (1986).

To estimate the models we follow the idea proposed in Dean et al. (2006) where N labeled observations and M unlabeled observations are available and all of them are used in the mixture model estimation. EM (see Dempster et al. 1977) and CEM (see Celeux and Govaert 1992) algorithms, which respectively maximize the likelihood and the complete-data likelihood, are used here. In addition to the different set of adopted models, the main contribution of our proposal is given by the choice of the model restrictions. While in Dean et al. (2006) the structure is chosen *ex-post* (the model estimation) using the BIC, we propose to choose *ex-ante* the structure using only the labeled data. Note that we can apply the test only on the labeled data since they can be assumed to be independent.

2.1 *Examples of Homometroscedasticity and Heteroscedasticity*

In this section two real examples of EV and VE configurations are presented. The first example is the famous crab data set (genus *Leptograpsus*) also considered in Campbell and Mahon (1974) and Ripley (1996). Here, following the setting of Peel

Table 1 Model, covariance restrictions, nomenclature and number of covariance parameters for each considered model (see Greselin et al. 2011)

Model	$\boldsymbol{\Lambda}_j$	$\boldsymbol{\Gamma}_j$	Nomenclature	# of covariance parameters
EE	Equal	Equal	Homoscedasticity	$p\frac{p+1}{2}$
EV	Equal	Variable	Homometroscedasticity	$Kp\frac{p+1}{2} - (K-1)\,p$
VE	Variable	Equal	Homotroposcedasticity	$p\frac{p+1}{2} + Kp$
VV	Variable	Variable	Heteroscedasticity	$Kp\frac{p+1}{2}$

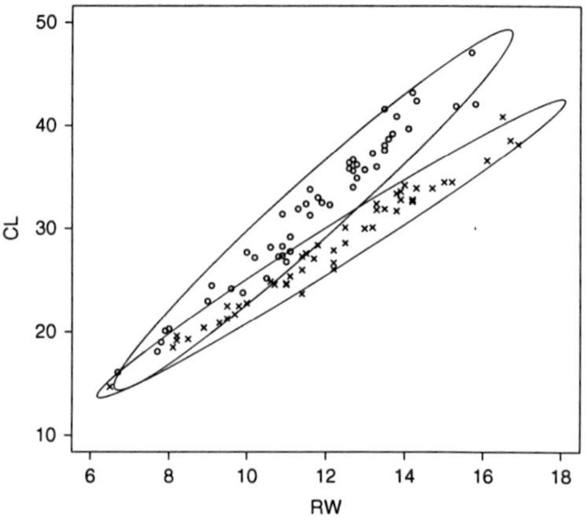

Fig. 1 Scatter plot of variables RW and CL for $n_1 = 50$ males and $n_2 = 50$ females blue crabs (**o** denotes male and **×** female). The ellipses of equal (95%) concentration are also superimposed

and McLachlan (2000), the attention is focused on the sample of $n = 100$ blue crabs, there being $n_1 = 50$ males (group 1) and $n_2 = 50$ females (group 2), each specimen having $p = 2$ measurements (in millimeters) on the rear width (RW) and the length along the midline (CL) of the carapace. In Fig. 1 the scatter plot of RW versus CL, in both groups, is shown jointly with the ellipses of equal (95%) concentration (arising from estimation under the assumption of heteroscedasticity). Here, the homometroscedasticity structure stands out since the group scatters are similar in size and shape but different in orientation. As shown in Greselin et al. (2011) this conjecture is confirmed also by applying the multiple testing procedure.

The second real example comes from the Swiss 1,000-franc bank notes, also considered in Flury and Riedwyl (1983), on which the following two variables are measured

$$\text{LEFT} = \text{width of bank note, measured on left side,}$$

$$\text{RIGHT} = \text{width of bank note, measured on right side.}$$

There are $k = 2$ groups of bills, genuine (group 1) and forged (group 2), each of them consisting of $n_j = 100$, $j = 1, 2$, observations. In Fig. 2a, b the scatter plots of

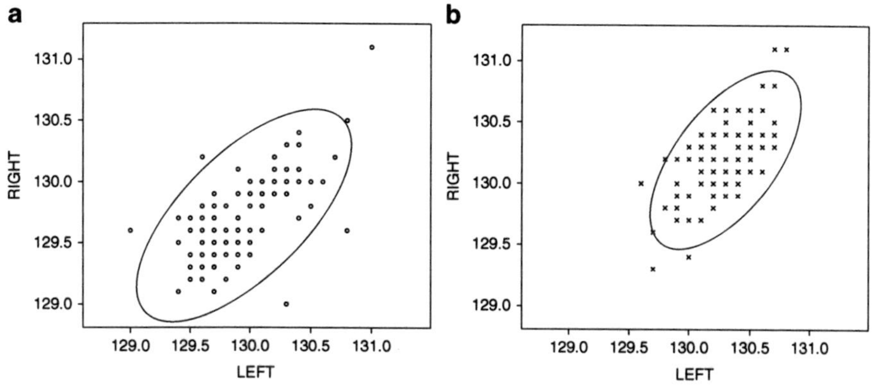

Fig. 2 Scatter plots, and related ellipses of equal (95%) concentration, of variables LEFT and RIGHT in two groups of Swiss bank notes. Coinciding points are marked by single symbol only. (**a**) Genuine, (**b**) Forged

LEFT versus RIGHT, for both groups, are jointly reported with the ellipses of equal (95%) concentration arising from estimation under the assumption of heteroscedasticity. As confirmed by the multiple test, the structure of homotroposcedasticity can be assumed: size and shape are different, but the orientations agree.

2.2 *Homotroposcedasticity Versus Common Principal Components*

The definition of model VE has been never used in DA, while models EE, EV and VV are considered also in EDDA. The EDDA counterpart of homotroposcedasticity exploits the Common Principal Components (CPC) model (see Flury 1984). The latter is in principle equal to VE but with the difference that in the spectral decompositions of $\boldsymbol{\Sigma}_j$ the eigenvalues of $\boldsymbol{\Lambda}_j$, $j = 1, \ldots, K$, are not constrained to be in decreasing order. In CPC the matrices of normalized eigenvectors are equal between groups but the requirement of decreasing eigenvalues is abandoned. Thus, the family of CPC models contains the family of models VE. Even if both configurations reach the same parsimony in terms of parameters, VE could provide better discrimination if this configuration really holds. Starting from the estimation of CPC, an ad hoc procedure to estimate VE has been implemented.

The difference between homotroposcedasticity and CPC is graphically explained in Fig. 3 considering ellipses of equal concentration of two normal groups. Both the ellipses have the same axes, according to the CPC model, although they differ in orientation. This means that homotroposcedasticity does not hold.

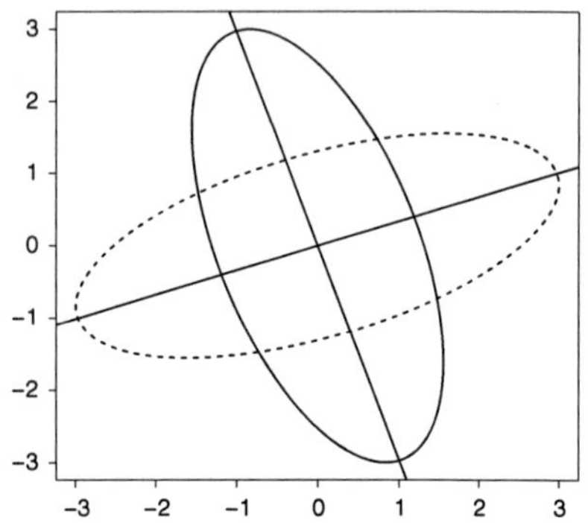

Fig. 3 Ellipses of equal concentration of two bivariate normal distributions, with the same principal axes but with different orientation

3 Simulation Study

In this section a simulation study is provided to evaluate the validity of the proposed procedure. The R code used to for both multiple testing procedure and mixture model estimation (EM and CEM) is available from the authors upon request. For sake of brevity only one scenario has been reported here and the results concern only the use of the EM algorithm. Similar results have been observed adopting the CEM algorithm. We have considered two normal groups ($K = 2$) in the bivariate context ($p = 2$). In each of the $M = 1{,}000$ replications, we have generated 90 data, 30 of which have been unlabeled and used to calculate misclassification errors. Data are generated so that the distance between centroids is equal to $\sqrt{2}$. Here, $\boldsymbol{\Sigma}_1$ has been randomly generated and its spectral decomposition has been computed:

$$\boldsymbol{\Sigma}_1 = \boldsymbol{\Gamma}_1 \boldsymbol{\Lambda}_1 \boldsymbol{\Gamma}_1' = \begin{pmatrix} \gamma_{11}^{(1)} & \gamma_{12}^{(1)} \\ \gamma_{21}^{(1)} & \gamma_{22}^{(1)} \end{pmatrix} \begin{pmatrix} \lambda_1^{(1)} & 0 \\ 0 & \lambda_2^{(1)} \end{pmatrix} \begin{pmatrix} \gamma_{11}^{(1)} & \gamma_{12}^{(1)} \\ \gamma_{21}^{(1)} & \gamma_{22}^{(1)} \end{pmatrix}',$$

with $\lambda_1^{(1)} \geq \lambda_2^{(1)}$. In order to simulate gradual departure from homoscedasticity, $\boldsymbol{\Sigma}_2$ has been generated in the following way

$$\boldsymbol{\Sigma}_2 = \boldsymbol{R}\boldsymbol{\Gamma}_1 \begin{pmatrix} d\lambda_1^{(1)} & 0 \\ 0 & \dfrac{\lambda_2^{(1)}}{d} \end{pmatrix} [\boldsymbol{R}\boldsymbol{\Gamma}_1]', \quad \text{with} \quad \boldsymbol{R} = \begin{pmatrix} \cos\vartheta & -\sin\vartheta \\ \sin\vartheta & \cos\vartheta \end{pmatrix},$$

where $\boldsymbol{R}$ is the rotation matrix of angle ϑ, and $d \geq 1$ is a sort of "deformation" constant: if $d = 1$, the concentration ellipsoids of $\boldsymbol{\Sigma}_1$ and $\boldsymbol{\Sigma}_2$ are homometroscedastic; the flattening of the concentration ellipsoids of $\boldsymbol{\Sigma}_2$ increases in line with d. Several values of ϑ and d have been considered: 4 values for ϑ (0, $\pi/6$, $\pi/3$, $\pi/2$ in radiants, i.e., 0°, 30°, 60°, 90°), and 3 values for d (1, 2 and 4). All combinations of these two factors have been taken into account and specified at the top of each subtable in Table 2. Subtables are arranged so that d increases moving

Table 2 Simulation results referred to misclassification errors for each choice of the couple (d, ϑ). In round brackets, the relative frequencies of the cases, over the 1,000 replications, in which the multiple testing procedure (with nominal level 0.05) has chosen each structure

(a) $\vartheta = 0°$, $d = 1$

	EE	EV	VE	VV
EE (0.965)	0.1998	0.2008	0.2027	0.2033
EV (0.006)	0.2000	0.1833	0.1889	0.1889
VE (0.028)	0.1821	0.1810	0.2012	0.2000
VV (0.001)	0.1333	0.1333	0.1333	0.1333
	0.1992	0.2001	0.2025	0.2030

(b) $\vartheta = 0°$ $d = 2$

	EE	EV	VE	VV
EE (0.546)	0.1831	0.1839	0.1742	0.1764
EV (0.005)	0.2867	0.2667	0.2667	0.2467
VE (0.439)	0.1861	0.1899	0.1774	0.1773
VV (0.010)	0.2333	0.2567	0.2367	0.2533
	0.1854	0.1877	0.1767	0.1779

(c) $\vartheta = 0°$ $d = 4$

	EE	EV	VE	VV
EE (0.009)	0.1000	0.1000	0.0778	0.0889
EV (0.000)	-	-	-	-
VE (0.967)	0.1742	0.1805	0.1389	0.1408
VV (0.024)	0.2306	0.2528	0.1944	0.1972
	0.1749	0.1815	0.1397	0.1417

(d) $\vartheta = 30°$ $d = 1$

	EE	EV	VE	VV
EE (0.255)	0.2541	0.2524	0.2621	0.2550
EV (0.712)	0.2234	0.1773	0.2224	0.1789
VE (0.007)	0.2476	0.2667	0.2619	0.2667
VV (0.026)	0.2141	0.1731	0.2167	0.1731
	0.2311	0.1970	0.2326	0.1988

(e) $\vartheta = 30°$ $d = 2$

	EE	EV	VE	VV
EE (0.069)	0.2725	0.2652	0.2551	0.2628
EV (0.390)	0.2368	0.1659	0.1955	0.1580
VE (0.083)	0.2514	0.2454	0.2285	0.2225
VV (0.458)	0.2336	0.1600	0.1810	0.1539
	0.2390	0.1766	0.1957	0.1687

(f) $\vartheta = 30°$ $d = 4$

	EE	EV	VE	VV
EE (0.002)	0.3000	0.2500	0.2500	0.2167
EV (0.006)	0.1944	0.1500	0.1222	0.0944
VE (0.116)	0.2451	0.2408	0.1997	0.1914
VV (0.876)	0.2320	0.1510	0.1428	0.1239
	0.2334	0.1616	0.1495	0.1317

(continued)

Table 2 (continued)

(g) $\vartheta = 60^\circ \quad d = 1$

	EE	EV	VE	VV
EE (0.149)	0.2729	0.2631	0.2718	0.2664
EV (0.805)	0.2566	0.1658	0.2000	0.1677
VE (0.009)	0.2222	0.1963	0.2630	0.2407
VV (0.037)	0.2658	0.1721	0.2117	0.1721
	0.2590	0.1808	0.2117	0.1832

(h) $\vartheta = 60^\circ \quad d = 2$

	EE	EV	VE	VV
EE (0.015)	0.3378	0.3022	0.2956	0.2933
EV (0.431)	0.2706	0.1514	0.1665	0.1448
VE (0.013)	0.2538	0.2436	0.2282	0.2154
VV (0.541)	0.2722	0.1518	0.1588	0.1402
	0.2723	0.1551	0.1651	0.1454

(i) $\vartheta = 60^\circ \quad d = 4$

	EE	EV	VE	VV
EE (0.000)	-	-	-	-
EV (0.006)	0.2833	0.1500	0.1333	0.1333
VE (0.025)	0.2853	0.2507	0.1893	0.1947
VV (0.969)	0.2798	0.1308	0.1234	0.1083
	0.2800	0.1339	0.1251	0.1106

(j) $\vartheta = 90^\circ \quad d = 1$

	EE	EV	VE	VV
EE (0.107)	0.2751	0.2748	0.2710	0.2738
EV (0.845)	0.2710	0.1630	0.1912	0.1632
VE (0.003)	0.3111	0.2889	0.2444	0.2667
VV (0.045)	0.2452	0.1511	0.1889	0.1600
	0.2704	0.1748	0.1998	0.1752

(k) $\vartheta = 90^\circ \quad d = 2$

	EE	EV	VE	VV
EE (0.005)	0.3200	0.3267	0.2733	0.2800
EV (0.453)	0.2795	0.1422	0.1493	0.1375
VE (0.006)	0.2944	0.2944	0.2222	0.2222
VV (0.536)	0.2805	0.1427	0.1557	0.1348
	0.2803	0.1443	0.1538	0.1373

(l) $\vartheta = 90^\circ \quad d = 4$

	EE	EV	VE	VV
EE (0.000)	-	-	-	-
EV (0.006)	0.3556	0.1222	0.1333	0.1278
VE (0.009)	0.2556	0.2889	0.1963	0.1963
VV (0.985)	0.2864	0.1209	0.1107	0.1011
	0.2866	0.1225	0.1116	0.1021

from left to right, while ϑ increases moving from top to bottom. In particular, while Table 2a is obtained under homoscedasticity, Table 2l has the greatest departure from homoscedasticity, both in terms of shape and orientation. Finally, it is interesting to note that the subtables d, g and j are related to homometroscedasticity, while the subtables b and c are referred to homotroposcedasticity.

Focusing on Table 2a we find, in round brackets, the relative frequencies of the cases, over the 1,000 replications, in which the multiple testing procedure (with nominal level 0.05) has chosen each structure. The last row reports the mean misclassification errors computed over the 1,000 replications, obtained by classifying the unlabeled data according to the rules based on the models EE, EV, VE and VV, respectively (specified in columns). The interpretation of the first row (EE) is the same of the one of the last row but the four averages are calculated only over the 965 replications in which the multiple test has chosen EE. The same reading holds for the other three rows EV, VE and VV. Thus we can consider the last row as a weighted mean, with weights reported in round brackets, of the averages misclassification rates of the corresponding column. The other subtables in Table 2 can be interpreted in the same way.

From Table 2a, and from the other subtables we can observe coherent results with the true underlying configurations, both in terms of the chosen structure and in terms of misclassification errors. For example, in Table 2c which reports a VE configuration, we can observe that in the 967 replications for which the multiple testing procedure has chosen this structure, the misclassification rate is the lowest one. Similar comments also hold by observing Table 2j and l which concern results under EV and VV configurations, respectively.

4 Conclusions and Further Developments

In this paper we propose a model-based classification rule based on a family of models recently introduced in Greselin et al. (2011). As model selection criterion we use the multiple testing procedure proposed by the same authors and the model is estimated by considering both labeled and unlabeled data. Simulations confirm the validity of the proposed procedure, both in terms of ability to identify the true underlying structure and in terms of classification. Further investigation is needed, maybe extending the multiple testing procedure in such a way that the best model structure could be chosen among a higher number of available patterns.

References

Bensmail, H., & Celeux, G. (1996). Regularized Gaussian discriminant analysis through eigenvalue decomposition. *Journal of the American statistical Association, 91*(436), 1743–1748.

Campbell, N. A., & Mahon, R. J. (1974). A multivariate study of variation in two species of rock crab of genus Leptograpsus. *Australian Journal of Zoology, 22*(3), 417–425.

Celeux, G., & Govaert, G. (1992). A classification em algorithm for clustering and two stochastic versions. *Computational Statistics & Data Analysis*, *14*(3), 315–332.

Dean, N., Murphy, T., & Downey, G. (2006). Using unlabelled data to update classification rules with applications in food authenticity studies. *Journal of the Royal Statistical Society: Series C (Applied Statistics)*, *55*(1), 1–14.

Dempster, A., Laird, N., & Rubin, D. (1977). Maximum likelihood from incomplete data via the em algorithm. *Journal of the Royal Statistical Society, Series B (Methodological)*, 1–38.

Flury, B. N. (1984). Common principal components in k groups. *Journal of the American Statistical Association*, *79*(388), 892–898.

Flury, B. N. (1986). Proportionality of k covariance matrices. *Statistics & Probability Letters*, *4*, 29–33.

Flury, B. N., & Gautschi, W. (1986). An algorithm for simultaneous orthogonal transformation of several positive definite matrices to nearly diagonal form. *SIAM Journal on Scientific and Statistical Computing*, *7*, 169–184.

Flury, B. N., & Riedwyl, H. (1983). *Angewandte multivariate statistik*. Gustav Fischer, Stuttgart, 112–124.

Friedman, J. (1989). Regularized discriminant analysis. *Journal of the American Statistical Association*, *84*(405), 165–175.

Greselin, F., Ingrassia, S., & Punzo, A. (2011). Assessing the pattern of covariance matrices via an augmentation multiple testing procedure. *Statistical Methods & Applications*, *20*(2), 141–170.

Hawkins, D. M. (1981). A new test for multivariate normality and homoscedasticity. *Technometrics*, *23*(1), 105–110.

Manly, R. F. J., & Rayner, J. C. W. (1987). The comparison of sample covariance matrices using likelihood ratio tests. *Biometrika*, *74*(4), 841–847.

McLachlan, G. (1992). *Discriminant analysis and statistical pattern recognition* (2nd printing). Hoboken, NJ: Wiley.

Peel, D., & McLachlan, G. J. (2000). Robust mixture modelling using the t distribution. *Statistics & Computing*, *10*(4), 339–348.

Ripley, B. (1996). *Pattern recognition and neural network*. Cambridge: Cambridge University Press.

Data Stream Summarization by Histograms Clustering

Antonio Balzanella, Lidia Rivoli, and Rosanna Verde

Abstract In this paper we introduce a new strategy for summarizing a fast changing data stream. Evolving data streams are generated by non stationary processes which require to adapt the knowledge discovery process to the new emerging concepts. To deal with this challenge we propose a clustering algorithm where each cluster is summarized by a histogram and data are allocated to clusters through a Wasserstein derived distance. Histograms are a well known graphical tool for representing the frequency distribution of data and are widely used in data stream mining, however, unlike to existing methods, we discover a set of histograms where each one represents a main concept in the data. In order to evaluate the performance of the method, we have performed extensive tests on simulated data.

1 Introduction

The increasing diffusion of efficient monitoring systems causes the recording of data streams generated continuously and at a very high data rates. Common example of these data include recordings of climatological variables, web logs, computer network traffic.

Querying and mining from data streams are challenging tasks since elements are on-line collected and the size of the dataset can be potentially unbounded so, often, data after processing are discarded or archived and become not easily available anymore (Babcock et al. 2002). Moreover, often, it is not possible to produce a

A. Balzanella (✉) · R. Verde
Second University of Naples, Naples, Italy
e-mail: antonio.balzanella@gmail.com; rosanna.verde@unina2.it

L. Rivoli
University of Naples Federico II, Naples, Italy
e-mail: lidia.rivoli@unina.it

P. Giudici et al. (eds.), *Statistical Models for Data Analysis*, Studies in Classification, Data Analysis, and Knowledge Organization, DOI 10.1007/978-3-319-00032-9_4,

new answer just when a new observation is seen, this is because the time required for computing answers can be bigger than the inter-arrival time among observations or due to blocking operators which require the recording of the whole set of data before to be able to provide the answer to the knowledge discovery task.

A further issue to take into account is that data streams are usually generated by non stationary processes which require to adapt the knowledge discovery process to new emerging concepts.

In order to deal with these constraints, techniques for data stream analysis should process the incoming observation in short and constant times using a very reduced amount of memory and performing a single scan of the data (Gama and Gaber 2007).

Data stream mining methods are often based on updating some summaries of data every time a new element or a batch of elements is collected and on extracting the knowledge starting from these summaries rather than directly from the observations. However, this way to process data enforces a trade-off between the accuracy of the knowledge discovery procedures and the computational and storage constraints. Thus, one of the aims is to develop summarizing methods which include the most possible information from data under the highlighted constraints, in order to provide results which approximate the ones from algorithms for data stored into traditional databases.

According to this premise, summarization is a main task in data stream processing and it has been widely dealt in literature. Main approaches are based on data sampling, synopses, data compression (Sebastiao 2007; Vitter 1985; Guha et al. 2001; Balzanella et al. 2010, 2011; Balzanella 2009).

In this paper we propose to use histograms as tool for data stream summarization. Histograms are a widely used tool to represent, in a concise way the frequency distribution of a dataset. They provide information about the general shape of the distribution, the variability and the location of data, the symmetry, the modality.

In data stream mining, histograms have found a wide use, for example in Sebastiao (2007), an on-line method for constructing the histogram of a data stream has been proposed and then it is used for monitoring the evolution of data by looking at how the on-line computed histogram changes over the time.

The novelty of our proposal is that we discover a set of histograms which summarize groups of data so that each histogram will represent a main concept in the data.

To reach this aim we introduce a clustering algorithm based on the CluStream in Aggarwal et al. (2003) extended to process histogram data.

It is an efficient clustering algorithm for data streams where the prototype of each cluster is a histogram and data are allocated to clusters through a Wasserstein derived distance for histogram data.

The method is based on an on-line step which performs a first summarization of the incoming data through a set of histogram micro-clusters and on an off-line step which reveals the final set of summaries using the information stored in the histogram micro-clusters.

2 On-Line Clustering for Histogram Data

In order to introduce our proposal, we recall, at first, the CluStream algorithm for data streams and then on this ground, we introduce the novelties of our method.

2.1 The CluStream Algorithm

CluStream is an one-pass algorithm ables to deal with evolving data streams which allows to get, as output, not only a summarization of the whole data stream but also the summarization of a part of it defined by the user at query time.

Let $Y = \{(y_1, t_1), \ldots, (y_j, t_j), \ldots, (y_\infty, t_\infty)\}$ be a data stream made by a set of real valued ordered observations y_j on a discrete time grid $T = \{t_1, \ldots, t_j, \ldots t_\infty\} \subseteq \Re$.

The CluStream is made by an on-line step and by an off-line step. The first one performs a first summarization of the incoming data and can be still divided into two phases that we call micro-clusters updating and snapshot recording; the second one performs the final summarization processing the output of the on-line step.

In the first phase of the on-line step, every time a new data item (y_j, t_j) is collected, it concurs to the updating of the statistics stored in one of the elements of the set $MC = [mC_1, \ldots, mC_k, \ldots, mC_K]$ appropriately selected through an allocation procedure. In particular, each element of MC is a micro-cluster $mC_k = [ssv_k, sv_k, sst_k, svt_k, n_k]$ which is a data structure collecting the following statistics:

- n_k is the number of data elements which belong to micro-cluster.
- $ssv_k = \sum_{l=1}^{n_k} y_l^2$.
- $sv_k = \sum_{l=1}^{n_k} y_l$.
- $sst_k = \sum_{l=1}^{n_k} t_l^2$.
- $st_k = \sum_{l=1}^{n_k} t_l$.

From the information stored in a micro cluster it is possible to compute its average and variance. As consequence, (y_j, t_j) is allocated to the micro-cluster $mC_k \in MC$ if the distance between y_j and the average value of mC_k is the lowest, and the increasing of variance in mC_k is not superior to a threshold value. If the second condition is not satisfied for any micro-cluster, a new one mC_{K+1} is started and if K reaches a threshold value which is set in order to keep under control the memory occupation, the two micro-clusters having the nearest average value are mixed into one.

The second part of the on-line step is named snapshots recording and consists in storing the set of micro-clusters MC on some available media at time stamps detected through a predefined temporal scheduling. With data flowing, this step makes available several snapshots of the micro-clusters which will be used for recovering the summarization of a user defined time period. For example, if the current time is t_{j*} and the user is interested to a time-horizon h, that is, the user wants to find the status of the micro-clusters formed in $[t_{j*-h}; t_{j*}]$, then it is sufficient to recover the snapshot temporally nearest to t_{j*-h} and the one corresponding to the current time and subtracting the values of the statistics stored in the snapshot related to t_{j*-h} to the corresponding ones of the current snapshot.

The off-line step of CluStream is a variation of the k-means algorithm where the input data points are the average value of each micro-cluster and the center of each cluster is computed as weighted average of the allocated data points.

2.2 CluStream Strategy for Histogram Data

The method we propose adapts the CluStream algorithm to the processing histogram data according to the following schema:

- **On-line phase**
  1. Splitting of incoming data into non overlapping batches.
  2. Representation of each data batch through an equi-frequency histogram.
  3. Allocation of the histograms to a micro-cluster through a Wasserstein derived distance function.
  4. Snapshot recording.
- **Off-line phase**
  1. Clustering of the histogram micro-cluster in order to obtain the summarization of the stream.

Following the previous schema, in the on-line phase the incoming data stream Y is split into batches Q_i such as $Q_i \bigcap Q_{i+1} = \emptyset, \forall i$. For each data batch Q_i we get an histogram H_i represented by a set of pair (I_{ij}, f_{ij}) as shown below:

$$H_i = \{(I_{i1}, f_{i1}), \ldots, (I_{ij}, f_{ij}), \ldots, (I_{iJ}, f_{iJ})\}, \tag{1}$$

where I_{ij} are intervals (or bins) obtained partiting the domain of Q_i and f_{ij} are empirical frequencies associated to I_{ij}. It noteworthy that we used equi-frequency histograms ($f_{ij} = f_{ij'}, \forall j \neq j'$ in (1)) and we set the number of bins equal to J for each histogram $H_i, \forall i$.

In our procedure, the histograms H_i are the data elements which concur to the updating of appropriate data structures that we call Histogram micro-Clusters (*HmC*). Similarly to CluStream, an histogram micro-cluster stores basic statistics about a set of histograms. Taking into account that each bin of the histogram H_i can be expressed as function of its center (mid point) and radius (half-width) that is $I_{ij} = c_{ij} + r_{ij}(2t - 1)$ for $0 \leq t \leq 1$ then, a Histogram micro-Cluster HmC_k (with $k = 1, \ldots, K$), summarizing a set of n histograms, stores the following information:

- n_k is the number of data elements which belong to the histogram micro-cluster.
- $\bar{c}_{kj}$ with $j = 1, \ldots, J$.
- $\bar{r}_{kj}$ with $j = 1, \ldots, J$.
- $sst_k = \sum_{l=1}^{n_k} t_l^2$.
- $st_k = \sum_{l=1}^{n_k} t_l$.

where $\bar{c}_{kj}$ and $\bar{r}_{kj}, j = 1, \ldots, J$ are, respectively, the centers and the radii of the bins of the histogram $\bar{H}_k$, which assumes the role of histogram micro-cluster prototype.

In the on-line step, every time a new batch of data Q_i is available, the correspondent histogram H_i has to be allocated to nearest micro-cluster HmC_k, selected on basis of the value of proximity between H_i and all $\bar{H}_k$, $k = 1, \ldots, K$. For this purpose, we need to introduce an appropriate metric for histogram data.

The comparison of histogram data can be considered as a particular case of distance between distribution functions. At this end, several distances have been presented in literature, as example we can cite the f-divergence based measures, the Kolmogorov (or Uniform metric), the Prokhorov–Levi distance. However in Irpino and Verde (2006), Verde and Irpino (2007), it is shown that the metric derived by the Wasserstein–Kantorovich–Monge–Gini metric and known as *Mallows' distance* (Mallows 1972) satisfies important proprieties in the case of histograms.

Let two histograms H_i and H_z and be F_i and F_z the (cumulative) distribution functions associated to them, their Mallows' distance can be computed as follows:

$$d_M(H_i, H_z) = \sqrt{\int_0^1 \left(F_i^{-1}(t) - F_z^{-1}(t)\right)^2 dt} \tag{2}$$

where F_i^{-1} and F_z^{-1} are the quantile functions corresponding to F_i and F_z. This expression has a serious defect because the distribution function is not always invertible. However, according Irpino and Verde (2006), the distance (2) between two histograms H_i and H_z can be written as:

$$d_M^2(H_i, H_z) = \sum_{l=1}^{m} \pi_l \left[(c_{il} - c_{zl})^2 + \frac{1}{3} (r_{il} - r_{zl})^2 \right], \tag{3}$$

where m is the number of uniformly dense intervals I_l determined on the base of the quantile functions associated to H_i and H_z and $\pi = [\pi_l]$ is the vector of the relative frequency associated to each bin I_l. In particular, when histograms are equi-frequency, $m = J$ and $\pi_l = \frac{1}{J}, \forall l$, that is π is the vector containing quantiles of the distribution functions.

The allocation of the histogram H_i, obtained from the data batch Q_i to the nearest histogram micro-cluster, is performed according to the proximity between H_i and $\bar{H}_k$ $k = 1, \ldots, K$ computed through the formula (3).

Once the allocation of H_i to HmC_k has been performed, the statistics of HmC_k are updated. The new prototype will be computed such to minimize the following sum of distances in the micro-cluster:

$$\sum_{i=1}^{n_k} d_M^2(H_i, \bar{H}_k) = \sum_{i=1}^{n_k}\sum_{j=1}^{m}\frac{1}{J}\left[\left(c_{ij} - \bar{c}_{kj}\right)^2 + \frac{1}{3}\left(r_{ij} - \bar{r}_{kj}\right)^2\right]. \tag{4}$$

According to Irpino and Verde (2006), it is sufficient to update the centers and radii of $\bar{H}_k$ through the following expression:

$$\bar{c}_{kj} = \frac{1}{n_k}\sum_{i=1}^{n_k} c_{ij} \quad ; \quad \bar{r}_{kj} = \frac{1}{n_k}\sum_{i=1}^{n_k} r_{ij}.$$

The updating of the remaining statistics of histogram micro-clusters and the snapshot recording are performed as in CluStream.

The off-line phase of our method has to consider that the prototype of each micro-cluster coming out from the on-line phase, is a histogram, thus, the k-means algorithm used in CluStream cannot be taken as it is. For this reason, we use the k-means like clustering algorithm introduced in Verde and Irpino (2007) which uses the distance described in (3) and which is characterized by prototypes which are histograms. It provides, as output, the final set of histograms which summarize the whole data stream.

3 Experimental Results

In order to evaluate the effectiveness of the proposed method in discovering concepts in data, we have performed several experimental tests on simulated data.

We have generated nine datasets composed by 110,000 temporally ordered observations according to two main concepts. Especially for all datasets, each concept have been kept for 50,000 consecutive and ordered time stamps and the observations are locally independent and identically distributed. A further transition concept made by 10,000 items has been introduced in the datasets, generating observations by a mixture of the two involved distributions with weights equal to 0.5. In all the cases, the main concepts are obtained by the random generation of values according to two Gaussian distributions having different parameters.

To discover if our strategy recognizes them, we have compared the prototype histograms obtained by the off-line clustering procedure with the histograms related to data generated by two Gaussian distributions. The comparison is based on the distance between two histograms H_i and H_j introduced in Verde and Irpino (2010):

$$d_M(H_i, H_j) = (\bar{x_i} - \bar{x_j})^2 + (\sigma_i - \sigma_j)^2 + 2\sigma_i\sigma_j(1 - \rho(H_i, H_j)) \tag{5}$$

where $\bar{x_i}$, $\bar{x_j}$, σ_i, σ_j and $\rho(x_i, x_j)$ are mean, variance and correlation of the quantile functions associated to two histograms. Through this distance function, we can evaluate the matching of two histograms in terms of location, size and shape. For this reason, we have distinguished three set of experiments. In the first one, the two distribution have the same *mean* μ and different *standard deviation* σ; in the second, they have different μ and same σ values and in the third, they have different μ and σ values. The dataset are generated according to the distributions shown in following table:

	First concept		Second concept	
Dataset Id	μ_1	σ_1	μ_2	σ_2
1	0	1	0	5
2	0	1	0	7
3	0	1	0	7
4	2	1	5	1
5	2	1	7	1
6	2	1	10	1
7	2	1	5	5
8	3	1	9	2
9	5	1	14	10

To run our procedure, we need to set the input parameters. The first input parameter is the size of each batch of data Q_i which, for our data, corresponds to 200 observations. According to the rule of the square root, we have set the number of bins of the histograms to $K = 15$. A further parameter to set is the boundary value δ. It is worth of noting that a too low value of δ involves that a lot of processed histograms will not be allocated to existing micro-clusters but they will start new ones. At the opposite, a too high value implies that histograms will be always allocated to some existing micro-cluster and it will be more difficult to capture the emerging concepts. Taking into account this and that we have considered maximum of 50 micro-clusters, a suitable choice is $\delta = 0.02$ for all dataset. Finally, the number of desired summaries has been set to two consistently with the number of main concepts in the generated data.

The results in Table 1 show that for all the datasets, the procedure has been able to catch the main concepts in the data. This emerges by looking at the values of the distance but also at the values of the single components of it. In particular, the values near to 0 for the first and second component, highlight that the obtained histograms have a good match to the original data in terms of average value and standard deviation while values near to 1, for the correlation component, show that there is also a very good match in terms of shape of the histograms.

Table 1 Results of the comparison between the histograms emerging from the proposed procedure and the generated data

Dataset	First concept				Second concept			
Id	$\mu - \mu_1$	$\sigma - \sigma_1$	ρ	d_M	$\mu - \mu_2$	$\sigma - \sigma_2$	ρ	d_M
1	−0.0496	−0.2422	0.9765	0.3590	−0.3480	−1.3024	0.9809	1.7161
2	−0.0802	−0.1118	0.9860	0.2405	−0.4457	−1.7096	0.9831	2.2537
3	−0.0877	−0.0550	0.9883	0.2101	−0.6223	−2.1930	0.9827	2.9110
4	0.1907	−0.1723	0.9715	0.3800	−0.0652	−0.2621	0.9812	0.3424
5	0.3404	0.0057	0.9656	0.4674	0.072	−0.2399	0.9826	0.3220
6	0.6190	0.1174	0.9304	0.8151	−0.0656	−0.2683	0.9804	0.3509
7	0.1676	−0.08	0.9797	0.3037	−0.2675	−1.4936	0.9786	1.9016
8	−0.1430	−0.5123	0.9811	0.6787	0.4075	0.0793	0.9618	0.5438
9	−0.7666	−2.6773	0.9789	3.5884	0.6431	0.3765	0.9592	0.8425

4 Concluding Remarks

In this paper, we have introduced a new strategy for summarizing a data stream and then, we have evaluated it on simulated data. Our approach provides, by a two-step clustering algorithm, a set of histograms which describes the main concepts emerging in a fast changing data stream. Unlike existent approaches in data streams literature, we use histograms to summarize the concepts emerging in data and to provide an intuitive graphic representation of them. Further contributions are the introduction of a Wasserstein derived distance to the contest of data stream mining and the proposal of a its more computationally efficient form for comparing equi-frequency histograms.

References

Aggarwal, C. C., Han, J., Wang, J., & Philip, S. (2003). A framework for clustering evolving data streams. In: *29th int. conf. on very large data bases*.

Babcock, B., Babu, S., Datar, M., Motwani, R., & Widom J. (2002). Models and issues in data stream systems. In: *21th ACM SIGMOD-SIGACT-SIGART symposium PODS '02* (pp. 1–16).

Balzanella, A. (2009). Clustering and summarizing massive data streams. PHD Thesis, http://www.fedoa.unina.it/4184(2009).

Balzanella, A., Irpino, A., & Verde, R. (2010). Dimensionality reduction techniques for streaming time series: a new symbolic approach. *Studies in classification, data analysis, and knowledge organization* (pp. 381–389). Heidelberg, Berlin: Springer.

Balzanella, A., Romano, E., & Verde, R. (2011). Summarizing and mining streaming data via a functional data approach. *Classification and multivariate analysis for complex data structures* (pp. 409–416). Heidelberg, Berlin: Springer

Gama, J., & Gaber, M. M. (2007). Learning from data stream. *Techniques in sensor networks*. Heidelberg, Berlin: Springer

Guha, S., Koudas, N., & Shim, K. (2001). Data-streams and histograms. In: *33th annual ACM symposium on theory of computing* (pp. 471–475). New York: ACM.

Irpino, A., & Verde, R. (2006). Dynamic clustering of histograms using Wasserstein metric. In A. Rizzi, & M. Vichi (Eds.) *COMPSTAT 2006 - Advances in computational statistics* (pp. 869–876). Heidelberg: Physica-Verlag.

Mallows, C. L. (1972). A note on asymptotic joint normality. *Annals of Mathematical Statistics, 43*(2), 508–515.

Sebastiao, R., & Gama, J. (2007). Change detection in learning histograms from data streams. *Progress in Artificial Intelligence*. Lecture Notes in Computer Science. Springer Berlin Heidelberg. ISBN: 978-3-540-77000-8

Verde, R., & Irpino, A. (2007). Dynamic clustering of histogram data: using the right metric. *Studies in Classification, Data Analysis, and Knowledge Organization*, Part I, 123–134, doi: 10.1007/978-3-540-73560-1 12.

Verde, R., & Irpino, A. (2010). Ordinary least squares for histogram data based on wasserstein distance. In Y. Lechevallier, & G. Saporta (Eds.) *COMPSTAT 2010* (pp. 581588). Berlin: PhysicaVerlag.

Vitter, J. S. (1985). Random sampling with a reservoir. *ACM Transactions on Mathematical Software, 11*(1), 37–57.

Web Panel Representativeness

Annamaria Bianchi and Silvia Biffignandi

Abstract Web panels are becoming more and more popular for data collection. They present specific problems and advantages with respect to the usual modes of collection. This paper analyzes possible re-weighting adjustments for non-response in panel data. Different weighting schemes are evaluated by means of a simulation study based on real data.

1 Introduction

In the last decade there has been a rapid growth of online panels for data collection purposes. Despite this great expansion, web panel representativeness is still an issue, both for volunteer panels and probability based panels. Refer to AAPOR (2010) and references therein for a description of different types of panels and related problems. Understanding how population representative estimates based on internet/web related data collection tools (both on-line panel and web surveys) can be obtained is still an open research issue. New methods and ideas need to be investigated.

When systematic differences in key background variables with respect to a population are present, not weighted results of surveys based on Internet panels are often misleading. Reweighting methods are generally used to reduce selection biases, thus increasing representativeness. The paper is focusing on the maximum entropy weighting technique. A simulation study is performed and weights based on different approaches are derived in a case where population panel and sampling data are available. Using calibration estimators based on the maximum entropy concept, alternative weights are derived; particular attention is paid to whether the use of

A. Bianchi (✉) • S. Biffignandi
DMSIA, University of Bergamo, via dei Caniana 2, 24127 Bergamo, Italy
e-mail: annamaria.bianchi@unibg.it; silvia.biffignandi@unibg.it

P. Giudici et al. (eds.), *Statistical Models for Data Analysis*, Studies in Classification, Data Analysis, and Knowledge Organization, DOI 10.1007/978-3-319-00032-9_5,

updated auxiliary information for late respondents may improve the estimates. A comparison of alternative weighting schemes is carried out in order to obtain some evidence on the impact of different weights on the estimates, especially for online collected data. The comparison is carried out by means of a simulation study based on real data; comparison is extended to Horvitz–Thompson, calibration based on Euclidean distance, and propensity scores estimates. The data used in the simulation are from the LISS panel which is part of the MESS project (Measurement and Experimentation in the Social Sciences) and it is administered by CentERdata (Tilburg University, The Netherlands).

The paper is organized as follows. Section 2 presents the dataset and the target variables. The MaxEnt approach and the proposed weighting schemes are introduced in Section 3. Section 4 provides results, conclusions and ideas for future work.

2 Data Source and Variables

The target population for the LISS panel is the Dutch-speaking population permanently residing in the Netherlands. The sampling and survey units are the independent, private households. The sampling frame is the nationwide address frame of Statistics Netherlands. From this sampling frame a simple random sample was drawn. Hence, this panel should adequately represent the general population. In order to reduce the coverage error due to the Non-Internet Population, households who were not yet online are loaned equipment to provide access to the Internet. Recruitment of the sampled households was done from May until December 2007 and, in order to cover the complete sample, households were approached in a traditional way: first, an announcement letter was sent and next respondents were contacted by an interviewer in a mixed mode design (CATI/CAPI).

In case a household takes part in the LISS panel, one of the household members answers a general questionnaire about some basic demographic characteristics of the household and the household members. As from then all household members are in the full LISS panel. Furthermore, all household members of age 16 and older indicate whether they want to participate in a monthly questionnaire or not. All persons of age 16 and older who have indicated that they want to answer the monthly questionnaires are participating members of the LISS panel. See Scherpenzeel (2009) for more details.

The demographics and other general background information on the households are updated monthly by the household contact person only. On the other hand, every month the participating members fill in a questionnaire. Notice that even if the panel is representative (please refer to Knoef and de Vos 2009 for details concerning representativeness of the LISS panel), if the response to the individual questionnaires differs among various household groups, the research results are not representative for the population. It is therefore necessary to investigate the representativeness of people who answered the questionnaire with respect to both the LISS panel and the target population. In this research we focus only on the representativeness with respect to the LISS panel and we consider results of the questionnaire on "Work and Schooling" (WS), which focuses on labour market participation, job characteristics, pensions, schooling and courses. This questionnaire was fielded in the panel in April 2008 and repeated in July 2008 for non respondents, using a remainder.

In order to evaluate the effects of reweighting corrections for non-response on the precision of estimators we perform a simulation study. As target variable we consider the personal gross monthly income in Euros. This variable is measured at the LISS level. Thus data on the target variable for both respondents and non-respondents to the WS questionnaire are available. The distribution of this variable is strongly skewed. A comparison of the means of this variable for respondents and non-respondents shows that they are significantly different. The mean for non-respondents is much higher than the one for respondents. As pointed out by Pérez-Duarte et al. (2010), this is a typical phenomenon in wealth surveys: unit non-response is unlikely to be a random phenomenon and it is likely to more severely affect wealthy households. As a consequence not weighted estimates are expected to be biased. Note that since general information and demographic data are available for the entire set of households belonging to the LISS panel, reweighting schemes can benefit from the knowledge of these variables collected on the entire household population, which is our target population. In order to choose the auxiliary variables, analyses of the non-response and the correlations between the target variable and the auxiliary variables were performed.

3 Proposed Weighting Approach and Experimental Design

The method employed here is a special case of calibration (Deville and Sarndal 1992) and it is based on the information theory concept of Maximum Entropy (MaxEnt). This weighting procedure allows to use the information that one has about the population of interest's observable covariates without making additional assumptions neither about the distribution of weights nor about the choice of the dissimilarity. A similar proposal is the Empirical Likelihood (EL) method introduced in Chen and Qin (1993) and Chen and Sitter (1999). Proper theoretical and empirical comparison of the proposed method with EL will be included in a subsequent research paper.

Consider now a random sample s of size n drawn from a population U of size N. Denote r the subset of respondents in the sample and y the variable of interest. The objective is to estimate its mean $\mu_y = N^{-1} \sum_{i \in U} y_i$ starting from the observed data $(y_i)_{i \in r}$. Assume that a sampling design is given and let π_i denote the inclusion probabilities. A common estimator for μ_y is the Horvitz–Thompson estimator (Sarndal et al. 1992), which has the good property of being unbiased in case of complete response. In the presence of non-response, this estimator in general turns out to be biased. Re-weighting techniques are then used to reduce this bias.

Suppose that an auxiliary vector variable x is given and its mean μ_x is known at the population level (either exactly or an estimate of it computed on the full sample). This additional information can be exploited by calibration to reduce the non-response bias. Originally calibration was introduced as a method to obtain a set of weights w_i calibrated on known population means of the auxiliary variables ($N^{-1}\sum_i w_i x_i = \mu_x$) with a minimum modification of the survey weights $d_i = 1/\pi_i$ with respect to some distance criterion. Calibration may also be used to adjust for non-response (see Kott 2006; Sarndal and Lundstrom 2005). In this case it is expected to be effective in reducing the non-response bias if the auxiliary variables are related either to the inverse of the response probabilities or to the variable of interest. On the other hand, precision is linked to the auxiliary information: to increase precision, population level information and a good linear relationship with the variable of interest are needed (see Fuller et al. 1994; Kott and Chang 2010).

In this research we focus on a specific kind of calibration technique called MaxEnt, which gives an interesting interpretation to it. In the MaxEnt approach the distance between the two distributions of weights is measured using Shannon's entropy measure. It is known that Shannon's entropy measure is used in information theory to measure the distance between a probability distribution and the uniform distribution (which corresponds to maximum uncertainty). Refer to Kapur and Kesavan (1992) and Jaynes (1957) for details about Shannon's entropy measure. In the following we show how we can match this framework by rescaling the variables of interest. Denote $\tilde{y}_i = nN^{-1}d_i y_i$, $\tilde{x}_i = nN^{-1}d_i x_i$, $p_i = \pi_i w_i n^{-1}$ and $p_i^0 = n^{-1}$. In this notation the Horvitz–Thompson estimator takes the form $\hat{\mu}_y^{HT} = N^{-1}\sum_{i\in r} d_i y_i = \sum_{i\in r} p_i^0 \tilde{y}_i$, i.e., it coincides with the mean of $\tilde{y}_i$ with respect to the uniform distribution on r. We aim at improving this estimator by deriving a probability measure $p = (p_i)_{i\in r}$ on r (and consequently weights $(w_i)_{i\in r}$) which is as close as possible to the uniform distribution $p^0 = \left(p_i^0\right)_{i\in r}$ while also respecting the set of constraints on the auxiliary variable, which can be rewritten as $\sum_{i\in r} p_i \tilde{x}_i = \mu_x$. It is therefore natural to measure the distance between p and p^0 using Shannon's measure of entropy $S(p) = -\sum_{i\in s} p_i \ln p_i$. It is not possible to derive an analytical solution for p_i. A solution must be found using an iterative search algorithm (Mattos and Viega 2004). Once a solution is found, the MaxEnt estimator is defined as

$$\hat{\mu}_y^M = \sum_{i\in r} \hat{p}_i \tilde{y}_i = N^{-1} \sum_{i\in r} w_i y_i,$$

where $w_i = nd_i \hat{p}_i$, $i \in r$.

To take advantage of specific characteristics of panel data, different weighting schemes are constructed. More specifically, these weighting schemes try to take advantage of the fact that demographic data are updated frequently. Indeed, a common situation in panel data is the following. A survey is carried out at time 1 and repeated at time 2 for non respondents, using a remainder. Further, updated background variables are available for the entire panel. Hence two different

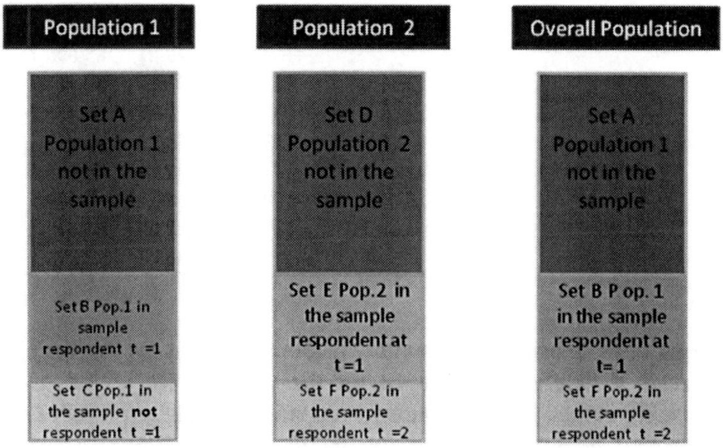

Fig. 1 Design of experiment on weights: population data sets

population databases are available: Population 1, which contains data available at time 1 and Population 2, based on background variables information updated at time 2. The full experimental design takes into account the cases represented in Fig. 1.

In Population 1 database sampled households are considered with the background characteristics detected at time 1 either they were respondent or not. In the Population 2 database—on the contrary—the background characteristics are the ones observed at time 2 either the response has been at time 1 or 2. In addition, a synthetic population is build up under the hypothesis that the population is fully described according to background variables at the time sampled households participate in the survey. As regards non-respondents, it is assumed that the population is that at time 1 (Set A). This population is called the Overall Population. As shown in Fig. 1, this population includes Set B of Population 1 and Set F of Population 2.

4 Results and Concluding Remarks

The performance of the proposed estimators is evaluated using a simulation study. First the WS survey data set is merged to the LISS panel, thus obtaining one data set containing both the sampled and the non-sampled units. Then 1,000 simple random samples of size $n = 3{,}000$ are drawn from the LISS panel. Non-respondents in the WS questionnaire are removed from the samples.

Table 1 Results of the simulation

Estimator	Bias	Variance	MSE	Bias rel
HT	−167.9	57,499.1	85,695.7	−5.981
MaxEnt.Ov	−138.1	61,309.7	80,374.3	−4.918
MaxEnt.1	−138.5	61,284.8	80,456.2	−4.932
MaxEnt.2	−143.0	61,399.0	81,853.3	−5.094
Euclidean.Ov	−137.8	61,171.4	80,156.4	−4.908
Euclidean.1	−138.2	61,144.1	80,239.5	−4.922
Euclidean.2	−142.4	61,273.5	81,548.3	−5.072
PS	−112.3	58,360.4	70,978.1	−4.001

The target variable is the gross monthly income. As far as the auxiliary variables are concerned, their choice is crucial for the performance of the reweighting adjustments. We choose gender, primary occupation and highest level of education (irrespective of diploma) of the household members. Indeed these variables appear to be predictors of the response indicator even though the correlation with the target variable is not high. In order to compute the weights, these variables are turned into a set of dummy variables (one dummy variable for each category). The dummy variables associated with each of these variables are used for the reweighting unless there are too few observations in that cell for the matrix to invert. Moreover an additional constraint is added to ensure that the sum of the final weights equals the number of households members in the LISS panel.

Estimates are based on the final survey data, which are from the Overall Population. The MaxEnt weights are computed using an R program based on the Minxent package (Asma 2012). In order to perform a proper comparison, also the classical calibration estimators based on Euclidean distance (Deville and Sarndal 1992) and the propensity score estimator are included in the simulation. The calibration estimators are computed using the "sampling" R package (Tillé and Matei 2012). The propensity score estimator is computed using weights adjusted for the individual response probabilities $\hat{\mu}_y^{PS} = N^{-1}\sum_{i \in r} d_i y_i/\hat{\theta}_i$, where $\hat{\theta}_i$ is the estimated conditional probability that an individual with given observed characteristics responds to the WS questionnaire. The response probabilities are estimated by means of a logit model (see Bethlehem and Biffignandi 2012 for more details).

The estimators that we compare are: Horwitz–Thompson (HT), MaxEnt based on the Overall Population (MaxEnt.Ov), MaxEnt based on Population 1 (MaxEnt.1), MaxEnt based on Population 2 (MaxEnt.2), calibration estimators based on Euclidean distance (Euclidean.Ov, Euclidean.1, Euclidean.2), and the Propensity score. Table 1 shows the results. The benchmark mean indicator is computed on Population 1.

The Horvitz–Thompson estimator shows a bias which reflects the difference between respondents and non-respondents. On the other hand, the proposed reweighting method has a positive impact on the bias with a small increase of the variance. The relative bias decreases of one percentage point with respect to HT. The mean squared error is smaller than the HT one. With regards to the different weighting schemes, a slightly better performance is shown by MaxEnt.Ov, even though it is not very different from MaxEnt.1, since the number of late respondents is low. The performance of MaxEnt.2 is worse than the other ones. However, we expect this technique to bring greater improvements over traditional estimates when the number of late respondents is higher. The propensity score estimator presents a strong bias reduction and a variance quite similar to the HT one. In this simulation the PS method seems to be the best solution. The results obtained with the MaxEnt estimators are not very different from those obtained with the calibration estimator based on the Euclidean distance. Leading to almost equal results, the MaxEnt approach presents the desirable characteristic of not requiring additional assumptions neither about the distribution of weights nor about the choice of the dissimilarity. Further insights about combining PS and MaxEnt approaches will be the subject of a forthcoming research paper.

The main conclusions are:

(a) Compared to Horvitz–Thompson, MaxEnt approach performs better in general and it is similar to the calibration estimator based on the Euclidean distance. In this case the PS approach performs even better. However, it should be advised that all the proposed methods are very sensitive to the choice of the auxiliary variables. The crucial point in empirical analyses is that no clear relationship exists between auxiliary variables models driving their inclusion in the estimation process and estimation performance.
(b) Online panels allow to update auxiliary information and to build up a synthetic population containing updated information for late respondents (the Overall Population). The empirical simulation presented in this paper shows that the use of updated socio-demographic information for late respondents could be a valuable approach for improving estimates. Greater improvements are expected depending on the portion of late respondents.

Acknowledgments The paper is supported by the ex 60 % University of Bergamo, Biffignandi grant and PAADEL project (Lombardy Region—University of Bergamo joint project). The authors are thankful to the anonymous referee for valuable comments and remarks. Paper within Cost Action is 1004 activities.

References

AAPOR. (2010). Research synthesis: AAPOR report on online panels. *Public Opinion Quarterly, 74*, 711–781. doi: 10.1093/poq/nfq048.

Asma, S. (2012). *Minxent: Entropy optimization distributions*. R package version 0.01. http://CRAN.R-project.org/package=minxent

Bethlehem, J., & Biffignandi, S. (2012). *Handbook of web surveys*. Hoboken: Wiley.

Chen, J., & Qin, J. (1993). Empirical likelihood estimation for finite population and the effective usage of auxiliary information. *Biometrika, 80*, 107–116.

Chen, J., & Sitter, R. R. (1999). A pseudo empirical likelihood approach to the effective use of auxiliary information in complex surveys. *Statistica Sinica, 9*, 385–406.

Deville, J. C., & Sarndal, C. E. (1992). Calibration estimators in survey sampling. *Journal of the American Statistical Association, 87*, 376–382.

Fuller, W. A., Loughin, M. M., & Baker, H. D. (1994). Regression weighting in the presence of nonresponse with application to the 1987–1988 Nationwide Food Consumption Survey. *Survey Methodology, 20*, 75–85.

Jaynes, E. T. (1957). Information theory and statistical mechanics. *Physical Review, 106*, 620–630.

Kapur, J. N., & Kesavan, H. K. (1992). *Entropy optimization principles with applications*. Boston: Academic.

Knoef, M., & de Vos, K. (2009). *The representativeness of LISS, an online probability panel*. http://www.lissdata.nl/lissdata/About_the_Panel/Composition_&_Response. https://www.lisspanel.nl/assets/uploaded/representativeness_LISS_panel.pdf

Kott, P. S. (2006). Using calibration to adjust for nonresponse and coverage errors. *Survey Methodology, 32*, 133–142.

Kott, P. S., & Chang, T. (2010). Using calibration weighting to adjust for nonignorable unit nonresponse. *Journal of the American Statistical Association, 105*, 1265–1275.

Mattos, R., & Viega, A. (2004). Entropy optimization: Computer implementation of the maxent and mixent principles, Working Paper.

Pérez-Duarte, S., Sanchez-Munos, C., & Tormalehto, V. M. (2010). *Re-weighting to reduce unit non-response bias in household wealth surveys: a cross-country comparative perspective illustrated by a case study*. http://www.ecb.int/home/pdf/research/hfcn/WealthSurveys.pdf?ee6a2f5e725b2ffcb7810e151fb7e304.

Sarndal, C. E., & Lundstrom, S. (2005). *Estimation in surveys with non response*. New York: Wiley.

Sarndal, C. E., Swensson, B., & Wretman, J. (1992). *Model assisted survey sampling*. New York: Springer.

Scherpenzeel, A. (2009). *Start of the LISS panel: Sample and recruitment of a probability-based Internet panel*. http://www.lissdata.nl/lissdata/About_the_Panel/Sample_&_Recruitment. http://www.lissdata.nl/assets/uploaded/Sample%20and%20Recruitment_1.pdf

Tillé, Y., & Matei, A. (2012). *Sampling: Survey sampling*. R package version 2.5. http://CRAN.R-project.org/package=sampling

Nonparametric Multivariate Inference Via Permutation Tests for CUB Models

Stefano Bonnini, Luigi Salmaso, and Francesca Solmi

Abstract A new approach for modelling discrete choices in rating or ranking problems is represented by a class of mixture models with covariates (Combination of Uniform and shifted Binomial distributions, CUB models), proposed by Piccolo (2003, *Quaderni di Statistica, 5*, 85–104), D'Elia & Piccolo (2005, *Computational Statistics & Data Analysis, 49*, 917–934), Piccolo (2006, *Quaderni di Statistica, 8*, 33–78) and Iannario (2010, *Metron, LXVIII*, 87–94). In case of a univariate response, a permutation solution to test for covariates effects has been discussed in Bonnini et al. (2012, *Communication in Statistics: Theory and Methods*), together with parametric inference. We propose an extension of this nonparametric test to deal with the multivariate case. The good performances of the method are showed trough a simulation study and the procedure is applied to real data regarding the evaluation of the Ski School of Sesto Pusteria (Italy).

1 Introduction

Usually in real applications the ordinal response in a rating or ranking problem depends on specific subjects' or objects' characteristics.

S. Bonnini (✉)
Economics and Management, University of Ferrara, Via Voltapaletto, 11-44121 Ferrara, Italy
e-mail: stefano.bonnini@unife.it

L. Salmaso
Department of Management and Engineering, University of Padova,
Stradella S. Nicola 3-36100 Vicenza, Italy
e-mail: luigi.salmaso@unipd.it

F. Solmi
Department of Statistical Sciences, University of Padova, Via C. Battisti, 241-35121 Padova, Italy
e-mail: solmi@stat.unipd.it

P. Giudici et al. (eds.), *Statistical Models for Data Analysis*, Studies in Classification, Data Analysis, and Knowledge Organization, DOI 10.1007/978-3-319-00032-9_6,

For defining the probability distribution of this ordinal response a new approach is represented by Piccolo (2003), D'Elia and Piccolo (2005), Piccolo (2006) and Iannario (2010) and generalized by Piccolo and D'Elia (2008), Iannario and Piccolo (2009) and Iannario and Piccolo (2012).

According to this approach, data are generated by a class of discrete probability distributions, that depend on two intrinsically continuous quantities (feeling and uncertainty) pertaining to the response. This is a mixture of a *shifted Binomial* and an *Uniform* random variable. Let us assume that n evaluators are rating a given item, hence the sample $\boldsymbol{y} = (y_1, y_2, \ldots, y_n)'$ is observed; moreover let $\boldsymbol{x}_i$ and $\boldsymbol{w}_i$, with $i = 1, \ldots, n$, be subjects' covariates for explaining feeling and uncertainty, respectively. Hence, the general formulation of a CUB (p, q) model (with p covariates to explain uncertainty and q covariates to explain feeling), is expressed by:

$$Pr(Y_i = y|\boldsymbol{x}_i, \boldsymbol{w}_i) = \pi_i \binom{m-1}{y-1}(1-\xi_i)^{y-1}\xi_i^{m-y} + (1-\pi_i)\left(\frac{1}{m}\right)$$

with $y = 1, 2, \ldots, m$, where $m > 3$ (Iannario and Piccolo 2012) is the known number of modalities for the rating survey, and $\pi_i = 1/(1 + e^{-\boldsymbol{x}_i \boldsymbol{\beta}})$ and $\xi_i = 1/(1 + e^{-\boldsymbol{w}_i \boldsymbol{\gamma}})$ where $\boldsymbol{\psi} = (\boldsymbol{\beta}', \boldsymbol{\gamma}')'$ is the vector of parameters associated to the covariates.

Inference on CUB models has been mainly developed in a parametric framework, via maximum likelihood and asymptotic theory (see Piccolo 2006 and Iannario and Piccolo 2009). A nonparametric solution to test for the effect of covariates in a CUB model has been recently proposed by Bonnini et al. (2012). It is a competitive alternative to the classical parametric test when the sample size is low.

In this work we extend the above presented permutation solution to the multivariate case. Nonparametric combination of dependent permutation tests (see Pesarin 2001 and Pesarin and Salmaso 2010) is used to end up with a global tool for comparing several nested CUB models at the same time on several aspects of the multivariate response. A simulation study is carried out in order to explore the performances of the multivariate test. The method is applied to real data regarding the evaluation of the Ski School of Sesto Pusteria in the Trentino Alto Adige region (Italy).

2 A Permutation Test for Multivariate Responses

Several types of permutation tests have been proposed in order to compare regression models (see Anderson and Ter Braak 2003 for a review). Permutation strategies can be divided in permutation of raw data and permutation of residuals. We propose a nonparametric solution for the test of significance for the coefficients of the covariates of the CUB model when a multivariate response is observed. Such solution performs permutations of raw data. In Bonnini et al. (2012) a nonparametric test is based on the constrained permutation of the tested covariates. Let us consider a test on the global influence of one or more covariates on a multivariate response. For each partial test (on the influence of the tested covariates on a single component of the multivariate response) the same set of covariates is taken into account. Hence the global alternative hypothesis is true when at least one of the partial null hypotheses is false.

Let $\mathbf{Y}' = (Y_1, \ldots, Y_c)$ be the multivariate response, and let a set of subjects' covariates $\left(\mathbf{x}', \mathbf{w}'\right) = \left(x_1, \ldots, x_p, w_1, \ldots, w_q\right)$ be observed on a set of n respondents. Formally the proposed solution works as follows:

i. Set the null and the alternative models that need to be compared (say $\mathscr{M}^{j,0}$ and $\mathscr{M}^{j,1}$ respectively), treating separately all the components of the multivariate response Y_j, $j = 1, \ldots, c$.
ii. For the j-th component of the multivariate response, $j = 1, \ldots, c$, consider the observed data $(y_i^j, \mathbf{x}_i', \mathbf{w}_i')$, with $i = 1, \ldots, n$ and perform one of the permutation tests proposed in Bonnini et al. (2012) to compare $\mathscr{M}^{j,0}$ and $\mathscr{M}^{j,1}$ (t^j say). In order to maintain the dependence due to the fact that for each test the responses come from the same n subjects, synchronized permutations have to be performed on the several tests.
iii. Consider the c separated tests t^j, $j = 1, \ldots, c$ and combine them into the global test t to test the global null hypothesis of interest, using the nonparametric combination of dependent permutation tests.
iv. If the global test t is significant, adjust the partial tests t^j, $j = 1, \ldots, c$, for multiplicity, using a closed testing nonparametric procedure (see Pesarin 2001 and Pesarin and Salmaso 2010), obtaining the adjusted p-values p_{adj}^j, $j = 1, \ldots, c$, and conclude that the tested covariates influence the response on those aspects where the adjusted p-values p_{adj}^j are significant. If the global test t is not significant, conclude that the tested covariates do not influence the multivariate response.

We remark that the permutations tests t^j can be chosen among the ones proposed for the univariate case (see Bonnini et al. 2012). Moreover measures of the significance of specific domain related models can also be tested.

3 Simulation Study

A Monte Carlo simulation study has been carried out in order to evaluate the performances of the proposed multivariate method.

As a multivariate version of the CUB model has not yet been defined, a first problem consists in how to simulate the data. Data were simulated assuming that the marginal components follow a CUB distribution. Moreover, to take into account for the dependence among the single components of the multivariate response the copula theory was used (see Nelsen 2006). The basic idea is to apply the probability integral transform to the single components and then specify the dependence among the resulting uniform random variables, instead of among the original ones. Copulas' theory does not share the same results for continuous and discrete data, and in particular identifiability issues arise in the case of discrete data. However copulas' models for discrete data keep on being valid constructions and, as suggested by Genest and Neslehová (2007), they are helpful in the context of simulation. The identifiability problem, indeed, concerns the estimation field and not the simulation one.

Hence the values of the cumulative distribution functions (c.d.f.s) were simulated from a copula and the response values were obtained inverting the c.d.f.s for a CUB model. Consider the general case of a c-dimensional response. The simulating procedure works according to the following two steps:

1. Simulate a sample from a multivariate copula, according to a pre-specified dependence degree, getting $\mathbf{u_i} = \left(u_i^1, \ldots, u_i^c\right)$, with $i = 1, \ldots, n$.
2. Consider each multivariate element of the simulated sample, $\mathbf{u_i}$, and transform it into the final element through the inverse c.d.f. of a CUB model, i.e. $\mathbf{y_i} = \left(y_i^1, \ldots, y_i^c\right) = \left(F_1^{-1}(u_i^1), \ldots, F_c^{-1}(u_i^c)\right)$.

We remark that at any combination (i, j), with $i = 1, \ldots, n$ and $j = 1, \ldots, c$, inverting the c.d.f. $F_j(y)$, a specific CUB model has to be considered according to the parameters values $\boldsymbol{\beta}^j$ and $\boldsymbol{\gamma}^j$ related to that component of the multivariate response and the values of the covariates. In general the c.d.f. of a CUB model, for the i-th subject, with uncertainty and feeling parameters $\pi_i^j = 1/\left(1 + e^{-\boldsymbol{x}_i \boldsymbol{\beta}^j}\right)$ and $\xi_i^j = 1/\left(1 + e^{-\boldsymbol{w}_i \boldsymbol{\gamma}^j}\right)$ can be derived as follows:

$$F_j(y) = \Pr(Y \leq y | \boldsymbol{x}_i, \boldsymbol{w}_i) = \sum_{l=1}^{y} \left[\pi_i^j \binom{m-1}{l-1} (1 - \xi_i^j)^{l-1} {\xi_i^j}^{m-l} + (1 - \pi_i^j) \left(\frac{1}{m}\right) \right]$$

$$= \pi_i^j \sum_{l=0}^{y-1} \left[\binom{m-1}{l} (1 - \xi_i^j)^l {\xi_i^j}^{((m-1)-l)} \right] + (1 - \pi_i^j) \left(\frac{y}{m}\right).$$

$F_j(y)$ corresponds to a mixture of two discrete distributions: a binomial $Bi(m-1, (1-\xi_i^j))$ calculated in $y - 1$ and a uniform distribution. The inversion of such function can be obtained using quantile functions of the c.d.f.s and the equation previously defined.

An Archimedean copula was used (see Nelsen 2006, p. 116), mainly because it allows to model the dependence with only one parameter (hereafter θ). The use of this copula family with discrete data can be found in Pfeifer and Neslehová (2004).

In the discrete case the copula alone cannot characterize the dependence between the several components of the multivariate response (see Genest and Neslehová 2007 for a more detailed discussion). Anyway one helpful property holds in the discrete case: the value of the parameter θ of the copula, increases with the dependence among the final discrete responses. In this simulation study the settings are similar to those used in Bonnini et al. (2012). Working in terms of rejection rates, the reliability of the method under the null hypothesis has been verified, and its power under alternative hypotheses has been studied. Only two test statistics for the permutation test were taken into account (hereafter t_{lrt}, t_{wald}, see Bonnini et al. 2012 for a description of the different tests), and a parametric counterpart is not available for the multivariate case. Such permutation tests are constructed considering the permutation distribution of the likelihood ratio and Wald type tests. Only the CUB (0, 1) model was studied under the alternative hypothesis, since this model is much used in real applications. For the same reason only dichotomous covariates were considered (see Bonnini et al. 2012 for a detailed discussion).

As regards the simulation settings, the relation between the power and the "distance" between the two populations defined by the dichotomous covariate under H_1 has already been studied (see Bonnini et al. 2012). Such a distance could be represented by the difference between the parameters of feeling of the two populations (hereafter δ_ξ). Hence only one value of δ_ξ was taken into account. We considered different numbers of components of the multivariate response simulated under the partial alternative hypothesis.

Moreover several values for the dependence parameter θ were considered in order to study the behavior of the power functions for different degrees of dependence among the components of the multivariate response. The dependence among the univariate responses does not depend on the underlying copula alone and it is instead influenced by the marginal distributions as well. Therefore in our particular case some dependence is surely attributable to the covariate when we simulate all the components under the alternative hypothesis, i.e. by following a CUB distribution with the same significantly influent covariate. Hence, even if it is not possible to say which kind of dependence is represented by θ, high values of θ (letting fixed all other parameters in the simulation setting) correspond to high dependence among the univariate responses, no matter which is the impact, on the global dependence, from the introduction of a significant covariate.

Table 1 shows the considered simulation settings for $c = 2, 3$ dimensions of the multivariate response: the table must be read in terms of number of components simulated under the alternative hypothesis. Settings from 1 to 3 refer to a bivariate simulated response, while settings from 4 to 7 to the case $c = 3$. Hence in the first and in the fourth settings data are simulated under the global null hypothesis that the covariate does not influence any of the components. The value $\delta_\xi = 0.2$ was chosen under H_1. Two values for the sample size ($n = 50, 100$) and three values for the dependence parameter ($\theta = 0, 5, 10$) were considered. The feeling parameter was set to $\xi = 0.1$ for the components simulated under the null hypothesis (CUB $(0, 0)$ model), and to $\xi_{(0)} = 0.1$, $\xi_{(1)} = 0.3$, in the two sub-groups identified by the covariate, for the components simulated under the alternative hypothesis (CUB $(0, 1)$ model). The uncertainty parameter was set always equal to $\pi = 0.9$ (low uncertainty, very frequent in real applications).

Table 2 reports the estimated rejection probabilities of the compared tests (partial adjusted and global permutation tests) on the parameter γ_1 for $m = 7$ and at a

Table 1 Simulation settings for $c = 2, 3$, each cell indicating under which hypothesis the specific component is simulated for the specific setting. The symbol − indicates that the component is not considered in that setting

Setting	Y_1	Y_2	Y_3
1	H_0	H_0	−
2	H_1	H_0	−
3	H_1	H_1	−
4	H_0	H_0	H_0
5	H_1	H_0	H_0
6	H_1	H_1	H_0
7	H_1	H_1	H_1

Table 2 Estimated rejection probabilities for the partial adjusted permutation tests on the single component of the multivariate response ($t_{\text{lrt}}^{Y_c}$ and $t_{\text{wald}}^{Y_c}$, with $c = 1, 2, 3$) and of the global solution ($t_{\text{lrt}}^{\text{glob}}$ and $t_{\text{wald}}^{\text{glob}}$), settings 1 to 7 and $\theta = 0, 5, 10$. The results refer to sample size $n = 50$, nominal level of $\alpha = 0.05$, $B = 1,000$ permutations and $CMC = 1,000$ replications. Estimates in bold indicate quantities under the (partial or global) null hypotheses

	$\theta = 0$				$\theta = 5$				$\theta = 10$			
	t_{lrt}		t_{wald}		t_{lrt}		t_{wald}		t_{lrt}		t_{wald}	
Setting	$t_{\text{lrt}}^{Y_{1,2,3}}$	$t_{\text{lrt}}^{\text{glob}}$	$t_{\text{wald}}^{Y_{1,2,3}}$	$t_{\text{wald}}^{\text{glob}}$	$t_{\text{lrt}}^{Y_{1,2,3}}$	$t_{\text{lrt}}^{\text{glob}}$	$t_{\text{wald}}^{Y_{1,2,3}}$	$t_{\text{wald}}^{\text{glob}}$	$t_{\text{lrt}}^{Y_{1,2,3}}$	$t_{\text{lrt}}^{\text{glob}}$	$t_{\text{wald}}^{Y_{1,2,3}}$	$t_{\text{wald}}^{\text{glob}}$
1	**0.032**	**0.057**	**0.027**	**0.058**	**0.035**	**0.057**	**0.035**	**0.056**	**0.029**	**0.048**	**0.031**	**0.048**
	0.031		**0.031**		**0.032**		**0.029**		**0.030**		**0.035**	
2	0.923	0.923	0.912	0.923	0.918	0.921	0.916	0.921	0.914	0.919	0.905	0.917
	0.052		**0.052**		**0.050**		**0.025**		**0.046**		**0.050**	
3	0.949	0.995	0.946	0.995	0.947	0.984	0.936	0.985	0.941	0.973	0.934	0.974
	0.956		0.948		0.947		0.936		0.944		0.938	
4	**0.024**	**0.056**	**0.027**	**0.060**	**0.016**	**0.050**	**0.016**	**0.053**	**0.019**	**0.057**	**0.019**	**0.057**
	0.015		**0.015**		**0.022**		**0.014**		**0.025**		**0.023**	
	0.020		**0.019**		**0.021**		**0.023**		**0.031**		**0.027**	
5	0.905	0.907	0.889	0.910	0.918	0.923	0.907	0.924	0.910	0.918	0.887	0.917
	0.026		**0.026**		**0.030**		**0.025**		**0.032**		**0.026**	
	0.025		**0.024**		**0.025**		**0.026**		**0.044**		**0.038**	
6	0.923	0.994	0.913	0.994	0.935	0.980	0.929	0.980	0.928	0.978	0.917	0.978
	0.915		0.903		0.923		0.916		0.924		0.903	
	0.041		**0.045**		**0.040**		**0.045**		**0.056**		**0.050**	
7	0.948	0.999	0.942	0.999	0.956	0.991	0.949	0.991	0.940	0.980	0.930	0.979
	0.939		0.926		0.953		0.937		0.945		0.933	
	0.950		0.948		0.955		0.950		0.933		0.929	

nominal level of $\alpha = 0.05$ for $n = 50$. A number of $B = 1,000$ permutations and $CMC = 1,000$ conditional Monte Carlo iterations have been considered. The obtained results show how the global permutation test controls the type I error when the global null hypothesis is true. Such test also turns out to be a powerful solution as soon as one of the partial null hypothesis is not true. Therefore a power increase can also be registered while increasing the number of false partial null hypotheses (hence passing from setting 2 to 3 and from setting 5 to 7). It also has to be underlined that when more than one partial null hypothesis is false (hence in settings 3, 6 and 7), the power decreases as the copula's dependence parameter θ increases, again confirming an expected behavior (see Pesarin and Salmaso 2010). In the end the results related to higher sample size ($n = 100$, which are available under request) suggest that the power of the global test increases as the sample size increases, for all the considered settings, reaching the value one.

4 Real Case Application

The S.E.S.T.O. (Statistical Evaluation of a Skischool from Tourists' Opinions) is the first Italian survey on the evaluation (by parents) of ski courses for young children (up to 14 years old) and it is a pilot study performed in the Ski School of Sesto, in the Dolomites near Bolzano in the North of Italy. Several customer satisfaction variables towards different aspects of ski teaching have been evaluated on a rating scale 1–10. A multivariate response has been considered in the study, related to five aspects of the customer satisfaction: "Easy Learning", "Helpful Teacher", "Fun", "Involvement" and "General Satisfaction". Moreover the dichotomous covariate "First presence in Sesto" for parameter ξ has been included in the analysis, to verify if the families who were in Sesto for the first time presented a different feeling toward the ski courses than the others.

The global p-value lower than 0.001 leads to the rejection of the global null hypothesis at $\alpha = 0.05$ hence the tested covariate affects the feeling. According to the adjusted p-values of the partial tests, to be in Sesto for the first time has no influence on the easy of learning but it positively affects the feeling of the respondents toward the helpfulness of the teacher (adjusted p-value equal to 0.009), the fun and the involvement of the children (adjusted p-value equal to 0.013 and 0.015 respectively) and also the general satisfaction (adjusted p-value lower than 0.001).

5 Conclusions

In this paper an extension of a permutation solution to test for covariate influence on an ordinal response, working within the CUB model framework, is presented. The method basically works implementing the permutation solution proposed in Bonnini et al. (2012) separately on each component of the multivariate response, anyway taking into account for the dependence among variables performing synchronized permutations on the several components.

The method's performances have been studied through a simulation study were the cases $C = 2$ and $C = 3$ dimensions of the multivariate response have been considered. Several settings have been explored, which differ from each other in terms of number of partial components under the alternative hypothesis. The results have shown the very good behavior of the global permutation solution, which is reliable under the global null hypothesis and powerful under the alternative even for low sample sizes. Its power increases (reaching value one) as the sample size increases and it is a decreasing function of the dependence among the components of the multivariate outcome. Moreover a power increase can be observed while increasing the number of false partial null hypothesis.

The permutation test has also been applied to real data regarding the evaluation of the Ski School of Sesto Pusteria in the Trentino Alto Adige region (Italy). The influence of the covariate "First presence in Sesto" on several responses is suggested by the use of the method.

We can conclude that the proposed permutation solution is useful in order to test for the influence of one covariate on a multivariate ordinal response, while working in the CUB models framework. Other parametric solutions do not exist at the moment to solve the multivariate aspect of such a problem.

Acknowledgments Authors wish to thank the University of Padova (CPDA092350/09) and the Italian Ministry for University and Research MIUR project PRIN2008 -CUP number C91J10000000001 (2008WKHJPK/002) for providing the financial support for this research.

References

Anderson, M., & Ter Braak, C. (2003). Permutation tests for multi-factorial analysis of variance. *Journal of Statistical Computation and Simulation, 73*, 79–86.

Bonnini, S., Picccolo, D., Salmaso, L., & Solmi, F. (2012). Permutation inference for a class of mixture models. *Communication in Statistics: Theory and Methods, 41*, 16–17, 2879–2895.

D'Elia, A., & Piccolo, D. (2005). A mixture model for preference data analysis. *Computational Statistics & Data Analysis, 49*, 917–934.

Genest, C., & Neslehová, J. (2007), A primer on copulas for count data. *ASTIN Bulletin, 37*, 475–515.

Iannario, M. (2010). On the identifiability of a mixture model for ordinal data. *Metron, LXVIII*, 87–94.

Iannario, M., & Piccolo, D. (2009). A program in R for *CUB* models inference, Version 2.0, available at http://www.dipstat.unina.it/CUBmodels1/.

Iannario, M., & Piccolo, D. (2012). CUB models: statistical methods and empirical evidence. In R. S. Kenett, & S. Salini (Eds.) *Modern analysis of customer surveys: with applications using R* (pp. 231–258). Chichester: Wiley.

Nelsen, R. B. (2006). *An introduction to copulas* (2nd edn). New York: Springer.

Pesarin, F. (2001). *Multivariate permutation tests: with applications in biostatistics*. Chichester: Wiley.

Pesarin, F., & Salmaso, L. (2010). *Permutation tests for complex data: theory, applications and software*. Chichester: Wiley.

Pfeifer, D., & Neslehová, J. (2004). Modeling and generating dependent risk processes for IRM and DFA. *ASTIN Bulletin, 34*, 333–360.

Piccolo, D. (2003). On the moments of a mixture of uniform and shifted binomial random variables. *Quaderni di Statistica, 5*, 85–104.

Piccolo, D. (2006). Observed information matrix for mub models. *Quaderni di Statistica, 8*, 33–78.

Piccolo, D., & D'Elia, A. (2008). A new approach for modelling consumers' preferences. *Food Quality and Preference, 19*, 247–259.

Asymmetric Multidimensional Scaling Models for Seriation

Giuseppe Bove

Abstract Singular value decomposition (SVD) of skew-symmetric matrices was proposed to represent asymmetry of proximity data. Some authors considered the plane (*bimension* or *hedron*) determined by the first two singular vectors to detect orderings (*seriation*) for preference or dominance data. Following these approaches, in this paper some procedures of asymmetric multidimensional scaling useful for seriation are proposed focalizing on a model that is a particular case of *rank-2* SVD model. An application to Thurstone's paired comparison data on the relative seriousness of crime is also presented.

1 Introduction

Gower (1977) and Constantine and Gower (1978) remark that the decomposition of an asymmetric proximity matrix **P** in a symmetric part **M** and a skew-symmetric part **N**, with $\mathbf{P} = \mathbf{M} + \mathbf{N}$, gives an orthogonal breakdown of sum of squares (see also Critchley 1988; Zielman and Heiser 1996):

$$\|\mathbf{P}\|^2 = \|\mathbf{M}\|^2 + \|\mathbf{N}\|^2$$

which suggests a separate analysis (or *modelling*) of **M** and **N**.

- **M** can be analysed by symmetric MDS.
- **N** can be analysed by *Singular Value Decomposition*.

G. Bove (✉)
Dipartimento di Scienze dell'Educazione, Universitá degli Studi Roma Tre,
Via Milazzo 11 B, 00185 Rome, Italy
e-mail: bove@uniroma3.it

P. Giudici et al. (eds.), *Statistical Models for Data Analysis*, Studies in Classification, Data Analysis, and Knowledge Organization, DOI 10.1007/978-3-319-00032-9_7,

Analysis of the skew-symmetric component seems promising in the context of preference or dominance data (e.g. Freeman 1997; Brusco and Stahl 2005), where it is usually relevant to find an ordering of preference (or *seriation*) of the objects compared. *Rank-2* approximations provided by the SVD of **N** allow us to depict in a diagram orderings of objects useful to support seriation analyses. In Sect. 2 aspects of the seriation problem and the type of data considered will be recalled, in Sect. 3 a *strategy of analysis* based on SVD of skew-symmetry will be proposed, and finally in Sect. 4 main results obtained by application of the strategy to Thurstone's paired comparison data on the relative seriousness of crimes are presented.

2 Type of Data

One of the main goal in the analysis of a *dominance/preference* $n \times n$ data matrix is to find an *ordering of dominance/preference* (or *seriation*) for the n objects (e.g. persons, political parties, teams, crimes, etc.). Many authors considered seriation essentially a combinatorial problem (see e.g. Hubert and Golledge 1981; Hubert et al. 2001; Brusco and Stahl 2005). In this case a *measure of adequacy* (i.e., the value for some *objective function*) is assigned to each of the reorganizations of rows and columns of matrix corresponding to the $n!$ permutations of the set. Then the permutations that maximize (or minimize) the measure of adequacy are located.

One natural measure in this context is the *sum of the entries above the main diagonal* of matrix, but many other measures can be considered.

The matrix in Table 1 represents dominance relationships among five players, for any pair (i,j), the corresponding entry is the number of games player i beats player j.

The maximum of the sum of the entries above the main diagonal (the optimal value of the objective function) can be obtained by reorganizing rows and columns according to the permutation (4,3,1,2,5), as in Table 2.

So the permutation (4,3,1,2,5) provides the optimal dominance ordering for the five players in this seriation problem, from the strongest to the weakest player. However, computational requirements of complete enumeration ($n!$ permutations) become excessive quickly as n increases, and can be excessive even when n is not very large.

In the following we will restrict attention to the type of $n \times n$ data matrix **P**, whose main diagonal elements are zeros and all other elements satisfy $p_{ij} + p_{ji} = k$ for $i \neq j (k \geq 0)$. We note that for this type of matrix, the symmetric component (with entries $\frac{1}{2}(p_{ij} + p_{ji})$) has no practical interest because all off-diagonal elements are equal to $\frac{k}{2}$. For $k = 0$, $\mathbf{P} = \mathbf{N}$ is skew-symmetric ($p_{ij} = -p_{ji}$).

As an example, a graded paired-comparison matrix on seriousness of 15 crimes is partially reproduced in Table 3 from a larger study of Thurstone (1927). These data will be analysed more in detail in Sect. 4.

Table 1 Dominance relationships among five players

Players	1	2	3	4	5
1	–	2	0	0	5
2	0	–	0	0	1
3	3	6	–	0	8
4	7	9	4	–	10
5	0	0	0	0	–

Table 2 Matrix in Table 1 reorganized

Players	4	3	1	2	5
4	–	4	7	9	10
3	0	–	3	6	8
1	0	0	–	2	5
2	0	0	0	–	1
5	0	0	0	0	–

Table 3 Thurstone's paired comparison data partially reproduced from Hubert and Golledge (1981)

Crimes	1.	2.	3.	4.
1. Arson	–	0.348	0.563	0.716
2. Embezzlement	0.652	–	0.752	0.774
3. Kidnapping	0.437	0.248	–	0.595
4. Seduction	0.284	0.226	0.405	–

For any pair (i,j), p_{ij} is the proportion of subjects who judged crime in column j is more serious than crime in row i (so that: $p_{ij} + p_{ji} = 1$, $(i \neq j)$).

We remark that any skew-symmetric matrix contains two types of information: size (or magnitude) and sign (or directionality) of skew-symmetry (e.g. Hubert et al. (2001) and Brusco and Stahl (2005) in combinatorial optimization; Bove (1989) in asymmetric multidimensional scaling). Thus, matrix **N** can be written

$$\mathbf{N} = \mathbf{T} \circ \boldsymbol{\Gamma}$$

where: $\mathbf{T} = \{t_{ij}\} = \{|n_{ij}|\}$, $\boldsymbol{\Gamma} = \{\gamma_{ij}\} = \{sign(n_{ij})\}$ with $sign(n_{ij}) = 1$ if $n_{ij} > 0$, $sign(n_{ij}) = -1$ if $n_{ij} < 0$ and $sign(n_{ij}) = 0$ if $n_{ij} = 0$, and $\circ$ is the Hadamard product. So, in the case of graded paired comparison data, orderings of the rows and columns of **N** can be formulated on the basis of both **T** and $\boldsymbol{\Gamma}$.

3 Two Complementary Approaches to the Seriation Problem

Brusco and Stahl (2005) proposed a bicriterion dynamic programming method based on three steps.

STEP 1 Choose an objective function for matrix $\boldsymbol{\Gamma}$ (e.g. *sum of the entries above the main diagonal* or similar functions).

STEP 2 Choose an objective function for matrix $\mathbf{T}$ (e.g. $L = \sum_{i=1}^{n-1} \sum_{j=i+1}^{n} (t_{ij} - |x_i - x_j|)^2$, where coordinates x_i, x_j are opportunely obtained from a permutation of the objects).

STEP 3 Look for a permutation of the n objects optimizing one (or more) opportune linear combination of the previous objective functions.

As a result an ordered list (or more ordered lists) of the n objects is obtained, sometimes not easy to interpret (for an application of this method to Thurstone's paired comparison data see Brusco and Stahl (2005, p. 338).

Now we show how data visualization based on SVD of skew-symmetry can help to support or find solutions to the seriation problem. First, we remark that the singular value decomposition of an $n \times n$ skew-symmetric matrix $\mathbf{N}$ is

$$\mathbf{N} = \mathbf{U}\boldsymbol{\Delta}\mathbf{V}' = \mathbf{U}\boldsymbol{\Delta}\mathbf{J}\mathbf{U}' \tag{1}$$

where $\Delta = diag(\delta_1, \delta_1, \delta_2, \delta_2, \ldots)$ and $\mathbf{J}$ is a block diagonal matrix with 2×2 matrices

$$\begin{bmatrix} 0 & 1 \\ -1 & 0 \end{bmatrix}$$

along the diagonal and, if n is odd, the last diagonal element is conventionally set to one. *Rank-2* approximation obtained by (1) is, in scalar form,

$$n_{ij} = (u_i v_j - u_j v_i) + e_{ij} \tag{2}$$

by which diagrams (usually named *Gower diagrams*) in a plane (*bimension* or *hedron*) are obtained. Many authors (e.g. Constantine and Gower 1978; Okada and Imaizumi 1987; Bove and Critchley 1993; Saito and Yadohisa 2005) studied the following linear particular case of model (2), that we can call *radius model*,

$$n_{ij} = (r_i - r_j) + e_{ij} \tag{3}$$

that is the simplest form of skew-symmetry. A least squares solution for the r_i' s in (3) was provided by Mosteller (1951) in the context of Thurstone case V scaling. Bove and Critchley (1989) provided the solution for the weighted least square problem.

Now we propose a *strategy of analysis* in three steps, to show how models (2) and (3) can be applied in different ways to matrices $\mathbf{T}$ and $\boldsymbol{\Gamma}$ to support the definition of an ordering of the n objects compared in the data matrices.

STEP 1 Find a unidimensional scaling of the n objects by applying to $\boldsymbol{\Gamma}$ the radius model (3) and visualize the ordering in a Gower diagram by model (2).

STEP 2 Display entries of matrix **T** with distances by symmetric Multidimensional Scaling.

STEP 3 Join the graphical analyses of matrix $\boldsymbol{\Gamma}$ and matrix **T** drawing circles around the points in the diagram obtained in STEP 2. Estimates of the radii of the circles can be based on the estimates of the radius model parameters (the row means of matrix $\boldsymbol{\Gamma}$), made non negative by adding an opportune constant (so that $\min(\hat{r}_i) = 0$) and normalized in order to be comparable with distances obtained in STEP 2.

4 Application

Hubert and Golledge (1981) analysed Thurstone's paired comparison data concerning the perceived severity of $n = 15$ criminal offences, reported in a 15×15 asymmetric proximity matrix. The 15 crimes are: Abortion (ABO), Adultery (ADU), Arson (ARS), Assault and Battery (ASB), Burglary (BUR), Counterfeiting (COU), Embezzlement (EMB), Forgery (FOR), Homicide (HOM), Kidnapping (KID), Larceny (LAR), Libel (LIB), Perjury (PER), Rape (RAP) and Seduction (SED). As already explained in Sect. 3 for Table 3, entries of the matrix represent the proportion of respondents who considered the column offence to be more serious than the row offence. Because the symmetric component of the data matrix contains values of 0.5 for all off-diagonal elements, it would be of no practical interest. For this reason, our attention will focus on the skew-symmetric component.

In the first step of the proposed strategy, the application of model (3) provides the following estimates of the scaling parameters $\hat{r}_i$, reported in parenthesis for each crime in non decreasing order: HOM (−0.9333), RAP (−0.8000), ABO (−0.6667), SED (−0.5333), ADU (−0.4000), KID (−0.2667), ARS (−0.1333), EMB (0.1333), COU (0.2667), PER (0.2667), BUR (0.4000), FOR (0.4000), ASB (0.5333), LAR (0.8000), LIB (0.9333).

So we obtain the following ordering from the most to the less serious crime: HOM, RAP, ABO, SED, ADU, KID, ARS, EMB, COU-PER, BUR-FOR, ASB, LAR, LIB, with the two ties COU-PER and BUR-FOR. The application of model (2) provides the Gower diagram depicted in Fig. 1. This diagram (that accounts for 81 % of the $\boldsymbol{\Gamma}$ total sum of squares) allows to depict the ordering, to solve for ties and to check for circular triads. By going from the top-right to the bottom-left we find the same ordering obtained by model (3). Besides, the diagram helps to complete the ordering because we see that PER $\succ$ COU and FOR $\succ$ BUR, thus the complete ordering will be: HOM, RAP, ABO, SED, ADU, KID, ARS, EMB, PER, COU, FOR, BUR, ASB, LAR, LIB, that is the same ordering obtained maximizing the sum of the entries above the main diagonal by Brusco and Stahl (2005, Table 3). All points are in only half plane so we do not have evidence for the presence of circular triads.

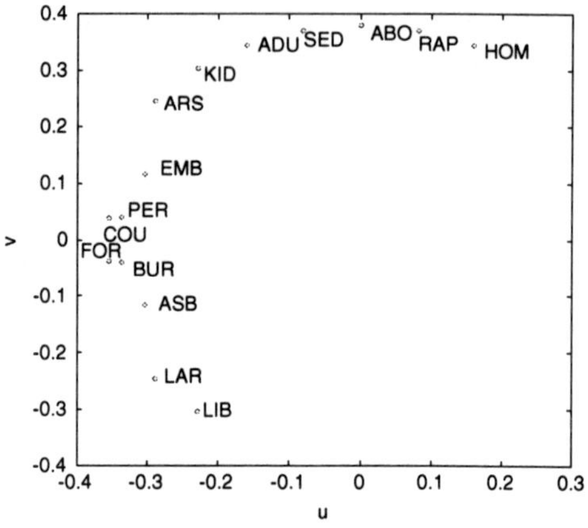

Fig. 1 Gower diagram for Thurstone's paired comparison crime data

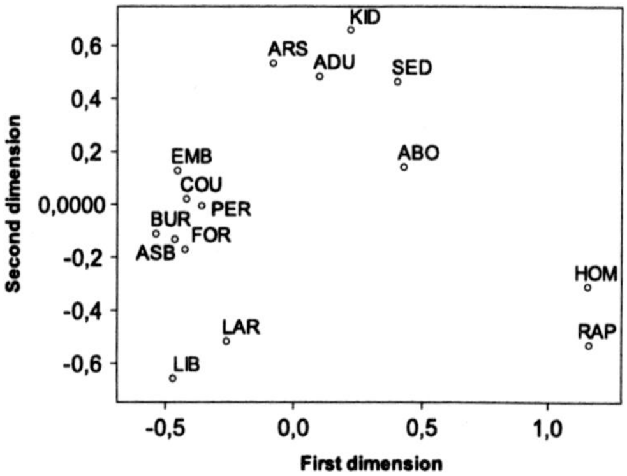

Fig. 2 Symmetric multidimensional scaling of the size of skew-symmetry

In the second step of the proposed strategy, we can display entries of matrix **T** by applying symmetric Multidimensional Scaling. Figure 2 shows the obtained configuration in two dimensions ($Stress\text{-}I = 0.11$). The size of skew-symmetry is represented by the distance between points, so we see for example that the pairs PER-COU and FOR-BUR have small skew-symmetry, while HOM-LIB and RAP-LAR have large skew-symmetry.

Finally, in the third step we can join the graphical analyses of matrix $\mathbf{\Gamma}$ and matrix **T** drawing circles around the points of the diagram obtained in STEP 2. Estimates of the radii of the circles can be based on the radius model parameters, as described in STEP 3 of the proposed strategy. The normalization factor for the radii was opportunely fixed to 7.9. Figure 3 represents the final result of the strategy. As in the Gower diagram, in Fig. 3 we can easily detect seriousness of crime by circles areas (crimes are less serious as larger is the circle area). So we see, for example, that Homicide and Rape are judged much more serious than Libel and Larceny.

Besides, a graduation of seriousness is provided that we cannot obtain by a dynamic programming approach. Crimes are depicted in four different groups: a

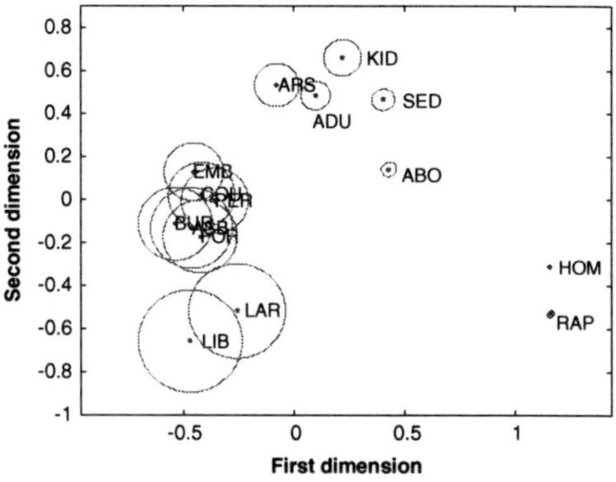

Fig. 3 Joint representation of size and direction of skew-symmetry

group of less serious crimes (Larceny, Libel), a group of "light" crimes (Embezzlement, Perjury, Counterfeiting, Forgery, Burglary, Assault and Battery), a group of serious crimes (Seduction, Adultery, Kidnapping, Arson), a group of more serious crimes (Homicide, Rape), with Abortion in a position between the third and the fourth group. So, we think that this last diagram can add further information and help a better interpretation of the results obtained for these data by combinatorial optimization methods (e.g. Hubert and Golledge 1981; Brusco and Stahl 2005).

References

Bove, G. (1989). *New methods of representation of proximity data*. Doctoral thesis, Dipartimento di Statistica, Probabilità e Statistiche Applicate, Universita'La Sapienza, Roma (in Italian).

Bove, G., & Critchley, F. (1989). On the representation of asymmetricproximities. *Atti delle Giornate di studio del Gruppo Italiano aderenti IFCS*, Erice 24–25 Ottobre 1988. Società Italiana di Statistica, Ila Palma Ed., Palermo (in Italian).

Bove, G., & Critchley, F. (1993). Metric multidimensional scaling for asymmetric proximities when the asymmetry is one-dimensional. In R. Steyer et al. (Eds.), *Psychometric methodology* (pp. 55–60). Stuttgart: Gustav Fischer Verlag.

Brusco, M. J., & Stahl, S. (2005). Bicriterion seriation methods for skew-symmetric matrices. *The British Journal of Mathematical and Statistical Psychology, 58*, 333–343.

Constantine, A. G., & Gower, J. C. (1978). Graphical representation of asymmetric matrices. *Applied Statistics, 27*, 297–304.

Critchley F. (1988). *On characterization of the inner products of the space of square matrices of given order which render orthogonal the symmetric and the skew-symmetric subspaces*. Warwick Statistics Research Report 171. Coventry: University of Warwick.

Freeman, L. C. (1997). Uncovering organizational hierarchies. *Computational and Mathematical Organization Theory, 3*(1), 5–18.

Gower, J. C. (1977). The analysis of asymmetry and orthogonality. In J. R. Barra et al. (Eds.), *Recent developments in statistics* (pp. 109–123). Amsterdam: North Holland.

Hubert, L. J., Arabie, P., & Meulman, J. (2001). *Combinatorial data analysis: Optimization by dynamic programming*. Philadelphia: SIAM.

Hubert, L. J., & Golledge, R. G. (1981). Matrix reorganization and dynamic programming: Applications to paired comparisons and unidimensional seriation. *Psychometrika, 46*(4), 429–441.

Mosteller, F. (1951). Remarks on the method of pair comparisons: I. The least squares solution assuming equal standard deviation and equal correlations. *Psychometrika, 16*, 3–9.

Okada, A., & Imaizumi, T. (1987). Non metric multidimensional scaling of asymmetric similarities. *Behaviormetrika, 21*, 81–96.

Saito, T., & Yadohisa, H. (2005). *Data analysis of asymmetric structures. Advanced approaches in computational statistics*. New York: Marcel Dekker.

Thurstone, L. L. (1927). The method of paired comparisons for social values. *Journal of Abnormal and Social Psychology, 31*, 384–400.

Zielman, B., & Heiser, W. J. (1996). Models for asymmetric proximities. *The British Journal of Mathematical and Statistical Psychology, 49*, 127–146.

An Approach to Ranking the Hedge Fund Industry

Riccardo Bramante

Abstract Due to the complexity and heterogeneity of hedge fund strategies, the evaluation of their performance and risk is a challenging task. Starting from the standard mutual fund industry, the literature has evolved in the direction of refining traditional measures (e.g. the Sharpe Ratio) or introducing new ones. This paper develops an approach, based on the Principal Component Analysis, to uncover the relevant information for performance measurement and combine it into a unique rank.

1 Introduction

In this paper the problem of performance assessment within the hedge fund industry is investigated. By combining commonly used and newly developed statistical performance indicators, a unique ranking of the industry is produced. A set of 18 indicators is firstly identified and then combined into a rank by a principal component analysis. This allows to detect the underlying drivers of the performance. Finally, a feature of the hedge funds that is not captured by this combination of indicators, namely the capability of raising the portfolio efficiency, is investigated. This is done by computing the upside gained by a balanced portfolio through the inclusion of a share of hedge funds.

Why should we not solely rely on the traditional risk adjusted measures to assess hedge funds performance? As a matter of fact, in the traditional mutual fund industry, performance is typically based on a version of the Sharpe Ratio (see for instance Morningstar and the well known star attribution system). Subscribing to the views expressed by Fung and Hsieh (1998, 1999), the transfer of this

R. Bramante (✉)
Department of Statistical Sciences, Catholic University of Milan, Milan, Italy
e-mail: riccardo.bramante@unicatt.it

P. Giudici et al. (eds.), *Statistical Models for Data Analysis*, Studies in Classification, Data Analysis, and Knowledge Organization, DOI 10.1007/978-3-319-00032-9_8,

methodology to the hedge fund industry could be arduous, since hedge funds differ from mutual funds mainly because of the variety of strategies adopted. This leads to widely different returns and volatilities depending on the particular fund. Thus, some hedge funds may be non-directional and less volatile than traditional bond and equity markets, while others may be fully directional and display higher volatility. As widely discussed by Ineichen (2003), hedge fund differences from mutual funds hinge on their relationship with the broader environment of financial markets. Dynamic reallocation of portfolios can create non-linear patterns with respect to the market (Agarwal and Naik 2004). In addition, hedge funds structurally use leverage. As discussed in Brealey and Kaplanis (2001), even if the fund remains exposed to the same market, variation in leverage, obtained by changing the net exposure between short and long position, may introduce further non-linearities. Finally, hedge-funds may invest in non-traditional financial assets such as derivatives. Some of these instruments display non-linearities because of their implicit features (Mitchell and Pulvino 2001). The structural characteristics of hedge-funds mentioned above, strongly affect the standard techniques of evaluating mutual funds.

If the risk adjusted return insufficiently portrays the hedge fund performance, how may we satisfactorily assess it? An answer to this question must entail the investor point of view, which ultimately defines the characteristics of hedge funds. Indeed, a hedge fund portfolio manager, in order to charge higher fees than mutual funds, makes every effort to provide the fund with some superior characteristics which can be summarized as follows:

- The capability of generating appealing absolute returns.
- The capability of protecting capital.
- The capability of raising the efficiency of a portfolio.

Generating appealing returns is attained by participating to the upside of a traditional investment, while protecting the capital may be achieved through an accurate hedging of the downside. These investment objectives cannot be reached through a linear exposure to the markets. By contrast, an asymmetric payoff resembling a call option is more suitable. In addition, the management of a hedge fund should result in a low correlation with "traditional" financial markets, thus increasing its appeal as a portfolio diversificator.

To address the problem of performance assessment, statistical indicators—chosen among widely used indicators—have been selected in accordance with the three identified features.

2 The Statistical Indicators

A burgeoning literature is currently discussing the appropriate tools for hedge fund performance measurement (Géhin 2004; Lhabitant 2004). The main problem hinges on the non-linear market exposure that often results in characteristics that limit the

Table 1 Statistical indicators

Annualized return	Sharpe ratio
Median return	Cornish–Fisher sharpe ratio
Frequency of positive returns	Adjusted sharpe ratio
Annualized volatility	Sortino ratio
Negative semi-deviation	Kurtosis
Maximum drawdown	Skewness
Value at risk	Normality test
Cornish–Fisher percentile	Correlation with MSCI world index
Cornish–Fisher value at risk	Correlation with BEA index

applicability of the classical bi-dimensional, risk and return, measures. In particular, the return distribution patterns, measured by moments higher than two, becomes relevant and could result in an over or underestimation of the performance. On this basis, measures that either come from the traditional investment world or have been accepted as useful tools to overcome the typical asymmetries of the hedge fund industry[1] are selected (Table 1).

Besides browsing among all the indicators, we reserve some notes on the measures capable of capturing the asymmetries typical of returns as it will turn out to be relevant in our application. Apart from skewness and kurtosis, a common solution to capture the shape of return distribution is attained by considering only the returns that lie below the average mean (Markowitz 1959). A measure that goes in this direction and, coherently with Kahnemann and Tversky (1979), that describes investors' viewpoint in an appropriate way, is the negative semi deviation

$$\sigma_R^- = \sqrt{\frac{1}{T}\sum_{t=1}^{T} Min(R_t, 0)^2}$$

where R_t is the sequence of the hedge fund log-returns.

Together with negative semi deviation practitioners often analyze the downside exposure, through empirical measures like the maximum drawdown[2]

$$MD_R = max_{t\in[0,T]}(max_{s\in[0,t]} R_s - R_t)$$

or the value at risk

$$VaR_R(\alpha) = \mu_R - z_\alpha \cdot \sigma_R$$

[1]MSCI World Index and Barclays Euro Aggregate (BEA) are chosen to represent correlation with Equity and Bond market.

[2]It should be pointed out that maximum drawdown is an empirical measure, without any statistical consistency.

where μ_R and σ_R are respectively the hedge fund average return and volatility and z_α satisfies

$$\Phi^{-1}(\alpha) = z_\alpha$$

where α is the desired percentile and Φ is the *CDF* of the standard normal distribution.

The last indicator has the same previously outlined problems in case returns exhibit non-normal features. To accomplish this problem, it is convenient to approximate the quantiles of the distribution via Cornish–Fisher approximations, which corrects the distortion in the returns distribution of the fund by integrating the effect of the moments of order greater than two on the left tail.[3] According to Hill and Davis (1968), it is called Cornish–Fisher percentile at the α significance level the α standard normal percentile corrected by the skewness $Skew(R)$ and the kurtosis $Kurt(R)$ effect[4]:

$$\Omega_\alpha = z_\alpha + \frac{1}{6}(z_\alpha - 1)^2 \cdot Skew(R) + \frac{1}{24}({z_\alpha}^3 - 3z_\alpha) \cdot Kurt(R) - \frac{1}{36}({z_\alpha}^3 - 5z_\alpha) \cdot Skew(R)^2$$

The value at risk correction is then:

$$VaR_R(\alpha) = \mu_R - \Omega_\alpha \cdot \sigma_R$$

Another important group of indicators refers to risk-adjusted measures. Beyond the Sharpe ratio that, in the case of hedge funds, provides a consistent under/over estimation of the risk adjusted performance, the two selected measures are the Sortino ratio

$$SO(R) = \frac{\mu_R - R_F}{\sigma_R^-}$$

and the Cornish–Fisher Sharpe ratio

$$S(R) - CF(R) = \frac{\mu_R - R_F}{\sigma_R \Omega_\alpha}$$

where R_F is the risk-free return.

[3] A different solution can be found in Bramante and Zappa (2011).

[4] Kurtosis are translated to zero.

3 Ranking the Hedge Fund Industry

To uncover most of the relevant information scattered across the statistical indicators adopted we have used the Principal Component Analysis (PCA). It may help to identify the implicit factors that significantly explain the overall variability, included in the variance and covariance matrix. Each of these factors consists of a linear combination of the original indicators. The proportion of variability explained by each of the factors is a natural way to estimate their relative strength. Moreover the components turn out to be uncorrelated among themselves so that they may be aggregated into a unique performance rank.

To explore whether the proposed ranking method can be applied within the hedge fund industry two different index families—*CS* (Credit Suisse/Tremont) and *HFR* (Hedge Fund Research)—are considered.[5] The data consist of monthly returns (from January 1994 to July 2011) of the universe of the two types of indices (55 for the *CS* and 32 for the *HFR* type respectively) and the ranking exercise is performed three times in July of the last 3 years.[6] The principal component technique, applied to the set of indicators previously normalized in the interval [0,10], extracts four factors that explain on average the 89 % of the total industry variability (Table 2).

To characterize the four components, it is convenient to analyze the correlations of the rotated component matrix.[7] In Table 3 results obtained in the 2009 analysis are reported[8] where correlations, between factors and observed variables, in absolute value greater than 0.7 are shown in italics. The component structure is amenable to interpretation:

- Factor 1: *Appealing returns and Capital Protection*. Represents absolute returns and captures the fund downside exposure (Annualized and Median return, Frequency of positive returns; Annualized volatility, Negative semi-deviation; Maximum drawdown and Value at Risk).
- Factor 2: *Asymmetry*. Measures the ability that the fund payoff resembles a call: consistent right tailed distribution, i.e. positive skewness/negative kurtosis (Skewness, Kurtosis, Normality and Cornish Fisher quantile).
- Factor 3: *Risk Adjusted Performance*. Captures the fund capability of balancing risk against reward (Sharpe ratio; Sortino; Adjusted and Cornish Fisher Sharpe ratio).

[5] These are the two widely recognized hedge fund index providers in the industry.

[6] Since the ability of this method in summarizing common patters depends on whether data contain strongly correlated variables, average partial correlation between variables was computed across the three considered years. Above all, the largest ones are between the three "Cornish Fisher" indicators and within the risk variables (Annualized Volatility, Negative Semi Deviation and Value at Risk 5 %).

[7] A varimax rotation was performed.

[8] Similar results, referred to the remaining two scenarios, are omitted.

Table 2 Total variance explained

	2009		2010		2011	
	% of variance	Cumulative %	% of variance	Cumulative %	% of variance	Cumulative %
Factor 1	28.87	28.87	26.28	26.28	34.77	34.77
Factor 2	27.50	56.37	25.86	52.13	25.42	60.19
Factor 3	22.48	78.85	25.41	77.54	14.47	74.66
Factor 4	13.96	92.82	13.76	91.30	8.26	82.92

Table 3 Rotated component matrix correlations

	Factor 1	Factor 2	Factor 3	Factor 4
Annualized return	*0.7345*	0.3235	0.5260	−0.2070
Frequency positive returns	*0.7397*	−0.1967	−0.0202	0.4901
Annualized volatility	*0.9319*	0.0498	−0.2341	−0.1201
Negative semi deviation	*0.9605*	0.2182	0.0147	−0.0833
Max drowdown	*0.8334*	0.2421	0.4144	−0.0148
Value at risk 5 %	*0.9793*	0.0782	−0.1189	0.0122
Sharpe ratio	0.0980	0.2817	*0.9448*	−0.0633
Cornish Fisher sharpe ratio	−0.0474	−0.1200	*0.9561*	−0.0409
Adjusted sharpe ratio	0.1098	−0.5585	*0.7896*	−0.0042
Sortino ratio	−0.1385	0.1227	*0.9285*	−0.1105
Kurtosis	0.1033	*0.9668*	0.0395	−0.0647
Skewness	0.2211	*0.8179*	0.3316	−0.0930
Normality test	0.0557	*−0.9637*	0.1084	0.0483
Cornish Fisher perc. 5 %	−0.0674	*−0.9867*	−0.0117	0.0679
MSCI world correlation	0.3217	0.0346	0.1745	*−0.8676*
BEA correlation	−0.1141	−0.2427	−0.2130	*0.8497*

- Factor 4: *Market correlation*. This component is self-explanatory and is approximated by the considered market proxies.

Since the components are by construction uncorrelated, the intuition suggests to assemble a ranking index, by linearly combining for each hedge fund the score of each factor with its weight, given by the percentage of variance explained. To facilitate the construction of a ranking grid, the score is replaced with its rank. In Table 4 the final ranking for the first ten positions of the *HFR* and *CS* indexes in 2009 is reported for explanatory purposes.

As a final test, we used the Spearman Rank correlation to compare previous and subsequent classifications. Our results indicate a strong correlation if 1 year lagged rankings are compared whereas—if a 2-year lagged ranking is considered—the null hypothesis of no persistence can be accepted with at least 95 % confidence in all the two types of the considered indices. Moreover, in the *HFR* analysis, the 2-year lagged rankings were found to be of opposite sign, though these results lacked statistical significance.

Table 4 *HFR* and *CS* indices final ranking

HFR ranking	*CS* ranking
FOF: CONSERVATIVE	EQTY MKT.NTL Y
EH: SHORT BIAS	EQTY MKT.NTL $
RV: MULTI- STRATEGY	EQTY MKT.NTL E
EH: EQUITY MARKET NEUTRAL	MULT STRATEGY SF
RV: YIELD ALTERNATIVES	MULT STRATEGY E
EMG MKTS: GLOBAL	CONV ARBITRAGE Y
RV: FIXED INC.- CONV.ARB.	MULT STRATEGY Y
FOF: DIVERSIFIED	DEDICATED SHT Y
FUND OF FUNDS COMPOSITE	HEDGE FUND SF
RV: FIXED INC.- CORPORATE	DEDICATED SHT SF

4 Raising Portfolio Efficiency

Assessing the ability of hedge funds to raise portfolio efficiency requires a separate analysis. The methodology originally developed by Modigliani and Modigliani (1997) is adapted to compare a balanced portfolio with another where 30 %[9] of the traditional investments are replaced by hedge funds. The efficiency spread is measured by the Modigliani–Modigliani index, that is the difference between the potential return of the (de)leveraged portfolio containing hedge funds and the return of the balanced portfolio. The former is (de)leveraged to the point where its volatility is equal to the balanced portfolio volatility. From this perspective, the Modigliani–Modigliani index can be interpreted as the return spread at the same level of volatility. The return spread is expected to be higher, the lower the correlation with traditional investments. It is important to bear in mind that the correlation effect is more important than the risk/return profile, since it acts directly on the volatility by reducing the systematic portfolio risk. Even if the hedge fund (*HF*), as in Fig. 1, is Pareto dominated by the balanced portfolio (*P*) it can happen that the combined portfolio (P^+) shares a sufficiently limited volatility to have a positive return spread.

Formally:

$$M = \frac{\mu_P^+ - r_f}{\sigma_P^+} \cdot \sigma_P - \frac{\mu_P - r_f}{\sigma_P} \cdot \sigma_P = \sigma_P \cdot (SR_{P^+} - SR_P)$$

A classification based on the Modigliani–Modigliani index may stress which are the hedge funds that contribute to portfolio diversification. We have computed, for each balanced portfolio obtained by replacing the 30 % with hedge funds, the Modigliani–Modigliani index.

[9] 30 % is arbitrarily chosen. However, empirical simulations show that 30 % of asset allocation in hedge funds seems to be closed to the optimum, in terms of the distance from the efficient frontier.

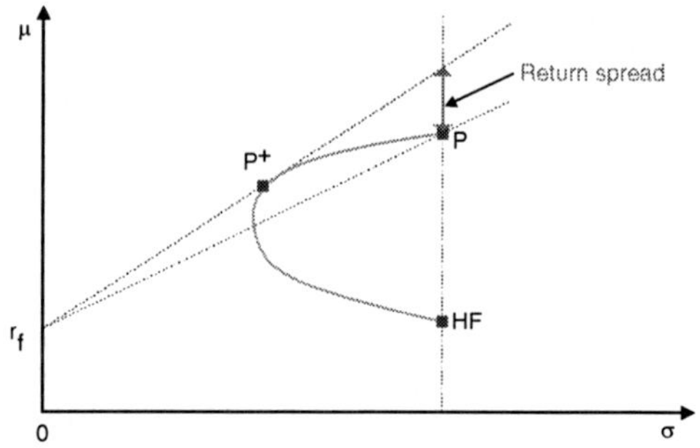

Fig. 1 Return spread and Modigliani–Modigliani index

Table 5 *HFR* indices Modigliani–Modigliani final ranking

Index name	Raising portfolio's efficiency	Raising performance
EH: SHORT BIAS	1	2
MACRO: SYST. DIVERSIFIED	2	32
ED: MERGER ARBITRAGE	3	17
FOF: MKT DEFENSIVE	4	24
MACRO (TOTAL)	5	28
RV: FIXED INC.- CONV.ARB.	6	7
EH: SECTOR-TECH/HEALTHCARE	7	29
RELATIVE VALUE (TOTAL)	8	13
EH: EQUITY MARKET NEUTRAL	9	4
EMG MKTS: ASIA EX-JAPAN	10	25

Table 6 *CS* indices Modigliani–Modigliani final ranking

Index name	Raising portfolio's efficiency	Raising performance
MANAGED FUT E	1	48
MANAGED FUT $	2	49
MANAGED FUT Y	3	46
RISK ARBITRAGE $	4	52
DEDI SHORT BIAS$	5	34
MANAGED FUT $	6	50
GLOBAL MACRO $	7	54
EVENT DRVNMSTRT$	8	47
CONV ARBITRAGE $	9	41
EVENT DRIVEN $	10	45

In Tables 5 and 6 rankings, only for the first ten positions, obtained according to the Modigliani–Modigliani index respectively for the *HFR* and *CS* indices are reported and compared to the ones retrieved from the previous analysis. It seems, as one may grasp from the results, that this technique points out that portfolio diversification is independent from the properties that the funds share as pure hedge funds.

Are hedge funds that have appealing performances alone also good contributor to portfolio diversification? Intuition may suggest that this cannot be true: appealing performance profile results from a non linear exposition to the market and this in some cases may affect portfolio risk. Moreover, the analysis of ranking permutations, accomplished by Spearman Rho, confirms that few average performing hedge funds display a consistent advantage in diversification: all the coefficients, albeit positive, are below 0.3 and indicate that the correlation is weak in all of the 3 years considered in the analysis.

Acknowledgments The development of this paper benefited significantly from the input and support of Alessandro Cipollini and Antonio Manzini.

References

Agarwal, V., & Naik, N. Y. (2004). Risk and portfolio decisions involving hedge funds. *Review of Financial Studies, 17*(1), 63–98.

Bramante R., & Zappa, D. (2011). Value at risk estimation in a mixture normality framework. In Proceedings of the Eighteenth International Conference Forecasting Financial Markets—Advances for Exchange Rates, Interest Rates and Asset Management, Marseille.

Brealey, R. A., & Kaplanis, E. (2001). Hedge funds and financial stability: An analysis of their factor exposures. *International Finance, 4*(2), 161–187.

Fung, W., & Hsieh, D. A. (1998). *Performance attribution and style analysis: From mutual funds to hedge funds*. Working paper, Duke University, Fuqua School of Business.

Fung, W., & Hsieh, D. A. (1999). A primer on hedge funds. *Journal of Empirical Finance, 6*, 309–331.

Géhin, W. (2004). *A survey of the literature on hedge fund performance*. Nice: EDHEC Risk and Asset Management Research Centre, EDHEC Business School Lille.

Hill, G. W., & Davis, A. W. (1968). Generalized asymptotic expansions of Cornish–Fisher type. *Annals of Mathematical Statistics, 39*, 1264–1273.

Ineichen, A. M. (2003). *Absolute returns: The risk and opportunities of hedge fund investing*. Hoboken: Wiley.

Kahnemann, D., & Tversky, A. (1979). Prospect theory: An analysis of decision under risk. *Econometrica, 47*(2), 263–292.

Lhabitant, F. S. (2004). *Hedge funds, quantitative insights*. Chichester: Wiley.

Markowitz, H. (1959). Portfolio selection. *Journal of Finance, 17*(1), 77–91.

Mitchell, M., & Pulvino, T. (2001). Characteristics of risk in risk arbitrage. *Journal of Finance, 56*, 2135–2175.

Modigliani, F., & Modigliani, L. (1997). Risk-adjusted performance—how to measure it and why. *Journal of Portfolio Management, 23*(2), 45–54.

Correction of Incoherences in Statistical Matching

Andrea Capotorti and Barbara Vantaggi

Abstract Statistical matching is studied inside a coherent setting, by focusing on the problem of removing inconsistencies. When structural zeros among involved variables are present, incoherencies on the parameter estimations can arise. The aim is to compare different methods to remove such incoherences based on specific pseudo-distances. The comparison is given through an exemplifying example of 100 simulations from a known population with three categorical variables, that carries out to the light peculiarities of the statistical matching problem.

1 Introduction

In several applications different data sources need to be integrated, in particular, we will deal with the so-called statistical matching problem (D'Orazio et al. 2006; Paass 1986; Rässler 2002), that can be represented by the following simple situation: two different sources of information, A and B, on a common population with some overlapping variables X and some other variables Y, Z collected only in one source A or B, respectively. To cope with these problems the available data are combined with assumptions strong enough to point-identify the joint probability distribution (see Rässler 2002 and references within), as those based on conditional independence of the variables Y and Z given X. However, in several situations, the independence assumption is not adequate (see e.g. D'Orazio et al. 2006). Actually, since there are many distributions on (X, Y, Z) compatible with the available partial

A. Capotorti (✉)
Dip. Matematica e Informatica, Universitá di Perugia, Italy
e-mail: capot@dmi.unipg.it

B. Vantaggi
Dip. Scienze di Base e Applicate per l'Ingegneria, Universitá La Sapienza, Roma, Italy
e-mail: barbara.vantaggi@sbai.uniroma1.it

P. Giudici et al. (eds.), *Statistical Models for Data Analysis*, Studies in Classification, Data Analysis, and Knowledge Organization, DOI 10.1007/978-3-319-00032-9_9,

information on (X, Y) and (X, Z), it is too restrictive to consider just one of the compatible joint distributions, obtained perhaps by taking a specific assumption (as noted in D'Orazio et al. 2006, Kadane 2001, Rubin 1986).

This problem has been already faced in a coherent conditional probability setting (Vantaggi 2008): coherence allows to check the compatibility of heterogenous partial (conditional) estimations, e.g. coming from field experts or different data sets, and to draw inferences by considering all the compatible joint distributions. A further remarkable advantage is related to structural zeros. In particular, in Vantaggi (2008) it is proved that when there is no structural zero, coherence is always satisfied, even by requiring conditional independence. On the other hand, when structural zeros are present, it is necessary to check global coherence of the relevant partial estimations drawn from the different sources and, when coherence does not hold, inconsistencies need to be removed.

Our contribution focuses on incoherences for categorical variables and the aim is to look for the coherent estimates as "close" as possible to the given one, with respect to different distances ($L1$, $L2$, Kulback–Leibler divergence, discrepancy). These (pseudo)distances need to be suitably tailored for partial conditional assessments. To properly deal with statistical matching, we introduce a specific adjustment of a discrepancy, originally introduced for conditional probability assessment (Capotorti et al. 2010): this shows the advantage of unsupervised localization of the sub-domains where incoherence must be removed.

Then, once coherence is restored, it is possible to draw inference, that means to extend the estimated coherent conditional probabilities (see e.g. Coletti and Scozzafava 2002). Actually, our proposal is in the same line with Rässler (2002), Rubin (1986)—based on multiple imputation in the multi-normal setting—and with D'Orazio et al. (2006)—based on maximum likelihood approach.

To show the advantages and drawbacks, and in particular to compare the different pseudo-distances, in Sect. 4 an empirical example is provided. This example is based on 100 sample couples, A with cardinality $n_A = 1,148$ and B with $n_B = 1,165$, randomly drawn from a finite population with three categorical variables. Among the 100 unconstrained maximum likelihood estimates of the marginal distribution for the common random variable X and of the conditional distributions of $Y|X$ and $Z|X$, we obtained 57 incoherent estimates. Once such estimates have been corrected, they induce the so called "credal sets" of joint distributions on (X, Y, Z) compatible with them. By performing standard χ^2 "goodness-of-fit" tests, it turns out that the 57 credal sets induced by the corrected estimate over-perform those induced by the 43 originally coherent ones.

2 Statistical Matching in a Coherent Setting

Let (X, Y, Z) be categorical variables with respectively I, J, and K categories and denote by $(X_1, Y_1), \ldots, (X_{n_A}, Y_{n_A})$ and by $(X_{n_A+1}, Z_{n_A+1}), \ldots, (X_{n_A+n_B}, Z_{n_A+n_B})$ two random samples (related to two sources A and B) concerning the same population and drawn according to the same sampling scheme. Thus, $(X_1, Y_1), \ldots, (X_{n_A}, Y_{n_A})$ (analogously $(X_{n_A+1}, Z_{n_A+1}), \ldots, (X_{n_A+n_B}, Z_{n_A+n_B})$), as well as the sequence $X_1, \ldots, X_{n_A}, X_{n_A+1}, \ldots, X_{n_A+n_B}$ can be regarded as exchangeable.

From the two files the relevant population parameters representing the following (conditional) probability values can be estimated: from file A the probability that the next unit has $Y = y_j$ on the hypothesis that $(X = x_i)$ (for any $i \in I$)

$$\mathbf{y}_{j|i} = P_{Y|(X=x_i)}(Y = y_j), \tag{1}$$

and analogously from file B

$$\mathbf{z}_{k|i} = P_{Z|(X=x_i)}(Z = z_k). \tag{2}$$

Moreover, by collecting data from both files we can evaluate

$$\mathbf{x}_i = P_X(X = x_i). \tag{3}$$

Such estimations are usually performed through the (unconstrained) partial maximum likelihood evaluations, that coincide with the following frequencies:

$$\mathbf{y}_{j|i} = \frac{n_A^{ij}}{n_A^{i\cdot}} \quad , \quad \mathbf{z}_{k|i} = \frac{n_B^{ik}}{n_B^{i\cdot}} \quad , \quad \mathbf{x}_i = \frac{(n_A + n_B)^{i\cdot\cdot}}{n_A + n_B} \quad , \tag{4}$$

where $n_A^{i\cdot}$, $n_B^{i\cdot}$ and $(n_A + n_B)^{i\cdot\cdot}$ represent the number of units expressing $(X = x_i)$ in samples A and B, while n_A^{ij} stands for the number of units in A with $(X = x_i) \wedge (Y = y_j)$ and n_B^{ik} the number of units in B with $(X = x_i) \wedge (Z = z_k)$.

Now, we should deal with the whole assessment $(\mathscr{E}, \mathbf{p})$ with

$$\mathscr{E} = \left\{ \begin{array}{r} (X = x_i),\ (Y = y_j)|(X = x_i),\ (Z = z_k)|(X = x_i) \\ \text{for any } x_i, y_j, z_k \end{array} \right\}, \tag{5}$$

$$\mathbf{p} = \{\mathbf{x}_i, \mathbf{y}_{j|i}, \mathbf{z}_{k|i}\}_{i,j,k} \quad .$$

Then, first of all we need to check its coherence, that means the compatibility of **p** with a full conditional probability (de Finetti 1972). Such compatibility is equivalent to the existence of a suitable class of joint probability distributions $\boldsymbol{\alpha}_1, \ldots, \boldsymbol{\alpha}_l$ agreeing with **p** (see for more details Coletti and Scozzafava 2002). Note that coherence is crucial being a prerequisite for a sound inference, that means extension of **p** to any new conditional event.

In Vantaggi (2008) it has been proved that when there is no structural zeros between Y and Z for any given value x_i of X (i.e. for any $i \in I$, if $(X = x_i) \wedge (Y = y_j) \neq \emptyset$ and $(X = x) \wedge (Z = z_k) \neq \emptyset$, then $(X = x) \wedge (Y = y_j) \wedge (Z = z_k) \neq \emptyset$ for all $j \in J$ and $k \in K$) coherence is assured. On the other hand, in the same paper it is proved that when there is some structural zero among the variables Y and Z, the coherence of the whole assessment (5) is not assured by coherence of the single assessments (1)–(3) and that, whenever present, incoherences can be localized among conditional events with the same conditioning event ($X = x_i$). Notice that the need of managing structural zeros, such as incompatibilities or implications among events generated by the random variables, arises from the practical applications and can be deduced from the structure of the problem and so are objectively known to the field experts (D'Orazio et al. 2006). For example in D'Orazio et al. (2006) Italian law imposes incompatibility of age "less than 17" with professional status "manager" or with educational level "degree".

Hence, the check of coherence of the whole assessments (5) can be reduced to the check of coherence for sub-assessments

$$\left\{\mathbf{y}_{j|i}, \mathbf{z}_{k|i} : \text{ for given } i \text{ and any } j,k\right\}. \tag{6}$$

Since the check of coherence is in general a NP-hard problem, its segmentation in several subproblems is a great advantage.

Thus, it is possible to proceed in two ways: a supervised procedure, where incoherent sub-assessments of type (6) are detected and attentions are focused only on them; or a unsupervised approach that adjusts the whole assessment (5).

In any case, adjustment can be performed by finding coherent estimates that derive from the minimization of some pseudo-distance, as shown in the next section.

3 Removing Inconsistencies in Statistical Matching

Estimate correction has been already studied (e.g. see Lindley and Tversky 1979), but this approach does not seem suitable in the context of statistical matching because of the lack of information due to the fact that Y and Z are not jointly observed, so the prior distribution cannot be updated and the likelihood function has a flat ridge (as already noted in Rubin 1986).

Our aim is to find coherent estimates as "close" as possible to the available information formed by the whole assessment (5). This implies the choice of some (pseudo)distance such as Euclidean distance, Kulback–Leibler divergence, Csiszár f-divergences. Some of them can be applied only among unconditional probabilities; while others could be applied also for partial conditional probability assessments.

Given two conditional probability estimates $\mathbf{p} = [p_1, \ldots, p_n]$ and $\mathbf{q} = [q_1, \ldots, q_n]$ on $\mathcal{E}$, the most widely adopted divergencies among them are $L1(\mathbf{p}, \mathbf{q}) = \sum_{i=1}^{n} |q_i - p_i|$, $L2(\mathbf{p}, \mathbf{q}) = \sum_{i=1}^{n} (q_i - p_i)^2$, $KL(\mathbf{p}, \mathbf{q}) = \sum_{i=1}^{n} (q_i \ln(q_i / p_i) - q_i + p_i)$.

$L1$ and $L2$ are usual metric distances, endowed with all their geometric properties, but until now remain without an intuitive probabilistic interpretation for conditional assessments. Moreover, their use in conditional context could lead to numerical troubles due to non-convexity of coherent assessments (see e.g. Biazzo and Gilio 2005).

KL is the so-called logarithmic Bregman divergence and, in the unconditional case, it is deeply used for its information theoretic properties. In fact, such divergence generalizes the well known Kulback–Leibler divergence (Kulback 1957) to partial assessments, however in some cases it presents some unpleasant situation due to the evaluation of just occurring events, without considering those turning out to be false. Moreover, $L1$, $L2$ and KL work on the whole domain $\mathscr{E}$, so their minimizations would induce changes also on the marginal estimate $\{\mathbf{x}_i\}_{i\in I}$ that, being coherent, could be valuable to avoid any change. Recently, to encompass the need of considering conditional probability assessments a discrepancy suitable for statistical matching has been introduced (for more details see Capotorti and Vantaggi 2011):

$$\Delta(\mathbf{p}, \{\boldsymbol{\alpha}_i\}_i) = \sum_i \mathbf{x}_i \left[\sum_j \left(q^{\alpha_i}_{j|i} \ln \frac{q^{\alpha_i}_{j|i}}{\mathbf{y}_{j|i}} + (1 - q^{\alpha_i}_{j|i}) \ln \frac{(1 - q^{\alpha_i}_{j|i})}{(1 - \mathbf{y}_{j|i})} \right) + \right.$$
$$\left. + \sum_k \left(q^{\alpha_i}_{k|i} \ln \frac{q^{\alpha_i}_{k|i}}{\mathbf{z}_{k|i}} + (1 - q^{\alpha_i}_{k|i}) \ln \frac{(1 - q^{\alpha_i}_{k|i})}{(1 - \mathbf{z}_{k|i})} \right) \right]. \qquad (7)$$

Each distribution α_i defined on the sample space spanned by $(Y = y_j)|(X = x_i)$ and $(Z = z_k)|(X = x_i)$, should fulfill the normalizing condition $\alpha_i(X = x_i) = \mathbf{x}_i$, and generates the conditional probabilities

$$q^{\alpha_i}_{j|i} = \frac{\alpha_i(Y = y_j)}{\alpha_i(X = x_i)} \quad \text{and} \quad q^{\alpha_i}_{k|i} = \frac{\alpha_i(Z = z_k)}{\alpha_i(X = x_i)}. \qquad (8)$$

Note that the generated estimate $\mathbf{q} = \{\mathbf{x}_i, q^{\alpha_i}_{j|i}, q^{\alpha_i}_{k|i}\}_{i,j,k}$ is coherent (see e.g. Vantaggi 2008).

In order to correct an estimate $\mathbf{p}$ we need to look for the assessment $\mathbf{q}_\mathbf{p}$, that is solution of the following nonlinear optimization program, with $\delta(\mathbf{p}, \mathbf{q})$ any pseudo-distance (if $\delta \equiv \Delta$ then $\mathbf{q}$ are those induced by $\{\boldsymbol{\alpha}_i\}_i$ as in (8)):

$$\min_{\mathbf{q}} \delta(\mathbf{p}, \mathbf{q}). \qquad (9)$$

Note that the discrepancy $\Delta(\mathbf{p}, \{\boldsymbol{\alpha}_i\}_i)$ profits from the already mentioned *segmentation* of the possible inconsistencies. In fact, it works separately on scenarios $(X = x_i)$ and its use in an optimization program like (9) allows to adjust only the values inside sub-domains where incoherences appear, without any other change. This feature distinguishes the specialized discrepancy (7) from the original formulation (Capotorti et al. 2010) so that, even being used in an unsupervised approach, it gives results as in a supervised one.

Another criterion for restoring coherence is based on the constrained maximum likelihood criterion (Little and Rubin 1983):
constrained maximum likelihood estimates of $\boldsymbol{\theta} = \big(P(X{=}x_i),\, P_{Y|(X=x_i)}(Y = y_j),\, P_{Z|(X=x_i)}(Z = z_k)\big)_{i,j,k}$ are the values of the parameters $\left(\hat{\theta}_i, \hat{\theta}_{j|i}, \hat{\theta}_{z|i}\right)_{i,j,k}$ solution of the program

$$\max_{\boldsymbol{\theta}} L(\boldsymbol{\theta}|n_A, n_B) = \prod_{i,j}(\theta_{j|i}\theta_i)^{n_A^{ij}} \prod_{i,k}(\theta_{k|i}\theta_i)^{n_B^{ik}} \tag{10}$$

under the constraint that $\boldsymbol{\theta}$ is a coherent conditional probability assessment over $\mathscr{E}$. Even in this situation we have an optimization problem with the observed data likelihood $L(\boldsymbol{\theta}|n_A, n_B)$ as non-linear objective function and a set of linear constraints characterizing the coherence of $\boldsymbol{\theta}$.

4 Corrections Comparison

In order to show the different behaviors of corrections presented in the previous section, we give a simulation study. We simulated 100 sample couples, with cardinality $n_A = 1,148$ and $n_B = 1,165$, respectively, drawn randomly from a finite population of three categorical variables (X, Y, Z), with $I = \{1,2\}$, $J = \{1,2,3\}$, $K = \{1,2,3\}$, distributed as described in Table 1, where the $(-)$ represent the structural zeros implied by the logical constraints

$$(Z = z_1) \wedge ((Y = y_1) \vee (Y = y_2)) = \emptyset \quad , \quad (Z = z_2) \wedge (Y = y_1) = \emptyset. \tag{11}$$

With a couple of samples A and B as in Sect. 2 we can obtain an estimate $\mathbf{p}$ of the conditional probability $\boldsymbol{\pi}$ of Table 2. Over the 100 estimates (4) of frequencies we observed 57 incoherent, as can be seen by computing e.g. $L2$ distances between $\boldsymbol{\pi}$ and the 100 estimates $\mathbf{p}$, so that $L2(\mathbf{p}, \boldsymbol{\pi}) = 0$ corresponds to coherent frequencies (see Fig. 1). Inconsistencies are mainly localized on $(X = x_2)$ and in particular due to the violation of the numerical bound $\mathbf{y}_{1|2} + \mathbf{y}_{2|2} + \mathbf{z}_{1|2} \leq 1$ implied by the first logical constraint in (11).

By means of the minimization (9) of pseudo-distances $L1, L2, KL, \Delta$ and the constrained likelihood maximization (10), for the 57 incoherent estimates over the whole domain $\mathscr{E}$ (hence with unsupervised procedures), we obtain five different data-sets with coherent corrections. To compare the performances we evaluate, through chi-squared goodness-of-fit test, the adequacy of the (credal) set of joint probability distribution compatible with each estimate with respect the joint distribution of the population. Results are in Fig. 2: there are box-plots of minimal χ^2 statistics associated to the five data-sets of corrections and that associated to the 43 coherent estimates obtained by frequencies (4).

Table 1 Finite population with (X, Y, Z) endowed with structural zeros $(-)$

	Z	z_1	z_2	z_3
X	Y			
x_1	y_1	(−)	(−)	116
	y_2	(−)	26	5
	y_3	54	108	25
x_2	y_1	(−)	(−)	277
	y_2	(−)	65	1
	y_3	321	1	1

Table 2 Conditional probabilities based on the population of Table 1

$\mathscr{E}$	π
$X = x_1$	0.3407
$X = x_2$	0.6593
$Y = y_1 \mid X = x_1$	0.1856
$Y = y_2 \mid X = x_1$	0.3763
$Y = y_3 \mid X = x_1$	0.4381
$Z = z_1 \mid X = x_1$	0.4903
$Z = z_2 \mid X = x_1$	0.0965
$Z = z_3 \mid X = x_1$	0.4131
$Y = y_1 \mid X = x_2$	0.3551
$Y = y_2 \mid X = x_2$	0.0783
$Y = y_3 \mid X = x_2$	0.5666
$Z = z_1 \mid X = x_2$	0.4105
$Z = z_2 \mid X = x_2$	0.0980
$Z = z_3 \mid X = x_2$	0.4915

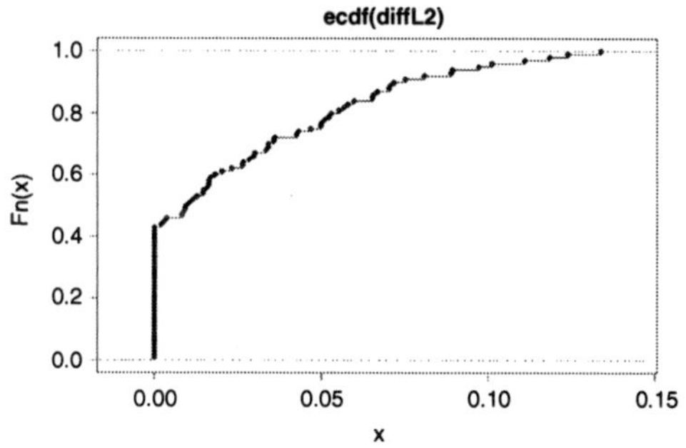

Fig. 1 Empirical cumulative distribution function of L2 distances between original probabilities π and simulated estimates $\mathbf{p}$.

Notice the over-performance of the corrected estimates with respect to the not changed ones and the best behavior of the minimization of $L2$ and Δ with respect to the other pseudo-distances. Then, it seems that the correction produces an information merging that maximum likelihood estimation does not capture. Among the different possible corrections, $L2$ and Δ minimizations seems to better preserve the original information, we privilege Δ for the automatic localization of the sub-domains of $\mathscr{E}$ where the changes are needed.

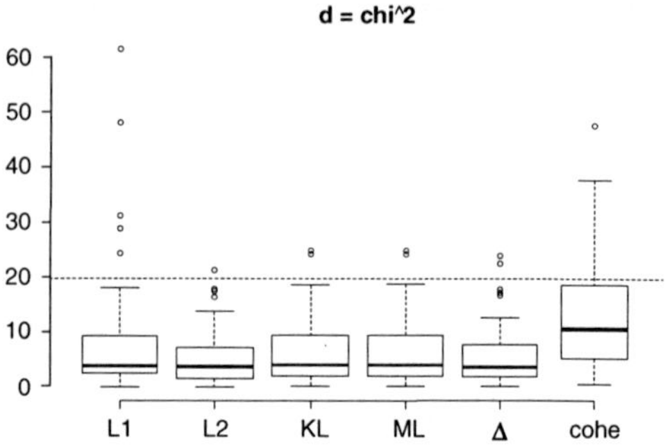

Fig. 2 Minimal χ^2 "goodness-of-fit" for credal sets induced by pseudo-distances minimizations (labels "L1", "L2", "KL", "Δ"), constrained maximum likelihoods (label "ML") and coherent frequencies (label "cohe") estimates. The *dotted line* corresponds to the 95% confidence threshold

References

Biazzo, V., & Gilio, A. (2005). Some theoretical properties of conditional probability assessments. *Lecture Notes in Computer Science LNAI, 3571*, 775–787.

Capotorti, A., Regoli, G., & Vattari, F. (2010). Correction of incoherent conditional probability assessments. *International Journal of Approximate Reasoning, 51*(6), 718–727.

Capotorti, A., & Vantaggi, B. (2011). In *Proc. of 7th int. symp. on imprecise probability: theories and applications - ISIPTA'11*. Incoherence correction strategies in statistical matching (pp. 109–1118). Innsbruck (Austria).

Coletti, G., & Scozzafava, R. (2002). *Probabilistic logic in a coherent setting*, Series "Trends in Logic". Dordrecht: Kluwer Academic.

de Finetti, B. (1972). Sull'impostazione assiomatica del calcolo delle probabilità. *Probability, induction, statistics*. London: Wiley.

D'Orazio, M., Di Zio, M., & Scanu, M. (2006). Statistical matching for categorical data: displaying uncertainty and using logical constraints. *Journal of Official Statistics, 22*, 137–157.

D'Orazio, M., Di Zio, M., & Scanu, M. (2006). *Statistical matching: theory and practice*. London: Wiley.

Kadane, J. B. (2001). Some statistical problems in merging data files. *Journal of Official Statistics, 17*, 423–433.

Kulback, S. (1957). *Information theory and statistics*. New York: Wiley.

Lindley, D. V., Tversky, A., & Brown, R. V. (1979). On the reconciliation of probability assessments. *Journal of Royal Statistical Society A, 142*(2), 146–180.

Little, R. J. A., & Rubin, D. B. (1983). On jointly estimating parameters and missing data by maximising the complete-data likelihood. *The American Statistician, 37*, 218–220.

Paass, G. (1986). In G. H. Orcutt & H. Quinke (Eds.) *Microanalytic simulation models to support social and financial policy*. Statistical match: evaluation of existing procedures and improvements by using additional information (pp. 401–422). Amsterdam: Elsevier Science.

Rässler, S. (2002). Statistical matching: a frequentist theory, practical applications and alternative Bayesian approaches. *Lecture notes in statistics*. New York: Springer.

Rubin, D. B. (1986). Statistical matching using file concatenation with adjusted weights and multiple imputations. *Journal of Business & Economic Statistics, 2*(1), 87–94.

Vantaggi, B. (2008). Statistical matching of multiple sources: A look through coherence. *International Journal of Approximate Reasoning, 49*(3), 701–711.

The Analysis of Network Additionality in the Context of Territorial Innovation Policy: The Case of Italian Technological Districts

Carlo Capuano, Domenico De Stefano, Alfredo Del Monte, Maria Rosaria D'Esposito, and Maria Prosperina Vitale

Abstract Evidence from economic literature suggests that innovative activities based on extensive interactions between industry, universities and local government can yield high levels of economic performance. In many countries, therefore, steps have been taken at an institutional level to set up innovation networks and, in particular, regional technological districts. Our paper deals with Italian Technological Districts: we aim to analyse the network additionality for territorial innovation determined by district policy. The analysis is based on a priori structural regional characteristics and on Social Network Analysis techniques.

1 Introduction

The evaluation of R&D policies is based on the concept of additionality of results considered in terms of inputs, outputs and firms' behaviour. Economic literature has recently criticized input or output additionality, as these concepts consider the firm as a black box and do not capture the impact of public intervention on the funded organizations' R&D behaviour. Behavioural issues are important if we are to understand the performance of policies, and they may be profitably employed to evaluate innovation policies according to the systemic-evolutionary foundations of innovation policy. Hence Behavioral Additionality (BA) (the change in the way a

C. Capuano (✉) · A. Del Monte
University of Naples Federico II, Via Cinthia 26, 80125 Napoli, Italy
e-mail: carlo.capuano@unina.it; delmonte@unina.it

D. De Stefano
University of Trieste, Piazzale Europa 1, 34127 Trieste, Italy
e-mail: domenico.destefano@econ.units.it

M.R. D'Esposito · M.P. Vitale
University of Salerno, Via Ponte don Melillo, 84084 Fisciano (SA), Italy
e-mail: mdesposi@unisa.it; mvitale@unisa.it

P. Giudici et al. (eds.), *Statistical Models for Data Analysis*, Studies in Classification, Data Analysis, and Knowledge Organization, DOI 10.1007/978-3-319-00032-9_10,

company undertakes R&D that can be attributed to policy actions) emerges as an important aspect of additionality to complement traditional evaluation approaches and there is a growing interest in how best to measure it. In order to tackle the BA measurement issue and to prove the BA of public subsidies, its multidimensional character needs to be taken into account. Since a prominent role is here ascribed to the collaborative and networking capabilities fostered by public interventions, Network Additionality (NA) must first be considered.

Given that a key role here is played by the concept of relationship, Social Network Analysis (SNA) techniques can be fruitfully used to measure the extent of occurrence, outspread and stabilization of a relationship. SNA may be used to integrate the analysis of the effects on the single components with the analysis of the effects on the system as a whole (Antonioli and Marzucchi 2010). Other methodologies used to analyse the effect of R&D policy principally concern input and output additionality and therefore are not strictly comparable with the SNA approach. This is the case of counterfactual analysis. If one of the effects of R&D policy is cooperation, it would, in fact, be very difficult to find two groups of organizations, those that cooperate and get supported and those that cooperate and are not supported, sharing the same characteristics except cooperation (Antonioli and Marzucchi 2010).

In this paper we focus attention both on the public policies undertaken to foster the creation and growth of the so called Technological Districts (TDs) (Antonelli 2000)—which are innovation networks originated as result of endogenous factors—and on the use of the SNA approach for their analysis (Del Monte et al. 2011). The paper is organized as follows: in Sects. 2 and 3 the theoretical framework to describe the characteristics of innovation networks and TDs in Italy is presented; in Sects. 3.1 and 3.2 the study of four Italian TDs through a Network Analysis approach is briefly discussed. Section 4 presents some preliminary concluding remarks.

2 Innovation Networks and Technological Districts: A Theoretical Framework

A policy to create a TD will not necessary have a positive effect from a welfare point of view (Capuano and Del Monte 2011). This can be seen with the help of some simple models of social connections developed by Jackson and Wolinsky (1996), where n organizations (e.g., firms, research centers, etc.) create a network through collaboration agreements (e.g., participation in research projects). These direct links between two organizations make it possible to share knowledge about new technology, to avoid the duplication of fixed costs and to derive a cooperative advantage from the strong complementarity of R&D activities. The attractiveness and efficiency of a network configuration depends on the level of direct and indirect benefits, costs and subsidies. When the benefits of direct effects are higher than the costs to create a link, a Network Based Policy (NBP), such as the creation of TDs, will not increase welfare because it will not create additional links between firms and between firms and other institutions. We expect that this will happen in areas where there are many firms in the innovative sector and a high concentration of knowledge. When, on the other hand, the benefits are lower than the costs to create a link, the firms will have no incentive to create links and without subsidies a network will be not created. Therefore, when the number of firms and research centers in a given location is high, the possibility that networks may be created spontaneously is high, and we can conjecture that a NBP will not be additional. If the number of potential partners in a location is low, a NBP may not be efficient. In the intermediate situations a NBP may be useful.

In order to evaluate the welfare effects and the network configuration of a TD, Network Analysis tools (Wasserman and Faust 1994) may be fruitfully used, bearing in mind that the aim is to evaluate the additionality and the welfare effect of Italian TDs. For example, if the network is highly centralised, the spread of information among organizations is high but there is a risk that the network will break down if the central organizations fail (or exit from the network). If the degree of centralization is low and many organizations play a central role, the failure of one central actor will not affect the other parts of the network. An index that might better describe the importance of having several links (e.g., where two organizations are involved in more than one research project) in a network may be defined by the ratio between the sum of valued links and the number of effective links activated for each actor (the higher the value of this index, the lower the transfer cost and the more efficient the network). Furthermore, blockmodeling analysis can help to define clusters of equivalent organizations that present similar relational behaviours in the district.

3 The Italian Technological Districts

In Italy, since 2002, steps have been taken at an institutional level to set up innovation networks to stimulate cooperation in R&D projects between firms and research centers. The government has identified 27 geographical areas (14 in Central and Northern Italy and 13 in the South) as potential locations in which to create organizational entities recognized by the Italian Ministry for University and Research (MIUR). The firms and other research institutions, belonging to a TD through formal agreements, can apply for public grants to carry out R&D projects.

A typology of the technological clusters[1] can be built by means of a classification of their share of productive and innovative activity in one of the four technological areas (pharmaceutical, biomedical, aeronautical and ICT) (Table 1). By using the model developed by Jackson and Wolinsky (1996), the number of districts where we can expect that the policy creates additional innovation can be computed. Table 1 shows that seven out of nine TDs in the North are located in an area where a

[1]Recently a list of technological clusters in Italy has been published (Intesa Sanpaolo - Servizio Studi e Ricerche 2010), which identifies four technological areas (pharmaceutical, biomedical, aeronautical and ICT) based on the ATECO 2007 classification. The conditions to identify a technological cluster in an area are: *i)* number of employees > 500; *ii)* share of employees > 5%; *iii)* number of firms > 20. If two of these conditions are satisfied, the area is considered a technological cluster.

Table 1 A = location of the technological area; B = Existence of an innovation cluster in the area of location of the TD (Yes/No); C = Types of clusters (H-H = High intensity when the share is higher than 20% of the national value; L-L = low intensity under 5% of the national value); D = Expected additional links as a result of NBP policy

	B			*C*		*D*		
A	Yes	No	Total	H-H	L-L	None	High	Low
North	7	2	9	3	2	4	3	2
Center	2	3	5	1	0	3		2
South	3	10	13	0	1	11	1	1
Total	12	13	27	4	3	18	4	5

technological pole already exists, but in four cases we do not expect the policy to create network additionality. The main reason for this is that there are spontaneous forces determining network formation and a specific policy will not create new opportunities. In the Center, our conjecture is that two TDs have a low possibility of producing additional links. And in the South we feel that only in one TD there is a high possibility of additional links being created.

In the following, we consider four TDs that a have different possibility of producing network additionality. The first (TD_1) is located in an area where ICT and aeronautical sectors play an important role. The second (TD_2) is located in an area with large firms but a low concentration in the high-tech sector. The third (TD_3) is not related to any technological cluster but there are several academic and important research centers. The fourth (TD_4) is located in an area where there is an intermediate level of firms involved in the high-tech sector and good public research centers. The first three TDs present different contexts but they have a low possibility that positive results on network additionality will be achieved; whereas the fourth one has a high possibility of achieving positive results. These results obtained via economic theory are confirmed by the analysis performed using Network Analysis techniques.

3.1 A Network Analysis Approach of Italian TDs

The collaboration established between organizations in the four TDs will be reconstructed by considering data provided by public grants for R&D projects financially supported by MIUR or by European funds, from the start-up phase of the TDs until the last granted projects in June 2011[2] through network indices and blockmodeling analysis (Ziberna 2007). These techniques enable the description of collaboration patterns in each TD and the identification of network patterns according to the definition of groups of equivalent organizations.

To specify the size of the four networks, we consider the provisional list of associated members of each TD deemed to be in the network in the start-up

[2]The online information provided by the TDs' websites and the databases of the research projects have been integrated with information directly obtained from TD administrative staff.

Table 2 Characteristics of the four Italian TDs

Associated members		Ass. Members Typology: Firms	Research Centers	Administr. Instit.	Others	Project Partners	Research Projects	Project Typology (%): National	European
TD_1	22	27.27	31.82	18.18	22.73	49	10	100.00	0.00
TD_2	25	56.00	8.00	8.00	28.00	41	7	100.00	0.00
TD_3	23	39.13	47.83	0.00	13.04	41	11	45.45	54.55
TD_4	29	44.83	37.93	3.45	13.79	21	15	86.67	13.33

phase and then add new associated members and all partners involved in the research projects by means of the connections observed from this initial "core".[3] Table 2 shows the number of organizations involved in each district according to certain characteristics (associated member *vs* project partner, typology of associated member—firms, public or private research centers, regions, foundations, etc.) and the number and the typology of research projects. In particular, according to the typology of associated members, the presence of firms is relevant in TD_2 and TD_4 (56.00% and 44.83%, respectively), while an involvement of public and private research centers is particularly present in TD_3 (47.83%). Regarding the kind of research projects, for TD_1 and TD_2 the only source of funding are national projects, whereas for TD_3 there is noteworthy percentage of research projects financially supported by the European Commission (54.55%).

3.2 Network Characteristics

The network analysis approach is applied to the collaboration data[4] to describe structural characteristics of the network and to highlight both the role and the position of organizations.[5]

[3]For TD_3 and TD_4, the district is considered as an actor in the network because it participates as a partner in some research projects. Furthermore, for TD_4 the different departments of the same institution are considered as single nodes in the network.

[4]Collaboration data are extracted from the set of research projects and from the set of organizations arranged in four affiliation matrices. $\mathbf{A}$ $(n \times p)$ is the affiliation matrix with a_{ij} $(i = 1, \ldots, n;$ $j = 1, \ldots, p) = 1$ if the i-th member participates in the j-th research project, *0* otherwise. From $\mathbf{A}$ we derive an adjacency matrix $\mathbf{G}_w$ of size $(n \times n)$ that represents an undirected weighted adjacency matrix, whose entries are equal to *0* if two organizations have never collaborated in research projects, or to the number of research projects shared by pairs of organizations. The $\mathbf{G}_w$ matrix can be analysed after removing the diagonal entries (which represent the total number of research projects for each member) and setting all entries greater than zero to "1". The new $\mathbf{G}_b$ matrix is an undirected binary adjacency matrix, where only the *presence* of ties is taken into account.

[5]In the following the network indices, at both global and actor level, are computed starting from the four $\mathbf{G}_b$ matrices to explore the collaboration patterns among members in each district, while blockmodeling analysis is performed on the four $\mathbf{G}_w$ matrices to identify the main characteristics of network structures.

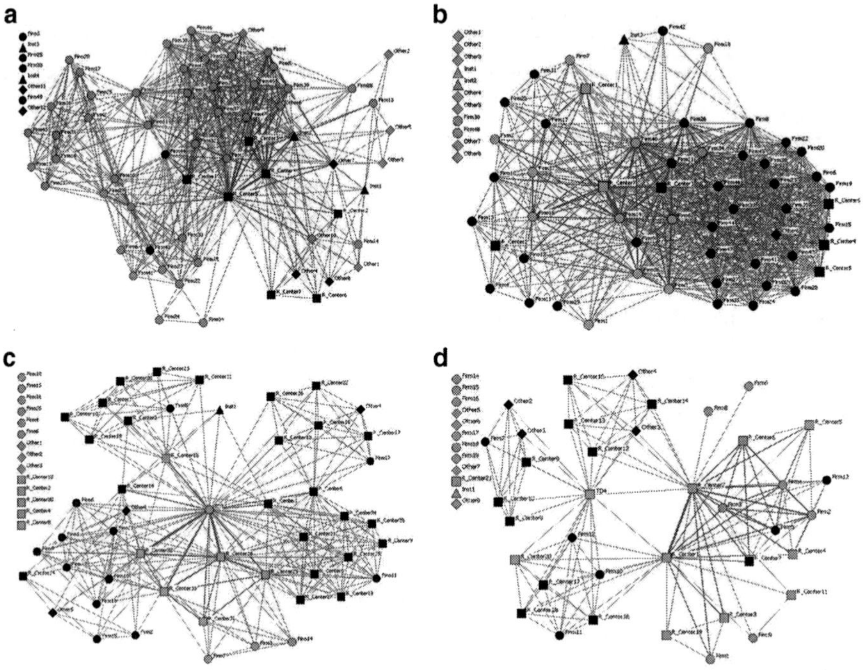

Fig. 1 Valued graphs showing the joint participation of pairs of organizations in research projects in the four TDs: (**a**) TD_1, (**b**) TD_2, (**c**) TD_3 and (**d**) TD_4. The colour of the nodes is related to associated members (*grey*) and project partner (*black*), whereas the shape shows the different member typology (*circle* = firms; *square* = research centers; *triangle* = institutions; *diamond* = other)

The network visualizations (Fig. 1) and the network indices (Table 3) show the presence of a relatively strong connectivity in all networks, especially for TD_2 and TD_1 (the density values are 0.38 and 0.25, respectively), and a high degree of network centralization for TD_3 (56.64%), due to the presence of few organizations in the network involved in several research collaboration agreements. The analysis of actor-level indexes, based on centrality measures (*degree* and *betweenness*), highlights the power of individual organizations arising from their relationships with others. The following characteristics can thus be traced:

- In TD_1 and TD_4 both firms and research centers play a central role in the network in terms of number of links (degree) activated with other organizations sharing research projects, whereas public research centers occupy advantageous positions in ensuring connections between organizations. In TD_4 the district also has a high betweenness centrality value.

Table 3 Network characteristics of the four Italian TDs

	Members	Isolated	Edges	Density	Avg. degree (*SD*)	Degree central.	Valued degree
TD_1	71	8	619	0.25	17.44 (*11.87*)	44.93	2.11
TD_2	66	11	824	0.38	24.97 (*16.37*)	46.06	2.22
TD_3	64	14	286	0.14	8.94 (*7.73*)	65.64	2.71
TD_4	50	12	146	0.12	5.84 (*5.30*)	38.61	2.50

- In TD_2 and TD_3 an opposed situation is observed: on the one hand TD_2 mainly presents firms in a central position but no organization plays a broker role; on the other TD_3 confirms the presence of highly active research centers with the district playing a strong broker role.

One quite important item of information for evaluating the network is the number of projects in which pairs of organizations participate, because if two organizations are involved in several projects there will be a high possibility that they are linked more than once. We then define an index that represents the average weight for the activated links with a value larger than one in the $\mathbf{G}_w$ matrix (Valued degree[6] in Table 3). Observation of this index for the four districts suggests that for TD_3 and TD_4 there could be a lower cost of information transfer and a higher network efficiency, given the greater number of projects shared by pairs of organizations.

The identification of groups of equivalent organizations is achieved by looking at the results obtained from a clustering procedure. The partition into three groups for TD_1 and TD_4, into four groups for TD_2 and into five groups for TD_3 derived from a hierarchical cluster analysis was further investigated by means of blockmodeling analysis.[7] According to the blockmodel types (Doreian et al. 2005) and the block composition, these initial results show that TD_1, TD_2 and TD_3 present a similar structure near to the *core-periphery model* with one core position internally cohesive and connected with other positions but also cohesive subgroups with intraposition ties. The core composition differs in the three TDs, with research centers and firms in TD_1, mainly firms in TD_2 and public research centers in TD_3. TD_4 presents *cohesive subgroups* and no ties between positions but just some light evidence versus core-periphery model. The core is composed of research centers and firms.

4 Conclusions

According to the theoretical framework to analyse the additionality of NBP policy, the network analysis results show, in terms of network size, a larger presence of associated members in TD_4 and partners in TD_1 with respect to the other two

[6]This index represents the average weight of activated links computed as the ratio of the sum of valued links compared to the number of activated links for each actor.

[7]Euclidean distance, Ward agglomerative method and blockmodeling analysis are performed on the valued adjacency matrix $\mathbf{G}_w$ using the blockmodeling package of R software (Ziberna 2007).

districts. On the other hand, TD_1 has the lower number of firms as associated members. This is an expected result because firms in the area where TD_1 is located have no incentive to enter into a bureaucratic structure like an Italian TD. They find it more convenient to establish links directly because it is easier for TD_1 to find partners for research projects since it is located in a highly developed economic environment. The same is observed for TD_2 which is located in an industrial area, whereas in TD_3 the high number of partners is mainly due to the presence of research centers in the area.

Then, if we evaluate the network size on the basis of associated members, our expectations about additionality are confirmed even though we cannot exclude the possibility that the NBPs in TD_2 and TD_3 have created new links. It is more difficult to analyse the effects from a welfare point of view, in terms of network configurations described by density and centralization degree results. In this respect, TD_4 presents both the lowest density and centralization values. Hence, the structure of TD_4 differs from the near core-periphery model observed for the other three districts and this result could be considered a positive characteristic. High valued degree scores suggest that TD_3 and TD_4, which are in less developed areas than the two other districts, could have a lower information transfer cost and a higher network efficiency, given the greater number of projects shared by pairs of organizations.

Acknowledgments Work supported by PRIN 2008 Network Theory, Evaluation of the technological districts and of the public policies for innovation.

References

Antonelli, C. (2000). Collective knowledge communication and innovation: The evidence of technological districts. *Regional Studies*, *34*, 535–547.

Antonioli, D., & Marzucchi, A. (2010). The behavioural additionality dimension in innovation policies: a review. Quaderni del Dipartimento di Economia Istituzioni Territorio, Universitá di Ferrara, *10* Available via DIALOG http://deit.economia.unife.it.

Capuano, C., & Del Monte, A. (2011). La politica per la costruzione di reti innovative e metodologia empirica. In A. Zazzaro (Ed.) *Reti d'imprese e territorio* (pp. 133–169). Bologna: Il Mulino.

Del Monte, A., D'Esposito, M. R., Giordano, G., & Vitale, M. P. (2011). Analysis of collaborative patterns in innovative networks. In S. Ingrassia, R. Rocci, & M. Vichi (Eds.) *New perspectives in statistical modeling and data analysis* (pp. 77–84). Heidelberg: Springer.

Doreian, P., Batagelj, V., & Ferligoj, A. (2005). *Generalized blockmodeling*. Cambridge: Cambridge University Press.

Intesa Sanpaolo - Servizio Studi e Ricerche (2010). Monitor dei distretti. Available via DIALOG http://www.osservatoriodistretti.org/sites/default/files/monitor-dei-distretti-dicembre-2010.pdf

Jackson, M. O., & Wolinsky, A. (1996). A strategic model of social and economic networks. *Journal of Economic Theory*, *71*, 44–74.

Ziberna, A. (2007). Generalized blockmodeling of valued networks. *Social Networks*, *29*, 105–126.

Wasserman, S., & Faust, K. (1994). *Social network analysis: methods and applications*. Cambridge: Cambridge University Press.

Clustering and Registration of Multidimensional Functional Data

M. Chiodi, G. Adelfio, A. D'Alessandro, and D. Luzio

Abstract In order to find similarity between multidimensional curves, we consider the application of a procedure that provides a simultaneous assignation to clusters and alignment of such functions. In particular we look for clusters of multivariate seismic waveforms based on EM-type procedure and functional data analysis tools.

1 Introduction

Looking for curve similarity is a main issue in many application fields, like graphics, computer vision, speech recognition and geographic information systems. Continuous transformations (alignment) are required to measure dissimilarity between curves in order to eliminate phase variability of functions; functional data analysis may provide useful tools to quantify these differences and explain the variability within and between functions.

We introduce an approach to find clusters from a set of individual seismic waveform, represented by seismograms and recorded by a seismic network, according to the functional nature of data, highlighting their common characteristics. A procedure for simultaneous clustering and alignment of sets of varying curves observed in time, such as seismic signals, is introduced: the alignment problem is handled with the introduction of a simple procedure, based on a similarity

M. Chiodi (✉) · G. Adelfio
Dipartimento di Scienze Economiche, Aziendali e Statistiche Università degli Studi di Palermo, viale delle Scienze ed. 13, 90128 Palermo, Italy
e-mail: marcello.chiodi@unipa.it; giada.adelfio@unipa.it

A. D'Alessandro
Istituto Nazionale di Geofisica e Vulcanologia, Centro Nazionale Terremoti, Italy

D. Luzio
Dipartimento di Scienza della Terra e del Mare, Università degli Studi di Palermo, Italy

P. Giudici et al. (eds.), *Statistical Models for Data Analysis*, Studies in Classification, Data Analysis, and Knowledge Organization, DOI 10.1007/978-3-319-00032-9_11,

measure between curves. In a different context, Chiodi (1989) proposed a method for clustering multivariate short time series, based on the similarities of shapes.

In Sect. 2 we introduce some notation of functional data, looking at the functional nature of waveforms and related continuous transformation of curves. The proposed method that aligns and assigns curves to clusters of waveforms according to an iterative EM-type procedure is proposed in Sect. 3. An application of the method to seismic waveform data is proposed in Sect. 4; Sect. 5 is devoted to discussion of results and some general conclusions.

2 Functional Curves

Let $\mathbf{x}_1, \mathbf{x}_2,..,\mathbf{x}_N$ be unsynchronized multidimensional curves with

$$\mathbf{x}_i(t) = \{x_{i1}(t), x_{i2}(t), \ldots, x_{id}(t)\},\ i = 1, \ldots, N$$

defined on closed real intervals such that $\mathbf{x}_i(t) : [0, T_i] \rightarrow \mathbb{R}^d$. In order to compare main features of curves $\mathbf{x}_i$ we have to look for a transformation of their domain, that is to use a registration procedure that optimizes a similarity criterion between curves and provides registered functions $\tilde{\mathbf{x}}_i$, $\tilde{\mathbf{x}}_i = \mathbf{x}_i \circ h_i,\ \forall\ i$, where $h_i(t)$ is a warping function, that is an invertible transformation of time t for each i. The registration or alignment of salient curves features requires the estimation of the time-warping transformations h_i of the argument t, by maximizing a similarity index between each curve and a reference curve (often named template).

From a seismological point of view we looked for simple transformation of waves, like the linear one, in order to slightly modify original waves. Furthermore we carried out a preliminary explorative analysis of the whole set of waves by following the approach suggested in James (2007) also fitting general nonlinear warping functions: from this analysis we observed that the individual optimal warping function of each wave was very close to the linear one.

3 Clustering and Registering of Curves

Clustering of curves can be seen as an issue of clustering of functional data and more generally it can be defined in the wide framework of partition type cluster analysis. Let $\{\mathbf{x}_1, \ldots, \mathbf{x}_N\}$ be N multivariate curves, where $\mathbf{x}_i,\ \forall i$ are functions in $\mathbb{R}^d, d \geq 1$. For example, $\mathbf{x}_i$ can be a four-dimensional seismic wave observed as a parametric curve in $\mathbb{R}^4$ and depending on a real parameter t. We seek a partition $\mathcal{P} = \{G_1, G_2, \ldots, G_k\}$ of N objects in k exhaustive clusters with strong internal homogeneity; as usual, we need to define a measure of internal homogeneity, or, alternatively, a measure of distance from a reference curve $\mathbf{C}$ defined in each group.

However some sort of curve registration procedure can improve the overall similarity of this kind of data. The problem of curves alignment has been studied in different fields: this is referred to as curve registration in statistics (Silverman 1995; Ramsey and Silverman 2006; Ramsey and Li 1998), time warping in engineering (Wang and Grasser 1997) and structural averaging in the context of computing an average curve (Kneip and Gasser 1992). As a first order approximation to the best warping function, a simple linear transform $f(\cdot)$ of the argument can be used for each of the d directions. The distance of a generic curve $\mathbf{x}_i$ from a group G_m is defined as the distance of the registered curve $\tilde{\mathbf{x}}_i$ from a reference curve $\mathbf{C}_m$ of G_m, in our implementation $\mathbf{C}_m$ will be taken as the average of the curves belonging to G_m, but other choices are of course possible, like median curves, medoids, etc.; see Sangalli et al. (2010). More formally this *registered* distance is defined as:

$$\tilde{\delta}(\mathbf{x}_i, \mathbf{C}_m) = \min_{\substack{a_{ims}, b_{ims} \\ s = 1, \ldots, d}} \delta(\tilde{\mathbf{x}}_i, \mathbf{C}_m) \tag{1}$$

where each registered curve $\tilde{\mathbf{x}}_i$ has d components $\tilde{x}_{is}$ $s = 1, \ldots, d$, obtained through a linear warping defined by $h_{ims} = a_{ims} + b_{ims}t$ with a_{ims} and b_{ims} $(m = 1, \ldots, k)$ cluster specific transform coefficients. For each of these distances $\tilde{\delta}(\mathbf{x}_i, \mathbf{C}_m)$, $2d$ parameters a_{ims}, b_{ims}, $s = 1, \ldots, d$ have to be estimated, and this can be done minimizing the distance with respect to a_{ims}, $b_{ims}, i = 1, \ldots, N$, $s = 1, \ldots, d$, $m = 1, \ldots, k$, separately for each direction.

Then, given a partition $\mathscr{P}$, we denote by $g(\cdot)$ an associated labelling function, such that $g(i) = m : \mathbf{x}_i \in G_m \; \forall i$ and define an overall measure of distances as the average of the N registered distances of the curves from their group template, that is:

$$\Delta_{\mathscr{P}} = \frac{1}{N} \sum_{i=1}^{N} \tilde{\delta}(\mathbf{x}_i, \mathbf{C}_{g(i)}) \tag{2}$$

Following the aim of the k-means method, and given a starting partition $\mathscr{P}^1 = \{G_1^1, G_2^1, \ldots, G_k^1\}$, the proposed clustering method deals with the simultaneous optimal clustering and warping of curves as in an iterative EM-type algorithm:

(a) At the v-th iteration, compute the reference curves $\mathbf{C}_1^v(t), \mathbf{C}_2^v(t), \ldots, \mathbf{C}_k^v(t)$ for each group.
(b) For each curve $\mathbf{x}_i^v(t)$, $i = 1, \ldots, N$, compute the registered distance from the reference curve $\mathbf{C}_m^v(t)$ of each group G_m^v of the current partition by means of the distance $\tilde{\delta}(\mathbf{x}_i^v(t), \mathbf{C}_m^v(t))$, defined in (1).
(c) For each curve $\mathbf{x}_i^v(t), i = 1, \ldots, N$ determine the labelling function $g^{v+1}(i)$ and then update the partition accordingly $\mathscr{P}^{v+1} = \{G_1^{v+1}, G_2^{v+1}, \ldots, G_k^{v+1}\}$.
(d) Apply to each curve $\mathbf{x}_i^v(t)$ the optimal warping functions (found at step (b)) of the G_i^*-th group and replace each curve with the registered curve $\mathbf{x}_i^{v+1}(t)$.
(e) Update v and repeat from step 1 until some stopping rule is satisfied.

At the step (d), the warping coefficients inside each group can be normalized in order to have a zero average shift and a unit average scale (see Sangalli et al. 2009). In our experience a reasonable starting partition is obtained cutting the tree of a hierarchical agglomerative procedure: this choice is usually considerably better than a random starting partition and is computationally acceptable when dealing with hundreds of multivariate waves.

4 Application to Seismograms

Earthquakes are usually generated by fracture processes that occur in Earth's lithosphere. The discontinuous strain fields associated with earthquakes are largely compatible with rock dislocation along faults. Dislocation causes partial release of the elastic strain energy stored by tectonic processes and the released energy is partially propagated away from its source as a wave-field. Waveforms, in particular, are the signals generated from the movement of waves in a physical medium. Both hypocenter and focal mechanism are determined by the analysis of the waveforms represented by seismograms recorded by a seismic network, that are a spatial sampling of the wave-field. Waveform correlation techniques have been introduced to characterize the degree of event similarity (Menke 1999) and in facilitating more accurate relative locations within similar event clusters by providing more precise timing of P and S arrivals (Phillips et al. 1997). Assuming that waveforms similarity implies the similarity of focal mechanisms, a procedure that finds clusters from a set of seismograms according to the functional nature of data is proposed in Adelfio et al. (2011), applying a variant of a k-means algorithm based on the principal component rotation of data.

4.1 The Sicilian Data

On September 6-th, 2002 at 01:21 GMT a shallow earthquake with $M_W = 5.9$ occurred near Palermo (Sicily); it was recorded by the Italian National Seismic Network operated by INGV and relocated by Giunta et al. (2004) in Southern Tyrrhenian Sea, some tens of kilometers offshore from the Northern Sicilian coast, followed by a sequence of other seismic events (Fig. 1). The Palermo aftershocks sequence is an interesting case of study because it constitutes a data set suitable for the detailed reconstruction of geodynamic and seismogenic models (Adelfio et al. 2006).

In the last years, the *Centro Nazionale Terremoti* of the INGV developed an own model of OBS/H (Ocean Bottom Seismometer with Hydrophone). In December 2009, about 7 years after the Palermo seismic crisis, the OBSLab deployed one OBS/H near the epicentral area of the mainshock at 1,500 m of depth. The OBS/H

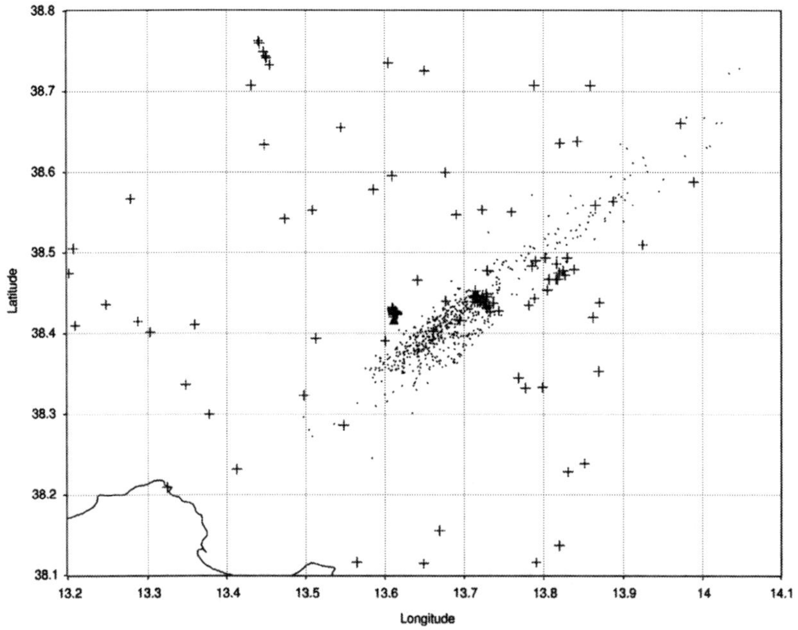

Fig. 1 Location of the Palermo 2002 seismic sequence epicenters (*points*) and local seismic activity recorded by the OBS/H in 2009-2010 (*crosses*). The *triangle* shows the OBS/H position

was recovered in July, after about 8 operative months. In this experiment the OBS/H was equipped by a 3-components velocity seismometer and one hydrophone. All the signals were sampled at 200 Hz. On the basis of the signal/noise ratio, 159 earthquakes (159×4 seismograms, i.e 3-space dimensions and pressure measurement) were selected from the initial dataset.

4.2 Warping-Clustering Results

The proposed simultaneous warping-clustering procedure is applied to the 159×4 signals relative to the selected events (for fixed $d_{max} = 0.6$) for different values of $k, k = 2, \dots, 12$.

We observed the sequence of the optimal values f_k^* of the target function (2) for $k = 2, \dots, 12$ and computed the simulated distribution of the target function from 100 random partitions of the 159 curves in k groups: for each value of k we compute the average ${}_fM_k$ and the standard deviation ${}_fS_k$ of the 100 simulated values and used them to standardize the sequence of f_k^*, such as $\bar{f}_k^* = \frac{f_k^* - {}_fM_k}{{}_fS_k}$ (Fig. 2). These values were plotted against k only for $k = 6, \dots, 12$, since values of $k < 6$ produced a large number of unclassified curves and also empty groups

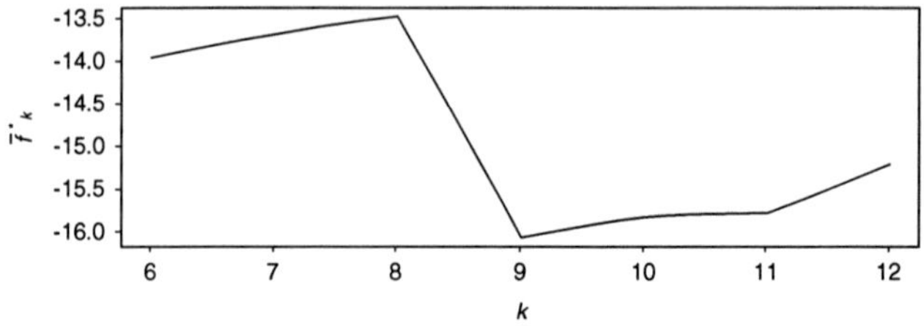

Fig. 2 Values of $\bar{f}_k^*$ against $k, k = 6, \ldots, 12$, for the choice of the number of groups

Table 1 Mean correlation and standard deviation of correlations with respect the cluster specific average curve and dimensions of the nine identified clusters

m	1	2	3	4	5	6	7	8	9
$\bar{r}_m$	0.931	0.889	0.971	0.695	0.958	0.712	0.801	0.932	0.802
$\bar{s}_m$	0.021	0.081	0.018	0.147	0.021	0.069	0.133	0.043	0.126
n_m	6	11	9	13	7	11	8	6	10

and therefore useless results. The results suggest the value $k = 9$, which showed a local minimum. For $k = 9$ the procedure stops at the third iteration (after there are no more changes in event positions), providing a value of the target function (2) that decreases from 0.202 to 0.147. When the clustering approach is applied to the unregistered curves, it returns decreasing values of the target function after two iterations from 0.227 to 0.218, suggesting the undeniable role of the warping procedure in order to find similarity between curves. The nine clusters contain signals with internal mean correlation ($\bar{r}_m = 1 - \Delta_m$, $m = 1, 2, .., 9$, with Δ_m the mean distance between curves of group m and their corresponding mean curve) greater than 0.6 relative to 78 of 159 earthquakes recorded. The $\bar{r}_m$ values and mean standard deviations $\bar{s}_m$ with respect the cluster-specific reference curves, together with cluster dimensions n_m, are reported in Table 1.

Figure 3 represents the distribution of the distances, according to the introduced measure in (1), between registered curves and their average for the nine identified clusters. Four of these nine groups (clusters 1, 3, 5 and 8) show a mean correlation inside clusters greater than 0.93 and a standard deviation less than 0.043; the events of these four clusters may be regarded as seismological multiplets, see also Carmona et al. (2009). As an example, the signals of events assigned to cluster 5 are shown in Fig. 4.

5 Concluding Remarks

We have presented a simple and computationally efficient procedure aimed to identify clusters of earthquakes with similar hypocentral parameters and focal mechanisms within the complex seismic activity of the investigated area. The

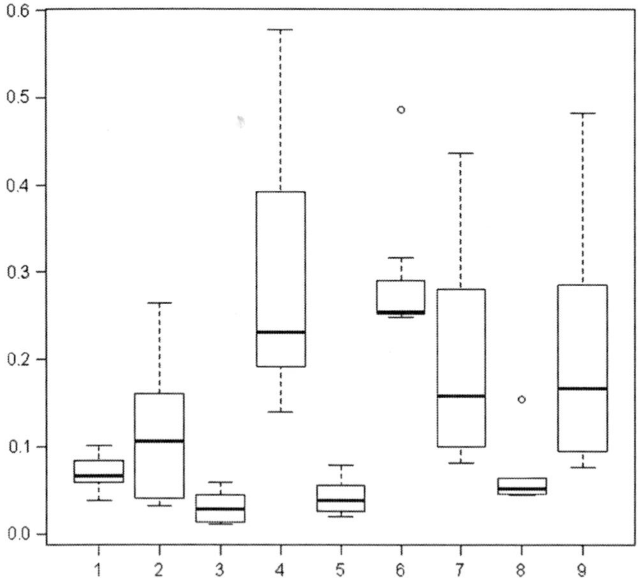

Fig. 3 Boxplot of distances between curves and their average for the nine clusters

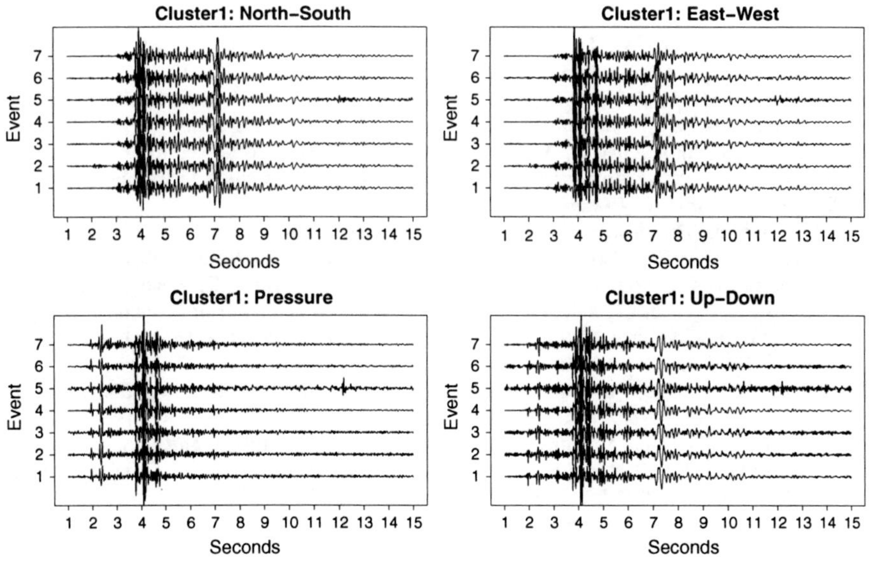

Fig. 4 Four dimensional signals of the events of cluster 5

proposed procedure is based on the assumption that seismic events which are similar with respect to the mentioned parameters generate similar wave fields. This procedure provides a simultaneous assignation and alignment of multivariate signals finding similar features, assumed as main characteristics of curves.

The application of this procedure to the small energy earthquakes recorded by the OBS/H showed its ability in identifying similarities between waveforms also for events with difference in magnitude greater than 1.5 and with difference in hypocentral distance greater than 10 km. We have identified four multiplets with very high mean correlation inside the cluster (>0.94) and very little standard deviation (<0.035). The signals relative to different multiplets are very dissimilar due to the different positions of their seismogenic volumes and their energy release. All the centers of seismic activity identified by the multiplets are clearly associated with the seismogenic volume of the seismic crisis of 2002 (Fig. 1). Given the high similarity of the signals, all the data related to the events of the multiplets can be used to determine the parameters of a single representative average earthquake. The data relative to the clusters characterized by lower values of average correlation coefficient will be used instead to make relative determinations of the parameters of each seismic event, since this kind of estimates is generally much more precise than the absolute estimates. The main advantage of this procedure is its capability to identify clusters of waveforms also when events have rather different hypocentral parameters, such as the local magnitude M_L, that depends on the logarithm of the maximum amplitude of a seismic waveforms, and the hypocentral distance.

References

Adelfio, G., Chiodi, M., De Luca, L., Luzio, D., & Vitale, M. (2006). Southern Tyrrhenian seismicity in space-time-magnitude domain. *Annals of Geophysics, 49*, 245–1257.

Adelfio, G., Chiodi, M., D'Alessandro, A., & Luzio, D. (2011). FPCA algorithm for waveform clustering. *Journal of Communication and Computer, 8*(6), 494–502. ISSN 1548–7709.

Carmona, E., Stich, D., Ibañez, M., & Saccorotti, G. (2009). Multiplet focal mechanisms from polarities and relative locations: the Izanjar swarm in Southern Spain. *Bulletin of the Seismological Society of America, 99*, 3421–3429.

Chiodi, M. (1989). The clustering of longitudinal data when time series are short. *The analysis of multiway data matrices*. Coppi R., & Bolasco, S. eds. Amsterdam: North-Holland, 445–453.

Giunta, G., Luzio, D., Tondi, E., De Luca, L., Giorgianni, A., D'Anna, G., Renda, P., Cello, G., Nigro, F., & Vitale, M. (2004). The Palermo (Siciliy) seismic cluster of Septermber 2002, in the seismotectonic framework of the Tyrrhenian Sea-Sicily border area. *Annals of Geophisics, 47*(6), 1755–1770.

Menke, W. (1999). Using waveform similarity to constrain earthquake locations. *Bulletin of the Seismological Society of America, 89*, 1143–1146.

James, G. M. (2007). Curve alignment by moments. *The Annals of Applied Statistics, 1*, 480–501.

Kneip, A., & Gasser, T. (1992). Statistical tools to analyze data representing a sample of curves. *Annals of Statistics, 20*, 1266–1305.

Phillips, W. S., House, L. S., & Feheler, J. (1997). Detailed joint structure in a geothermal reservoir from studies of induced microearthquake studies. *Journal of Geophysical Research, 102*, 745–763.

Ramsey, J. O., & Li, X. (1998). Curve registration. *Journal of the Royal Statistical Society, Series B, 60*, 351–363.

Ramsey, J. O., & Silverman, B. W. (2006). *Functional data analysis*. New York: Springer.

Sangalli, L. M., Secchi, P., Vantini, S., & Veneziani, A. (2009). A case study in explorative functional data analysis: geometrical features of the internal carotid artery. *Journal of the American Statistical Association, 104*(485), 37–48.

Sangalli, L. M., Secchi, P., Vantini, S., & Vitelli, V. (2010). Functional clustering and alignment methods with applications. *Communications in Applied and Industrial Mathematics, 1*(1), 205–224.

Silverman, B. W. (1995). Incorporating parametric effects into functional principal components analysis. *Journal of the Royal Statistical Society, Series B, 57*, 673–689.

Wang, K., & Grasser, T. (1997). Alignment of curves by dinamic time warping. *The Annals of Statistics, 25*, 1251–1276.

Classifying Tourism Destinations: An Application of Network Analysis

Rosario D'Agata and Venera Tomaselli

Abstract Tourism is basically a spatial phenomenon, which implies moving consumption within space. Starting from the assumption that the destinations are nodes of a network, we are able to reconstruct a spatial grid where each locality shows different grades and types of centrality. The analysis, focusing on the spatial dimension, shows clusters of locations. By shifting interest from single locations to destination networks, the study points out the structural features of each network. Employing traditional network analysis measures, we classify destinations considering the routes of a self-organized tourists sample that visited more than one destination in Sicily.

1 Tourism Mobility Among Destinations: A *NA* Approach

A tourism destination is a system composed of both a large amount of natural, cultural, artistic resources—also artificially built, such as museums, theme parks or sport complexes—and a network of groups of economic, non economic and institutional actors, whose prevalent activity is providing tourism related services to visitors and travellers. So, we can consider the concept of tourism destination as "a physical space in which a visitor spends at least one overnight. ... It has physical and administrative boundaries defining its management and images and perceptions defining its market competitiveness" (UNWTO 2002).

In the study of tourism mobility, the spatial distribution of tourists affects some areas at regional and/or sub-regional level where the economic and environmental impact of tourism is more concentrated.

R. D'Agata (✉) • V. Tomaselli
University of Catania, Vitt. Emanuele II, 8, Catania, Italy
e-mail: rodagata@unict.it; tomavene@unict.it

P. Giudici et al. (eds.), *Statistical Models for Data Analysis*, Studies in Classification, Data Analysis, and Knowledge Organization, DOI 10.1007/978-3-319-00032-9_12,

In our opinion, *Network Analysis* (*NA*) methods could be appropriate to define a territorial network of tourism demand by mapping the spatial distribution of tourism mobility. If the tourism destination pattern tends to share certain formal, informal and structural properties as a whole, by means of *NA*, we can classify destinations (termed by *nodes*) by a set of metrics able to measure the relationships (displayed as *links*) among tourism destinations.

Network analysis (*NA*) has been used both to explore how networks are formed in the relationship among different *nodes* and—taking the network structure as given—to analyse economic phenomena that are constrained by the structure (Wasserman and Faust 1994). *NA* also plays an important role in determining the effectiveness of the typology of the linkages. It takes into account the structure of links among nodes and their 'location' in the network, deriving consequences for both the nodes and the whole system (Hanneman and Riddle 2005). So, we can analyse the combination of the network components—*nodes* and their *links*—and how they are assembled according to their number and their distribution, referring to the static elements and to the dynamic processes that govern a network system.

Although the *NA* methods are quite well known and tourism is a network business, little has been done so far to apply these techniques to the study of the tourism sector. Some scholars show the usefulness and the effectiveness of this approach, with special regard to the analysis of the features of a tourism network (Scott et al. 2008).

An interesting paper shows the use of these methods to revise the organization of tourism facilities and services in each destination by measuring the structural features of routes taken by tourists in multi-destination trips (Shih 2006).

Here, we used *NA* methods to analyse a database drawn from a survey[1] about the features of tourism demand carried out in Sicily, in order to specify both the relevant and the marginal destinations by their centrality within the route network.

2 *NA* Measures for Tourism Destinations

According to graph theory, after collecting relational data and organising it into a matrix, we compute the network measures of *density*, *in-degree* and *out-degree centrality*, *betweenness centrality* and *closeness centrality*.

In our study, density is used to count the actual number of links as tourism routes among destinations as a ratio of the maximum number of potential connections in the Sicilian area. So, we use density as a property of the whole network. It describes the general level of linkage among the points in a graph. The more

[1]The survey is carried out as PRIN 2007–2009 "Socio-economic effects of behavior and motivations of real tourism in Sicily. Internal mobility and its economic effects" by University of Palermo, Catania, Sassari, Bologna. Selected data consist in face-to-face interviews submitted to tourists during their departures from most important Sicilian airports and ports.

points are connected to one another, the denser will the graph be. The *density* of a graph is defined as the number of lines incident with each node in a graph, expressed as the ratio of the number of relationships that exist compared with the total number of possible ties $g\,(g-1)/2$, if each member were tied to every other member (Wasserman and Faust 1994, p. 101). Density measure is calculated as:

$$\Delta = \frac{L}{\frac{g(g-1)}{2}} \tag{1}$$

where L is the number of lines present. This measure can vary from 0 to 1. So, the density of a complete graph[2] is 1, because all possible ties exist (Rowley 1997).

Afterward, we use centrality as basic measure to identify the most important nodes of the tourism destination network. So, we can recognize some central or main tourism destinations within the network, comparing the different centrality measures. *Degree centrality* is defined as the number of links incident upon a node (Opsahl et al. 2010). Since the network is directed, we specify the two separate measures of degree centrality, namely *in-degree* and *out-degree*: for a node, the first is the number of head endpoints adjacent to a node and *out-degree* is the number of tail endpoints that the node directs to others. The *in-degree* is denoted $deg^{-}(v)$ and the *out-degree* as $deg^{+}(v)$. A vertex with $deg^{-}(v) = 0$ is called a source, as it is the origin of each of its incident edges. Similarly, a vertex with $deg^{+}(v) = 0$ is called a sink. For a directed graph, the degree sum formula states that:

$$\sum_{v \in V} \deg^{+}(v) = \sum_{v \in V} \deg^{-}(v) = |A|. \tag{2}$$

In our research, in order to analyze the tourism destination network, we measure the *betweenness* as an indicator of the importance or influence of a single destination in a pattern. *Betweenness centrality* is a measure of node centrality in a network. Basically, the fraction of shortest paths between node pairs, from all vertices to all others, that pass through that node of interest. It is a more useful measure of the node importance into the network. So, the *betweenness centrality* of a node v is given by the expression:

$$g(v) = \sum_{s \neq v \neq t} \frac{\sigma_{st}(v)}{\sigma_{st}} \tag{3}$$

where σ_{st} is the total number of shortest paths from node s to node t and $\sigma_{st}(v)$ is the number of shortest paths from s to t that pass through a vertex v.

Another important measure of centrality is the *closeness*. In our study, we compute the *in-closeness* centrality to reveal the extent to which each tourism

[2] A complete graph is one in which all the points are adjacent to one another (Wasserman and Faust 1994, p. 102) and each point is connected directly to every other point.

destination as node of the network is reachable from every other destination. *Closeness* is defined as the mean geodesic distance (i.e., the shortest paths) between a vertex v or node and all other vertices reachable from it:

$$\frac{\sum_{t \in V \setminus v} d_G(v,t)}{n-1} \tag{4}$$

The *closeness* $C_C(v)$ for a vertex v is the reciprocal of the sum of geodesic distances to all other vertices of V. We use the reciprocal in order to count as 0 the vertices that are not reachable, because we want that higher values are taken by the most central vertices:

$$C_C(v) = \frac{1}{\sum_{t \in V \setminus v} d_G(v,t)} \tag{5}$$

3 Results

Analysis carried out on a sample of 3,182 self-organized tourists leaving from the airports of Palermo and Catania after they spent a holiday in Sicily visiting at least two destinations. Starting from collected data, we focused on destinations. So we constructed a 74 by 74 adjacency matrix. Then, we considered tourist destinations as network nodes (Fig. 1) connected by direct relationships.

Each of 74 selected destinations is represented by a square whose size is proportional to its network centrality (*indegree*): the higher the number of tourists who chose such destination, the larger the size of the square is.

In the first step, by means of UCINET 6.211 (Borgatti et al. 2002), we calculated some descriptive statistics of whole network cohesion (Scott 2000). The first measure to consider in order to describe the network is the *density* that shows a value of 0.324 (SD = 1.765). Since *density* varies from 0 to 1, the network does not seem to be very dense. In other words, a *density* of 0.324 indicates that the network includes the 32 % of all possible links. Another important network descriptive measure is *centrality*. Considering the aim of the analysis, we focus on *indegree* centrality that, to simplify, we could define as the attractive capability of destinations. An average *indegree* of 23.33 (SD = 42.42) indicates the presence of few attractive nodes.

In a second step, in order to describe every single network of most central nodes, employing an ego-networks analysis, we focused on the network of each destination. The analysis of ego-network (Table 1), reveals as Palermo shows a normalized value of *indegree* (*NrmInDeg*) equal to 5.871 followed by Siracusa, Catania and Agrigento, all with a value higher than 5. Palermo, thus, appears to be the most central destination of both inflow and outflow, as remarked by the value (7.449) of normalized *outdegree* (*NrmOutDeg*).

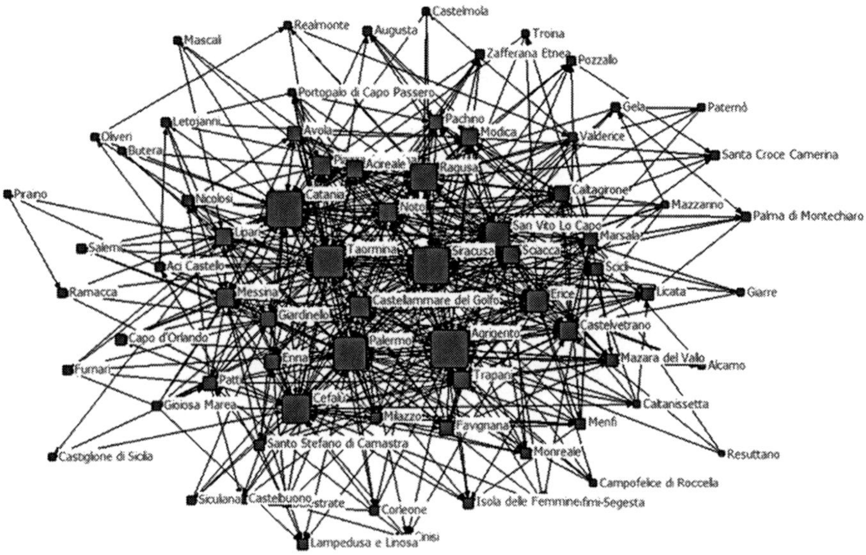

Fig. 1 Network graph of tourism destinations in Sicily

Table 1 Ego-network indicators of most central destinations in Sicily

Destination	NrmInDeg	NrmOutDeg	Size	Densit	nEgoBe
Palermo	5.871	7.449	50	18.29	24.37
Siracusa	5.492	5.840	46	19.18	23.04
Catania	5.240	4.735	44	20.51	20.79
Agrigento	5.019	4.198	48	18.66	20.17
Taormina	4.609	4.135	42	21.37	22.49

Furthermore, centrality can be seen as vicinity to other nodes. In this case we consider the *outcloseness* based on number of outward connections. *Outcloseness*, then, could be seen as proximity to a lot of destinations. The network shows an average value of *outcloseness* equal to 31.97 (SD = 3.3). Ego-network analysis (Table 1) confirms the central role of Palermo (40.91), followed by Taormina (38.71).

Finally the *betweenness* indicates the frequency with which each node is within the shorter route linking every other couple of nodes. It highlights the role of intermediary between two destinations. In the network we observe an average *betweeness* of 84.49. In other words, about the 84 % of links presents intermediary destinations.

In the last step, we aim to point out the destinations caught from the identical places and left towards the identical places. From a heuristic point of view, knowing these destinations could be useful in analyzing tourist routes and allows to detect clusters of places "playing" the identical role inside the network. In order to identify these clusters, we employ the concept of *Structural Equivalence* (Borgatti and

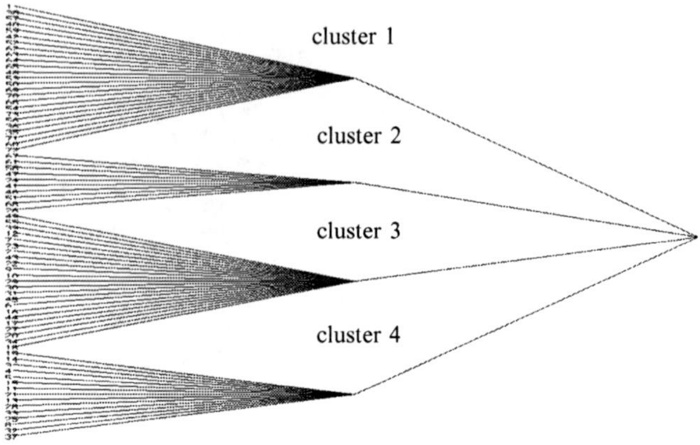

Fig. 2 Clusters of tourism destinations in Sicily

Everett 1992) considering the positions of the destinations in the network. "Two actors are structurally equivalent if and only if they have identical ties to and from identical other actors" (Wasserman and Faust 1994, p. 468). In our application actors are the destinations representing network's nodes. In the traditional field of network analysis, *Structural Equivalence* is considered a strongly limited method because of its severe assumptions and often other concepts of equivalence are preferred (e.g. *Automorphic Equivalence* or *Regular Equivalence*). In tourism studies, however, what is usually seen as a limit could represent an important knowledge element and a useful tourism-planning tool. Clustering the groups of structurally equivalent destinations, in fact, allows to identify alternative routes replacing a destination with an equivalent another one, helping tourism firms in order to differ territorial tourism supply.

Among the techniques proposed in literature to investigate the pattern of similarities about nodes' tie-profiles and to group the nodes in equivalence classes, we employ the procedure of *CONnvergence of iterated CORrelation* (CONCOR). Formalized by Breiger et al. (1975), the CONCOR algorithm is the most common *blockmodelling* method (Nunkesser and Sawitzki 2005; Schwartz 1977).

By means of *Structural Equivalence* approach, we group the 74 destinations of the network in 4 clusters (Fig. 2). The clusters 1 and 2 include destinations mainly located in the South and East of Sicily, whereas the clusters 3 and 4 appear chiefly to characterize the West side of Island and just few central destinations.

Actually geographic proximity plays an important role in determining *structural equivalence*. Therefore, it is probable that neighbouring destinations are structurally equivalent but, focusing on the destinations inside each cluster, we observe that it is not always like this. In the first cluster, for example, we find destinations as ACI CASTELLO, a seaside village in the Eastside of the Island and RESUTTANO, a small hilly village in the Westside and also in the other clusters we observe similar cases. It is important to remember that inside each cluster, every destination shares with each other the *in*-flow and the *out*-flow. In other words, tourists arrive

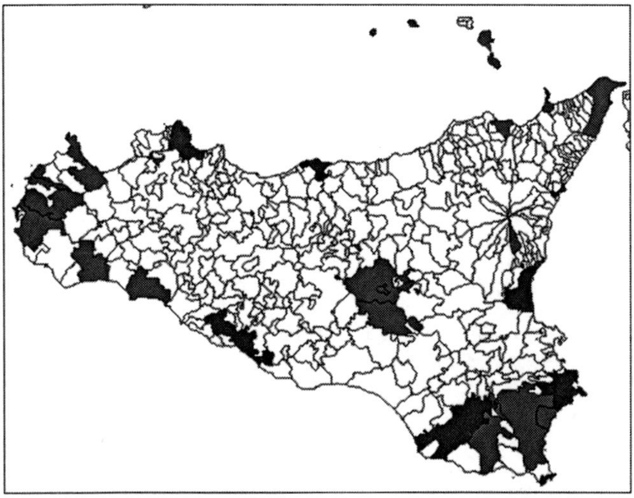

Fig. 3 The map of Lambda Set destinations in Sicily

in those destinations leaving from the identical destination and leave them towards the identical place. Understanding the reason of this *Structural Equivalence* could provide important information in order to develop tourism industry in Sicily.

Another important feature in tourism studies concerns the "bridge" role played by some destinations. Such bridging destinations represent an intermediate point along the route and investigating them could help the tourism routes planners to "re-draw" tourism paths.

In order to analyze the bridging destinations, let switch the attention from the nodes to the links and introduce the *lambda sets* (Borgatti et al. 1990). Quoting Wasserman and Faust (1994) we can say that considering "pairs of nodes in the subgraph G_s with node set N_s, the set N_s, is a lambda set if any pair of node in a lambda set has larger line connectivity than any pair of node consisting of one node from within the lambda set and a second node outside the lambda set". Comparing connectivity lines, so, it is possible to rank network links from the most important ones to the less important. The most important links represent the bridges without which the network might loose its cohesion.

Considering tourism destination in Sicily, we analyzed the lambda sets of the node ties aiming to point out the bridging destinations which most tourism routes follow. Figure 3 shows the map of lambda sets in Sicily, where in darker grey we find both core and bridge destinations, while in light grey just the bridge destinations. Except for two destinations localized in the central area of the Island (ENNA and PIAZZA ARMERINA), all the other bridge destinations are situated in the seaside, confirming so the exclusively bathing vocation of tourism in Sicily.

In conclusion, *NA* takes into account the structure of links among the tourism destinations and their position in the network, deriving effects both for the single destinations and the whole system.

The paper tries to provide a framework for placing tourism within the context of mobility. From individual routes, by means of *Network Analysis* measures, it is possible to classify tourism destinations.

Our proposal could be useful in understanding patterns of tourism flows and the territorial features of tourism market.

References

Borgatti, S. P., & Everett, M. G. (1992). Notions of positions in social network analysis. *Sociological Methodology, 22*, 1–35.

Borgatti, S. P., Everett, M. G., & Freeman, L. C. (2002). *Ucinet for windows: Software for social network analysis*. Harvard: Analytic Technologies.

Borgatti, S. P., Everett, M. G., & Shirey, P. (1990). LS sets, lambda sets and other cohesive subsets. *Social Networks, 12*, 337–357.

Breiger, R., Boorman, S., & Arabie, P. (1975). An algorithm for clustering relational data with applications to social network analysis and comparison with multidimensional scaling. *Journal of Mathematical Psychology, 12*, 328–383.

Hanneman, R. A., & Riddle M. (2005). *Introduction to social network methods*. Riverside: University of California. http://faculty.ucr.edu/~hanneman/nettext/Introduction_to_Social_Network_Methods.pdf.

Nunkesser, M., & Sawitzki, D. (2005). Blockmodels. In U. Brandes & T. Erlebach (Eds.), *Network analysis: Methodological foundations* (pp. 253–292). Hidelberg: Springer.

Opsahl, T., Agneessens, F., & Skvoretz, J. (2010). Node centrality in weighted networks: Generalizing degree and shortest paths. *Social Networks, 32*, 245–251.

Rowley, T. J. (1997). Moving beyond dyadic ties: A network theory of stakeholder influences. *The Academy of Management Review, 22*(4), 887–910.

Schwartz, J. E. (1977). An examination of CONCOR and related methods for blocking sociometric data. *Sociological Methodology, 7*, 255–282.

Scott, J. (2000). *Social network analysis: A handbook* (2nd ed.). London: Sage.

Scott, N., Baggio, R., & Cooper, C. (2008). *Network analysis and tourism. From theory to practice*. Clevedon: Channel View.

Shih, H. Y. (2006). Network characteristics of drive tourism destinations: An application of network analysis in tourism. *Tourism Management, 27*(5), 1029–1039.

UNWTO. (2002). *News release: WTO Think Tank enthusiastically reaches consensus on frameworks for tourism destination success*. http://destination.unwto.org/en/content/conceptual-framework-0. Accessed June 2011.

Wasserman, S., & Faust, K. (1994). *Social network analysis. Methods and applications*. Cambridge: Cambridge University Press.

On Two Classes of Weighted Rank Correlation Measures Deriving from the Spearman's ρ

Livia Dancelli, Marica Manisera, and Marika Vezzoli

Abstract Weighted Rank Correlation indices are useful for measuring the agreement of two rankings when the top ranks are considered more important than the lower ones. This paper investigates, from a descriptive perspective, the behaviour of (*i*) five existing indices that introduce suitable weights in the simplified formula of the Spearman's ρ and (*ii*) an additional five indices we derive using the same weights in the Pearson's product-moment correlation index between ranks. For their evaluation, we consider that a good Weighted Rank Correlation index should (1) differ from ρ, if computed on the same pair of rankings and (2) assume a broad variety of values in the range $[-1, +1]$, in order to better discriminate amongst different reorderings of the ranks. Results suggest that linear weights should be avoided and show that indices (*ii*) do not have equalities with ρ and are more sensitive.

1 Introduction

Weighted Rank Correlation (WRC) indices are a useful tool for measuring the agreement between rankings when the top ranks are considered more important than the lower ones. Examples can be found in Blest (2000), Dancelli et al. (2012), Pinto da Costa and Soares (2005), Quade and Salama (1992). This paper investigates, from a descriptive perspective, five existing WRC indices that introduce suitable weights in the simplified formula of the Spearman's ρ. In addition, we derive a

L. Dancelli · M. Manisera (✉)
Department of Economics and Management, University of Brescia, C.da S. Chiara, 50 - 25122 Brescia, Italy
e-mail: dancelli@eco.unibs.it; manisera@eco.unibs.it

M. Vezzoli
Department of Molecular and Traslational Medicine, University of Brescia, Viale Europa 11 - 25123 Brescia, Italy
e-mail: marika.vezzoli@med.unibs.it

P. Giudici et al. (eds.), *Statistical Models for Data Analysis*, Studies in Classification, Data Analysis, and Knowledge Organization, DOI 10.1007/978-3-319-00032-9_13,

new class of WRC measures introducing weights in the Pearson's product-moment correlation index between ranks.

Suppose that $A: a_1, a_2, \cdots, a_i, \cdots, a_n$ and $B: b_1, b_2, \cdots, b_i, \cdots, b_n$ are two rankings. For simplicity, no ties are allowed. Without loss of generality, $A: 1, 2, \cdots, i, \cdots, n$, where 1 is the "most important" rank (the top rank) and n is the "least important" one. In the Spearman's ρ (Spearman, 1904; 1906), ranks replace variables in the Pearson's product-moment correlation index

$$\rho = \frac{1}{n\sigma_a\sigma_b}\sum_{i=1}^{n}(a_i - \bar{a})(b_i - \bar{b}), \tag{1}$$

which is -1 in the case of total disagreement (when one ranking is the inverse of the other one) and $+1$ in the case of total agreement (when the two rankings coincide). Indices satisfying this condition are standardized. Index (1) is usually presented in the simplified form (see, for example, Zani 1994, p. 233), which, in the case of no ties, is

$$\rho = 1 - 2\frac{\sum_{i=1}^{n}(a_i - b_i)^2}{(n^3 - n)/3}. \tag{2}$$

According to some authors, the Spearman's ρ has an implicit weighting scheme (among others, Kendall and Gibbons 1990, Tarsitano 2009). On the contrary, the WRC measures introduce explicit weights in order to emphasize discrepancies on the top ranks. In this framework, some existing WRC measures were obtained by introducing weights in (2) and determining the maximum that makes the index standardized, that is

$$\rho_w = 1 - 2\frac{\sum_{i=1}^{n}(a_i - b_i)^2 w_i}{\max\sum_{i=1}^{n}(a_i - b_i)^2 w_i}. \tag{3}$$

The paper focuses on indices where weights w_i are functions of a_i and b_i and treat them symmetrically (symmetric weights, henceforth).

Another possible choice is to introduce weights directly in the unsimplified formula (1), obtaining

$$\rho_w^* = \frac{\sum_{i=1}^{n}(a_i - \bar{a}^*)(b_i - \bar{b}^*)w_i}{\sigma_a^*\sigma_b^*\sum_{i=1}^{n} w_i} \tag{4}$$

where $\bar{a}^*, \bar{b}^*, \sigma_a^*$ and σ_b^* are weighted means and weighted standard errors, with weights w_i (Dancelli et al. 2011). These indices are standardized by definition.

Generally, the two classes of indices lead to different results, because (2) follows from (1) with $\bar{a} = \bar{b} = (n+1)/2$ and $\sigma_a = \sigma_b = \sqrt{(n^2-1)/12}$ that hold only when A and B are unweighted rankings. The same holds for alternative formulas of ρ existing in the literature. Henceforth, we label the class (3) as $\text{WRC}_{S\rho}$ (WRC measures from the Simplified ρ) and the class (4) as $\text{WRC}_{U\rho}$ (WRC measures from the Unsimplified ρ).

The aim of this paper is to evaluate, from a descriptive perspective, the two classes of indices and to establish which weights perform better. To do this, our exploratory study considers two aspects: *first*, indices must provide values different from the Spearman's ρ when computed on the same pair of rankings (except for the extreme cases of total agreement or disagreement); *second*, indices must not assume the same repeated value at different pairs of rankings. We compare (*i*) five indices of class WRC$_{S\rho}$ already existing in the literature, and (*ii*) the corresponding five indices of class WRC$_{U\rho}$ we obtain by introducing the same weights in formula (4).

Section 2 introduces the 5 + 5 indices, along with the two criteria we use to evaluate their performance. The computational study and the results are described in Sect. 3. A brief discussion and the conclusions are given in Sect. 4.

2 Some WRC Measures and Their Evaluation

The five existing WRC$_{S\rho}$ indices under consideration are obtained by introducing the weights $sr_i = \frac{1}{a_i} + \frac{1}{b_i}$, $rp_i = \frac{1}{a_i \cdot b_i}$, $rs_i = \frac{1}{a_i + b_i}$, $l_i = (n - a_i + 1) + (n - b_i + 1)$, $q_i = [(n - a_i + 1) + (n - b_i + 1)]^2$ in formula (3):

$$\rho_{sr} = 1 - 2\,\frac{\sum_{i=1}^{n}(a_i - b_i)^2\left(\frac{1}{a_i} + \frac{1}{b_i}\right)}{(n+1)\sum_{i=1}^{n}\frac{[2i-(n+1)]^2}{i(n-i+1)}}$$

$$\rho_{rp} = 1 - 2\,\frac{\sum_{i=1}^{n}(a_i - b_i)^2\left(\frac{1}{a_i \cdot b_i}\right)}{2(n+1)\sum_{i=1}^{n}\frac{1}{i} - 4n}$$

$$\rho_{rs} = 1 - 2\,\frac{\sum_{i=1}^{n}(a_i - b_i)^2\left(\frac{1}{a_i + b_i}\right)}{n(n-1)/3}$$

$$\rho_{l} = 1 - 2\,\frac{\sum_{i=1}^{n}(a_i - b_i)^2[(n - a_i + 1) + (n - b_i + 1)]}{n(n^3 + n^2 - n - 1)/3}$$

$$\rho_{q} = 1 - 2\,\frac{\sum_{i=1}^{n}(a_i - b_i)^2[(n - a_i + 1) + (n - b_i + 1)]^2}{n(n^2 - 1)(n+1)^2/3}$$

In Tarsitano (2009), ρ_{sr} is attributed to Salama and Quade; ρ_{rp} was proposed in Salama and Quade (1982) and ρ_{rs} in Quade and Salama (1992). The attribution of ρ_l gave rise to a controversy: Pinto da Costa and Soares (2005) introduced it, but Genest and Plante (2007) pointed out that it was their symmetrized version of the Blest's coefficient (Blest 2000). Pinto da Costa and Soares (2007) claimed a different derivation and a simpler form, suitable for generalization. In particular, they suggested to square the weights in ρ_l, obtaining ρ_q. Since the maximum in ρ_q shown in Pinto da Costa and Soares (2007) is wrong, we report the right one we derived in Dancelli et al. (2011).

The counterparts of these measures are the five indices of class $\text{WRC}_{U\rho}$ denoted with ρ^*_{sr}, ρ^*_{rp}, ρ^*_{rs}, ρ^*_l, ρ^*_q that can be obtained by introducing the same weights sr_i, rp_i, rs_i, l_i, q_i in formula (4).

In order to investigate the behaviour of the $5+5$ indices, in this exploratory study we perform a simulation considering all the possible exchanges of ranks. Hence, we calculate the indices between the target permutation $A : 1, 2, \cdots, i, \cdots, n$ and the $n!$ permutations of the set $\{1, 2, \cdots, n\}$. To evaluate their goodness we consider the two criteria already mentioned in Sect. 1.

First, a WRC index must generally not assume the same value of the Spearman's ρ if computed on the same pair of rankings. Otherwise, the introduction of weights does not serve the purpose. We evaluate the performance of each index by computing how many values are equal to ρ.

To measure the dispersion of a WRC index around the Spearman's ρ, we also calculate the mean M_d of the absolute differences between the values of the two indices. WRC indices showing a higher dispersion around ρ are more able to emphasize top ranks.

Second, a good WRC index must generally assume a broad variety of values in the range $[-1, +1]$, in order to discriminate better amongst different reorderings of the ranks (with corresponding weights). To evaluate the discrimination power (sensitivity) of WRC indices, one must consider that they can assume repeated values at different pairs of rankings. Hence, we evaluate the sensitivity of an index by considering the number of its unique values. Since the non-unique values can be replicated few or many times, we also calculate the mean M_r of their frequencies. WRC indices with a lower mean have a higher sensitivity.

3 Results

The computational study[1] was performed:

- Assuming low values of n because the identification of the equalities with ρ (*first criterion*) and the unique values (*second criterion*) makes sense only with short

[1]Computations were obtained by the statistical software R 2.13.2 with full precision.

Table 1 Number of values equal to ρ, mean M_d of the absolute differences from ρ, number of unique values, and mean M_r of the frequencies of the repeated values for the 5 + 5 WRC indices

	Indices of class $\text{WRC}_{S\rho}$					Indices of class $\text{WRC}_{U\rho}$			
	Values = ρ	M_d	No. unique values	M_r		Values = ρ	M_d	No. unique values	M_r
ρ_{sr}	0 (0.00%)	0.13	316 (79.80%)	2.41	ρ^*_{sr}	0 (0.0%)	0.15	396 (100.00%)	-
ρ_{rp}	0 (0.00%)	0.14	271 (68.43%)	2.44	ρ^*_{rp}	0 (0.0%)	0.24	396 (100.00%)	-
ρ_{rs}	3 (0.76%)	0.08	369 (93.18%)	2.08	ρ^*_{rs}	0 (0%)	0.11	395 (99.75%)	2.00
ρ_l	74 (18.69%)	0.05	149 (37.63%)	3.35	ρ^*_l	0 (0.0%)	0.12	370 (93.43%)	2.20
ρ_q	9 (2.27%)	0.11	310 (78.28%)	2.21	ρ^*_q	0 (0.0%)	0.22	395 (99.75%)	2.00

rankings. Otherwise, the exchanges of ranks (even the top ones) have little effect on the values of the indices and their differences are only a matter of decimals.

- Discarding all the inverse permutations from the $n!$ permutations of the set $\{1, 2, \cdots, n\}$ to avoid unnecessary replications of values.[2] Hereafter, the permutations maintained in the simulation study are denoted as "reduced permutations".

We report in detail the results obtained for $n = 6$, giving rise to 396 reduced permutations. In addition, some selected results for $n = 7, 8, 9$ are provided.

First criterion (comparison of each WRC index with the Spearman's ρ).

The results reported in Table 1 show that ρ_l is the index with the highest number of values equal to ρ (18.69%). Indices ρ_{rs} and ρ_q have a negligible number of equalities with ρ, while for ρ_{sr} and ρ_{rp}, no equality was found. It is interesting to note that all the 5 indices of class $\text{WRC}_{U\rho}$ show no equalities with ρ.

ρ_l is also the index with the lowest mean M_d of the absolute differences between its values and ρ (0.05). This underlines its limited dispersion around ρ. When considering the same weights, indices of class $\text{WRC}_{U\rho}$ always have higher values of M_d than the corresponding indices of class $\text{WRC}_{S\rho}$, suggesting a better performance in weighting.

Second criterion (power of discrimination).

[2]An inverse permutation B^{inv} of B is obtained by substituting each number with the number of the place it occupies. A WRC index with symmetric weights must give the same result when computed between A and B and between B^{inv} and A, because the pairs (a_i, b_i) and (b_i^{inv}, a_i) are the same. Then, one of the two rankings, B or B^{inv}, must be excluded from the $n!$ permutations, unless B and B^{inv} coincide. For example, let $A : 1, 2, 3, 4, 5, 6$ and $B : 2, 3, 1, 5, 4, 6$. The inverse permutation of B is $B^{inv} : 3, 1, 2, 5, 4, 6$. It is evident that the pairs are the same if we rewrite B^{inv} in the natural order ($B^{inv'} : 1, 2, 3, 4, 5, 6$) and, consequently, rearrange A, obtaining $A' : 2, 3, 1, 5, 4, 6$.

Consider now the number of unique values and the mean M_r of the number of replications of the repeated values, also reported in Table 1.

The number of unique values of the Spearman's ρ is only 34 (8.59%). Results in Table 1 show that, as expected, all the 5 + 5 WRC measures better discriminate different rankings. In detail, indices of class $\text{WRC}_{U\rho}$ better distinguish cases than their counterparts of class $\text{WRC}_{S\rho}$ by showing a higher number of unique values. Note that the worst index of class $\text{WRC}_{U\rho}$ (ρ_l^*) and the best index of class $\text{WRC}_{S\rho}$ (ρ_{rs}) have a similar number of unique values (370 and 369, respectively). No indices of class $\text{WRC}_{S\rho}$ have 100% of unique values; percentages vary from 37.63% (ρ_l) to 93.18% (ρ_{rs}). Linear weights l_i lead to the smallest number of unique values for both classes. Otherwise, when nonlinear weights are used in formula (4), the percentage of unique values is virtually 100%.

The mean M_r of the frequencies of the repeated values for ρ is 11.65. The means of all the WRC indices are much lower, ranging from 2.08 (ρ_{rs}) to 3.35 (ρ_l) for the indices of class $\text{WRC}_{S\rho}$ and from 2.00 (ρ_{rs}^* and ρ_q^*) to 2.20 (ρ_l^*) for the indices of class $\text{WRC}_{U\rho}$. This underlines their higher sensitivity.

Focusing on ρ_l, which is the worst index with respect to both criteria, it is interesting to look in detail at some rankings involving relevant exchanges in the ranks. Measuring their agreement with the target permutation $A : 1, 2, 3, 4, 5, 6$, we always have $\rho_l = \rho$, meaning that weights are neutralized. Some selected cases follow.

$\rho = \rho_l = -0.7143$:
$[4, 5, 6, 2, 3, 1]$ $[4, 5, 6, 3, 1, 2]$ $[4, 6, 3, 5, 2, 1]$ $[5, 3, 6, 4, 2, 1]$ $[5, 6, 2, 3, 4, 1]$ $[6, 3, 5, 2, 4, 1]$

$\rho = \rho_l = -0.5429$:
$[4, 3, 6, 5, 2, 1]$ $[4, 5, 6, 1, 2, 3]$ $[6, 3, 2, 5, 4, 1]$

$\rho = \rho_l = -0.4286$:
$[3, 4, 5, 6, 2, 1]$ $[3, 5, 4, 6, 1, 2]$ $[3, 6, 4, 2, 5, 1]$ $[5, 2, 4, 6, 3, 1]$ $[6, 2, 3, 4, 5, 1]$

$\rho = \rho_l = +0.2000$:
$[2, 3, 6, 4, 1, 5]$ $[2, 4, 5, 1, 6, 3]$ $[2, 4, 6, 1, 3, 5]$ $[2, 6, 3, 1, 4, 5]$

$\rho = \rho_l = +0.6571$:
$[1, 3, 4, 5, 2, 6]$ $[2, 3, 1, 5, 6, 4]$ $[2, 3, 1, 6, 4, 5]$ $[2, 4, 1, 3, 6, 5]$

$\rho = \rho_l = +0.8857$:
$[1, 3, 2, 5, 4, 6]$ $[2, 1, 3, 4, 6, 5]$.

Note that the last ranking [2,1,3,4,6,5] combines the two rankings used by Quade and Salama (1992) for introducing the meaning of "weighted rank correlation".

Results for $n = 7, 8, 9$ confirm that as n increases, differences between the values of the indices are detected with difficulty. For example, looking again at the worst index ρ_l, the number of equalities with the Spearman's ρ are 8.88% ($n = 7$), 5.84% ($n = 8$) and 2.96% ($n = 9$). For the other indices of class $\text{WRC}_{S\rho}$, as n

increases, percentages of values equal to ρ tend to zero, and as expected, the same happens for indices of class $\text{WRC}_{U\rho}$, because such percentages were already zero for $n = 6$. Since the mean M_d remains substantially constant over the different n's, the conclusions drawn for $n = 6$ hold also when $n = 7, 8, 9$.

With reference to the discrimination power of indices, as n increases, the percentage of unique values tends to decrease for each index (even if the absolute values increase). Such a trend is much more evident for indices belonging to class $\text{WRC}_{S\rho}$: for example, for the worst index ρ_l, the percentage of unique values decreases from 37.63 ($n = 6$) to 14.20 ($n = 7$), then to 3.36 ($n = 8$) and 0.61 ($n = 9$), while for ρ_{rs} (the best among class $\text{WRC}_{S\rho}$ in Table 1), such a percentage moves from 93.18 ($n = 6$) to 86.75 ($n = 7$), 74.54 ($n = 8$) and 63.81 ($n = 9$). From this point of view, indices of class $\text{WRC}_{U\rho}$ also perform better than those in class $\text{WRC}_{S\rho}$ when n increases: except for ρ_l^* (56.46% of unique values when $n = 9$), all the other indices maintain percentages higher than 98% when $n = 9$. The mean M_d computed for $n = 7, 8, 9$ confirms the higher sensitivity of indices of class $\text{WRC}_{U\rho}$, with respect to their counterparts of class $\text{WRC}_{S\rho}$ and, generally, of WRC indices with respect to the Spearman's ρ.

4 Discussion and Conclusions

In the present study, we investigated, from a descriptive perspective, the performance of 5 + 5 WRC measures from two aspects: (1) their ability in weighting the top ranks and (2) their power of discriminating different pairs of rankings. Results suggest that linear weights, as in ρ_l and ρ_l^*, are not appropriate to emphasize the top ranks. This is due to compensations between the exchanges of the ranks and the linear weights (Dancelli et al. 2011). In addition, from the points of view (1) and (2), indices belonging to class $\text{WRC}_{U\rho}$ seem to be preferable. Besides, since these indices are standardized by definition, the analytical identification of the maximum in $\text{WRC}_{S\rho}$ is not necessary. This can be useful, for example, when other weights are chosen or when tied ranks are considered. In fact, real applications sometimes include ties, which may be handled by simple procedures (tied ranks are usually assigned the midpoint value that would result if the objects were ranked consecutively) or by more sophisticated techniques (Hájek and Sidák 1967, pp. 118–123).

In this study, we maintained low n because the focus was on the two particular aspects (1) and (2), which cannot be conveniently evaluated with high values of n. In fact, as n increases, the exchanges of ranks (even the top ones) have little effect on the final value of a WRC index and it is difficult to detect differences with the values of the Spearman's ρ (same paired rankings) or within the values of the WRC index itself (different paired rankings). In addition, in practice, only a few decimals are usually retained when computing rank correlation measures. With few decimals, differences are even more difficult to see, because the number of equalities of WRC indices with ρ increases while the discrimination power decreases. Actually, results with full precision do not differ much from results with few decimals when $n = 6$. Differences stand out when n increases, since the number of possible values of indices with a fixed limited number of decimals is small while the number $n!$ of permutations of $A : 1, 2, \cdots, i, \cdots, n$ rapidly increases, even if inverse permutations are removed. For example, with $n = 9$, the possible different values for whatever index with 4 decimals is $2 * 10^4 - 1 = 19{,}999$ (not including -1 and $+1$), while the number of reduced permutations is 185,152. Moreover, the choice of low n is consistent also with real applications, where long rankings are not very frequent.

However, simulation studies with higher values of n are certainly recommended when the focus is on other topics, such as the sampling properties of the indices.

The promising results obtained in this preliminary study encourage future research aimed at further refining guidelines helping the choice of the best weights to use in practice and deepening the study of the indices, for example, when measuring the agreement between "nonlinear rankings" (Tarsitano 2009).

Acknowledgments We wish to thank the anonymous referee for his/her comments that greatly improved the quality of the paper.

References

Blest, D. (2000). Rank correlation - an alternative measure. *Australian and New Zealand Journal of Statistics, 42*, 101–111.

Dancelli, L., Manisera, M., & Vezzoli, M. (2011). On some weighted rank correlation measures related to the Spearman's ρ. Working Paper n. 363, Department of Quantitative Methods, University of Brescia, Italy.

Dancelli, L., Manisera, M., Vezzoli, M. (2012), Weighted Rank Correlation measures in hierarchical cluster analysis, in *Book of short papers JCS - CLADAG 2012*, pp. 1–4.

Genest, C., & Plante, J. F. (2007). Letter to the editor. *Australian and New Zealand Journal of Statistics, 49*, 203–204.

Hajek, J., & Sidak, Z. (1967). *Theory of rank tests*. New York: Academic Press.

Kendall, M., & Gibbons, J. (1990). *Rank correlation methods* (5th edn). Oxford: Oxford University Press.

Pinto da Costa, J., & Soares, C. (2005). A weighted rank measure of correlation. *Australian and New Zealand Journal of Statistics, 47*, 515–529.

Pinto da Costa, J., & Soares, C. (2007). Rejoinder to letter to the editor. *Australian and New Zealand Journal of Statistics, 49*, 205–207.

Quade, D., & Salama, I. A. (1992). A survey of weighted rank correlation. In P. K. Sen, & I. A. Salama (Eds.) *Order statistics and nonparametrics* (pp. 213–224). Amsterdam: Elsevier.

Salama, I. A., & Quade, D. (1982). A nonparametric comparison of two multiple regressions by means of a weighted measure of correlation. *Communications in Statistics - Theory and Methods, 11*, 1185–1195.

Spearman, C. (1904). The proof and measurement of association between two things. *The American Journal of Psychology, 15*, 72–101.

Spearman, C. (1906). A footrule for measuring correlation. *British Journal of Psychology, 2*, 89.

Tarsitano, A. (2009). Comparing the effectiveness of rank correlation statistics. Working Paper n. 6, Department of Economics and Statistics, University of Calabria, Italy.

Zani, S. (1994). *Analisi dei dati statistici* (vol. I). Milano: Giuffrè.

Beanplot Data Analysis in a Temporal Framework

Carlo Drago, Carlo Lauro, and Germana Scepi

Abstract We propose in this work a new approach for modelling, forecasting and clustering beanplot financial time series. The beanplot time series like the histogram time series or the interval time series can be very useful to model the intra-period variability of the series. These types of new time series can be very useful with High Frequency financial data, data collected with often irregularly spaced observations.

1 Introduction

There are situations in which it is necessary to use some types of aggregation which can arise in a direct loss of information, for example in financial time series. In particular high frequency financial data shows some relevant characteristics (they are inequally spaced and contains errors) which suggest the use of some time series like intervals or histograms. Alternatives in the context of data types concretely used, are belong the field of the Symbolic Data Analysis (Billard Diday 2006; Arroyo Maté 2008). However, we propose the use of the beanplot data (Kampstra 2008) which summarize the initial data by returning the relevant data features. We have proposed a transformation of the original time series in a time series of beanplots to analyse the intra-day variability over time. Beanplot allows to keep the relevant information and data structure of the initial data which could be hidden by the aggregation. In this way we take into account the entire intra-period variation over time. Here we propose a new parameterization of the beanplots with aim of forecasting and clustering.

C. Drago (✉) · C. Lauro · G. Scepi
University of Naples "Federico II" Complesso Universitario Monte Sant'Angelo via Cinthia, Naples, Italy
e-mail: carlo.drago@unina.it; clauro@unina.it; scepi@unina.it

P. Giudici et al. (eds.), *Statistical Models for Data Analysis*, Studies in Classification, Data Analysis, and Knowledge Organization, DOI 10.1007/978-3-319-00032-9_14,

2 Beanplot Modeling, Forecasting and Clustering

The beanplot data represents the intra period variability of the initial series. In particular beanplot data show the centre by period like the aggregate data, then the size or the range of the variability, and finally the density trace represent the entire data structure. We start from a classical time series which generates a beanplot time series (Fig. 1), where a beanplot data at time t can be defined as:

$$\hat{f}_{h,t} = \frac{1}{nh}\sum_{n}^{i=1} K(\frac{x - x_i}{h}) \tag{1}$$

where K is a Kernel and a h is a smoothing parameter defined as a bandwidth. K can be a gaussian function with mean zero and variance 1. The Kernel is a non-negative and real-valued function $K(z)$ satisfying: $\int K(z)dz = 1, \int zK(z)dz = 0, \int z^2 K(z) = k_2 < \infty$ with the lower and upper limits of integration being $-\infty$ and $+\infty$. It is possible to use various Kernel functions: uniform, triangle, epanechnikov, quartic (biweight), tricube (triweight), gaussian and cosine. In the gaussian case the variance can be controlled through the parameter h:

$$K(\frac{x - x_i}{h}) = \frac{1}{\sqrt{2\pi}} e^{-\frac{x-x_i^2}{2h^2}} \tag{2}$$

Various methods was proposed in Literature to choose the bandwidth h. In particular Jones Marron Sheather (1996) reviews bandwidth choice methods. In the data visualization we use the Sheather-Jones criteria that defines the optimal h in a data-driven choice (Kampstra 2008). A Beanplot data $\{b_{Y_t}\}$ is a combination between a 1-d scatterplot and a density trace. In a beanplot we take in account both the interval between the minimum a_{L_t}, the maximum a_{U_t} and the density as the kernel nonparametric estimator (the density trace see Kampstra 2008). Every single observation y_{it} is represented on the one-dimensional scatterplot. This feature is useful to detect visually observations distant from the others. The beanline at time t is a central measure of the beanplot (and a measure of location).

2.1 Beanplot Internal Modeling

By assuming each temporal observation as a mixture of distributions, we estimate the parameters of each mixture distribution (Du 2002) and we obtain a set of parameters for each beanplot data:

$$A_t = [p_{1,t}, p_{2,t}, \ldots, p_{k,t}]' \tag{3}$$

We define this procedure internal modeling. For each model at time t we obtain a measure of goodness of fit I_t representing the quality of the representation. The parameters replace both the original data and the densities at time t as the

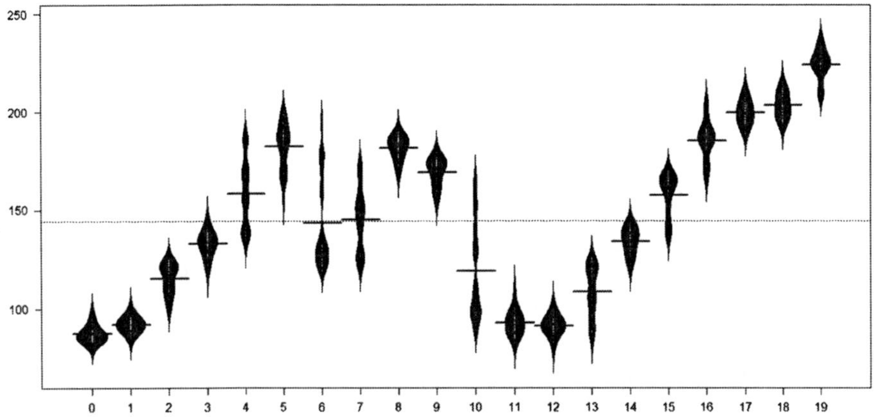

Fig. 1 Beanplot time series of the stock Apple for the period 2007–2010. Typically we can observe regular symmetric shape, where there is a structural change the shape modify in two distincts parts

models of the intra-period variability. The model parameters by each temporal interval t summarize the relevant data aspects of the densities as the location, the size, and the shape. The parameters give important information on the sequential process, in fact they show us the presence of change points at a time t.

2.2 *Detecting Structural Changes*

We take into account the parameter sequences for checking structural changes over time t. In particular for each parameter we consider the associated time series. So, we consider each parameter in A_t, $p_{1,t}, p_{2,t}, \ldots, p_{j,t}$ and we estimate:

$$p_{j,t} = \beta_0 + \sum_{q=1}^{Q} \beta_q \delta_q + \omega_j \tag{4}$$

where δ_q is a dummy variable representing a specific period or an interval period of time, in which the null hypothesis of no structural change is tested and ω_j is a residual. In presence of structural change $\beta_q \neq 0$. We return the dates of the structural changes for all the parameters in A_t

2.3 *Multiple Beanplot Forecasting*

We propose a forecasting method of the beanplot models based on the parameters, and a clustering procedure based on a distance of models (in the next paragraph). By obtaining the parameters, that represents the intra period variability over time, we can forecast the next beanplot observation by considering a time series forecasting method. The procedure is divided in two parts. Firstly, it is necessary a specific synthesis of the models by using the Time Series Factor Analysis (in particular Gilbert Meijer 2005). Starting from the trajectories obtained by each parameter time series we can estimate

$$p_{j,t} = \alpha + \beta\xi_t + \epsilon_t \tag{5}$$

With α as a vector of intercepts, β as a n,k matrix of factor loadings and ϵ is a n vector of random errors. So, by considering n parameters or observed processes $p_{j,t}$ with $i = 1..n$ and $t = 1..T$, in which we are searching from k the factors as unobserved processes $\xi_{i,t}$ with $t = 1..T$ and $i = 1..k$. In particular, for each time series, we can obtain a set of factors. To measure the factor, as well, we use:

$$\xi_t = (B^t \Psi^t_{-1} B)^{-1} B^t \Psi^{-1} B^t (z_t - \alpha_t) \tag{6}$$

In which $\psi_t = Cov(\epsilon_t)$. In particular the loadings, in the FA estimators (say ML) can be estimated by using the sample covariance of the error (Gilbert Meijer 2005). In this way it is possible to compute the factor time series for the trajectories defined by parameters. Each factorial time series represent the dynamics of the beanplot time series and allow to identify the external shocks which affect the beanplot dynamics. We can estimate a VAR or a VECM model based on the factorial time series considered. In the case the results is not completely satisfying we can use a combination of different forecasting methods, for example a VAR, VECM model and some univariate methods (for a review De Gooijer Hyndman 2006) with equal or different weights (eventually changing over time). By considering explicitly $f^1, f^2 \ldots f^m$ as different competing external models we have as external model forecasts combination f_t:

$$f_t = \gamma_1 f^1 + \gamma_2 f^2 + \ldots + \gamma_m f^m + \zeta \tag{7}$$

In this case ζ is an error term with zero mean and $\gamma_1 \ldots \gamma_n$ are different weights. In particular the weighted combination of the models allow to improve forecasting in a context of parameter drift and structural change. Poor results in in-sample forecasting can occur in a revision of the initial parameters considered (the kernel K used and the bandwidth h for example).

2.4 *Clustering Multiple Beanplot Time Series*

Finally for Clustering the different beanplot time series, we use the parameters obtained. In particular we use an adequate distance, like the distance from models in Romano Giordano Lauro (2006) where we are trying to consider the dynamics of the different factors. In this way it is possible to obtain the dissimilarity matrices related, and use the clustering method. So, in order to cluster multiple time series of beanplot $W_{v,t}$ with $v = 1 \ldots V$, we use a suitable distance between models that combines a convex function of the differences in model parameters with corresponding fitting indexes I (Romano Giordano Lauro 2006). The two pieces of information are combined to define the following measure IM (or intra-model distance). Following Signoriello (2008) we have:

$$IM(A_t, A_{t\prime}|\lambda) = \lambda IM_P + (1 - \lambda) IM_R \tag{8}$$

with $\lambda \in [0, 1]$. The λ is related to weight to apply both to the parameters and the model fit in the final distance. The *IM* measure is a sum of IM_P and IM_R, where IM_P is the L_2-norm between the parameters. In practice we are considering the structural differences between two beanplots by considering their parameters. It is important to stress the fact that the parameters of the models are related to the structural part of the beanplot so we are clustering not the initial beanplot but their models. So we have:

$$IM_P = \left[\sum\nolimits_{k=1}^{K-1} \left(p_{j_k} - p_{j'_k}\right)^2\right]^{\frac{1}{2}} \quad (j \neq j\prime) \tag{9}$$

and IM_R is the L_1-norm between the *Chisquare* related the different models (Du 2002). At the same time we need to consider the fit of the models to the real data and the difference between these indexes. In this case the fit of the model is related to the capability to the model to capture the initial data. Clearly an unsatisfactory fit can lead to different choices in the beanplot modelling. So we have in this case:

$$IM_R = \left|I_{j\,K} - I_{j\prime\,K}\right| \quad (j \neq j\prime). \tag{10}$$

In order to cluster multiple time series of beanplot $W_{v,t}$ with $v = 1 \ldots V$, we use the Romano Giordano Lauro distance of models by considering the entire series of the models. In this case the distance between the models need to take into account all the suberiods and the differences of the parameters in each subperiod.

$$IM_{P_T} = \sum_{t=1}^{T} \left[\sum\nolimits_{k=1}^{K-1} \left(p_{j\,k_t} - p_{j\prime k_t}\right)^2\right]^{\frac{1}{2}} \quad (j \neq j\prime) \tag{11}$$

At the same time we can consider the differences between the model fit.

$$IM_{R_T} = \sum_{t=1}^{T} \left|I_{j\,K_t} - I_{j\prime\,K_t}\right| \quad (j \neq j\prime). \tag{12}$$

The two pieces of information are combined to define the following distance in which we take into account both the structural aspects of the parameters (and their differences) and the differences of the model fit related to the initial data. So we finally have:

$$IM_T(p_{t_j}, p_{t_{j\prime}}, \ldots p_{T_j}, p_{T_{j\prime}}|\lambda) = \lambda IM_{P_T} + (1 - \lambda) IM_{R_T} \tag{13}$$

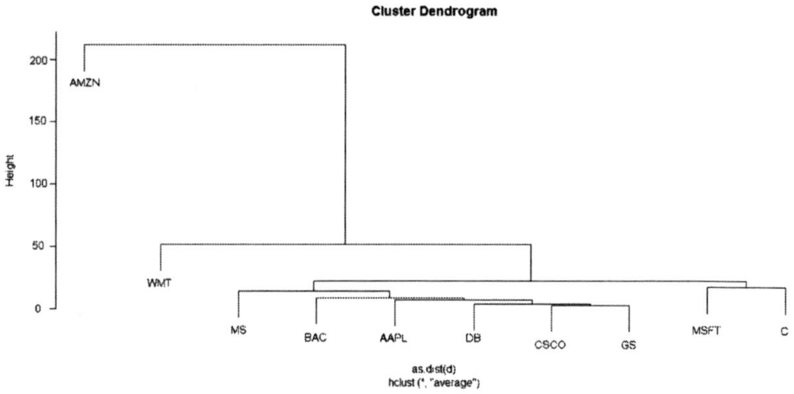

Fig. 2 Dow Jones Market: subperiod years 2007–2008

In this case we are considering in an direct way all the different subperiods and so we compare the beanplots time series by considering the entire beanplot dynamics and not a single subperiod.

2.5 Application on Financial Data 1990–2011

We consider the period from 1990 to 2011 yearly observations for the FCHI France and the GDAXI Germany Market. In this sense we transform the original data in beanplot time series. Then we extract the mixtures of the data and we compute the factorial time series from the data. The factorial time series seems to respond well to the same economic shocks affecting the different economies over time. Then we estimate a VAR considering the two factorial time series and obtain the 1 year forecast for the year 2010 characterized by known instability. The results overperform the univariate models for the Germany but show some instability due clearly to the crisis. At the same time by considering the predictions for the year 2010 the MAPE became: 27 (France) and 14 (Germany). For the clustering methods we consider various stocks of the US market for asset allocation purposes (years 2007–2011). It is interesting to observe the strong structural change before and after the crisis (Figs. 2 and 3). At the same time the financial stocks tend to behave similarly (due to the strong negative shocks related to the financial crisis) where in the market there was some not financial companies that perform differently (Apple and Amazon between others) which better sustained the financial crisis (Figs. 4 and 5).

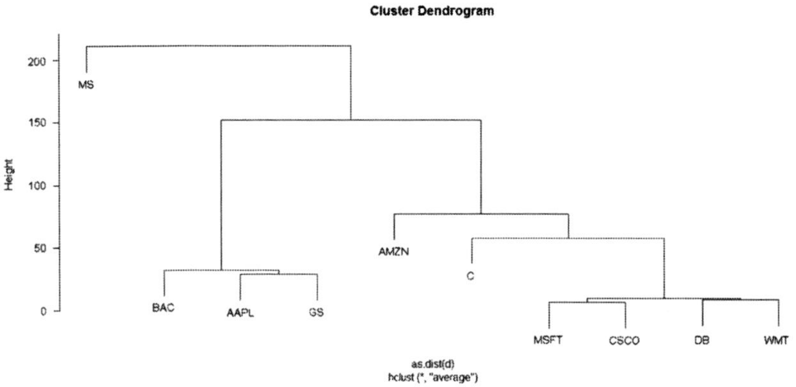

Fig. 3 Dow Jones Market: subperiod years 2007–2008

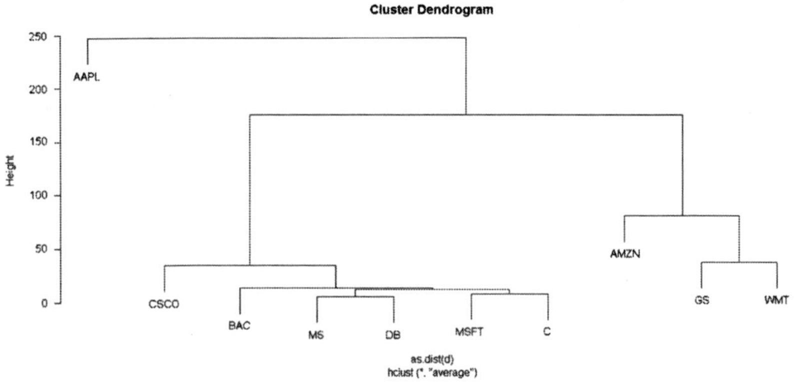

Fig. 4 Dow Jones Market: subperiod years 2009–2010

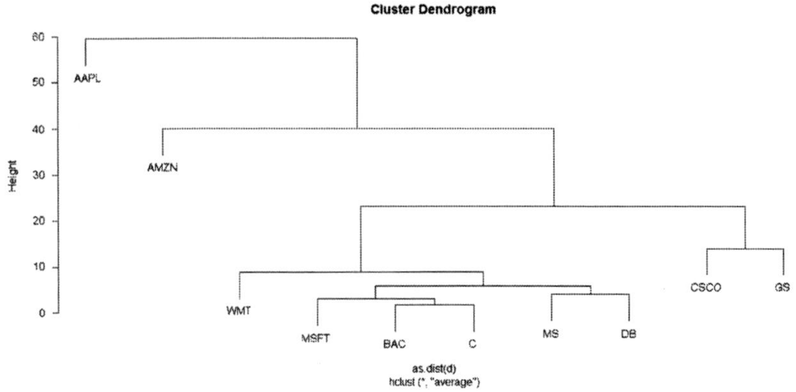

Fig. 5 Dow Jones Market: subperiod years 2010–2011

References

Arroyo, J., & Matè, C. (2008). Forecasting histogram time series with the K-neirest neighbour methods. *International Journal of Forecasting, 25*(2009), 192–207.

Billard, L., & Diday, E. (2006). *Symbolic data analysis: conceptual statistics and data mining*. Chichester: Wiley.

De Gooijer, J. G., & Hyndman, R. J. (2006). 25 years of time series forecasting. *International Journal of Forecasting, 22*(3), 443–473. Elsevier

Du, J. (2002). Combined algorithms for constrained estimation of finite mixture distributions with grouped data and conditional data. Department of Mathematics and Statistics, McMaster University.

Gilbert, P. D., & Meijer, E. (2005). Time series factor analysis with an application to measuring money, research report 05F10. University of Groningen, Research Institute SOM (Systems, Organisations and Management).

Jones, M. C., Marron, J. S., & Sheather, S. J. (1996). A brief survey of bandwidth selection for density estimation. *Journal of the American Statistical Association, 91*, 401–407.

Kampstra, P. (2008). Beanplot: a boxplot alternative for visual comparison of distributions. *Journal of Statistical Software*, Code Snippets *28*(1), 1–9. http://www.jstatsoft.org/v28/c01/.

Romano, E., Giordano, G., & Lauro, C. N. (2006). An inter model distance for clustering utility function, *Statistica Applicata, 18*(3), 521–533.

Signoriello, S. (2008). Contributions to symbolic data analysis: a model data approach. Ph.D. thesis, Department of Mathematics and Statistics, University of Naples Federico II.

Supervised Classification of Facial Expressions

S. Fontanella, C. Fusilli, and L. Ippoliti

Abstract Over the last decade, the statistical analysis of facial expressions has become an active research topic that finds potential applications in many areas. As the expression plays remarkable social interaction, the development of a system that accomplishes the task of automatic classification is challenging. In this work, we thus consider the problem of classifying facial expressions through shape variables represented by log-transformed Euclidean distances computed among a set of anatomical landmarks.

1 Introduction

Automatic facial expression analysis—FEA—started with the pioneering work of Mase and Pentland (1991). The reasons for the interest in FEA are multiple, but they are mainly due to the advancements accomplished in related research areas such as face detection, face tracking and face recognition.

Given the significant role of the face in our emotional and social lives, it is not surprising that the potential benefits from efforts to automate the analysis of facial signals are varied and numerous.

The analysis of facial expressions has a great relevance in sociological, medical and technological researches. For example, in social science, the relevance of this new research area is due to the growing importance of investigating the role of social intelligence in the interaction between human beings. In this context, an important research domain is the Social Signal Processing—SSP. The term "social intelligence" is referred here to the ability to express and recognize social signals produced during social interactions (e.g. agreement, politeness, empathy,

S. Fontanella (✉) · C. Fusilli · L. Ippoliti
University G. d'Annunzio, Chieti-Pescara, Italy
e-mail: ippoliti@unich.it

P. Giudici et al. (eds.), *Statistical Models for Data Analysis*, Studies in Classification, Data Analysis, and Knowledge Organization, DOI 10.1007/978-3-319-00032-9_15,

friendliness, conflict, etc), coupled with the ability to manage them in order to act wisely in human relations (Pantic et al. 2001). SSP is thus the new research and technological domain that aims at providing computers with the ability to sense and understand human social signals.

In the medical field, facial expressions are analysed in order to study deficits in emotional expressions and social cognition in neuropsychiatric disorders. In fact, researchers have shown that different neurologic and psychiatric disorders may present peculiarities in the facial expression. Thus, FEA provides the possibility of accelerating the process of diagnosis.

Another useful application of FEA is in security systems. Monitoring and interpreting facial signals can provide important information to lawyers, police, security, and intelligence agents regarding deception and attitude. Hence, here the aim is that of designing systems that are able of recognizing friendly, unfriendly or aggressive faces in order to support the law enforcement process.

In recent years, due to the availability of relatively cheap computational power, an automatic facial expression analysis has been investigated as facial pattern recognition using imaging techniques.

Facial expressions are generated by contractions of facial muscles, which results in temporally deformed facial features such as eye lids, eye brows, nose, lips and skin texture, often revealed by wrinkles and bulges. Thus, for example, an increase in distance between lip corners may indicate a smiling or happy face. In our work, this action is transformed into geometrical shape using landmark coordinates. Specifically, a face is coded by interactively locating the coordinates of specific landmarks. These landmarks are defined by biological features on the face and are usually referred to as anatomical landmarks. We thus use these landmarks to identify important features of the expressions in order to provide an accurate classifier for "shape" (expression) allocation.

The outline of the paper is as follows. In Sect. 2 we describe the data and how to present them in a form suitable for a classification procedure. Specifically, we compute distances between landmarks to provide partial shape information. In Sect. 3, we first review the basic of linear discriminant analysis (LDA) and then discuss the high dimension/small sample size problem to introduce a set of LDA-based classifiers to deal with the problem of multicollinear data. Then, in Sect. 4, we illustrate classification results for the FG-NET database while Sect. 5 concludes the paper with a discussion.

2 Notation and Data Description

In this paper we assume that there are multivariate data available for G groups (or classes). The vectors in the training set are denoted by a p-dimensional vector $\mathbf{z}$. There are also n_j observational units available in the j-th group. In general, considering the G groups, we have n statistical units. Finally, the estimated mean in the j-th group is denoted by $\tilde{\mathbf{x}}_j$ while $\mathbf{W}$ and $\mathbf{B}$ are the within-group and between-group covariance matrices, respectively.

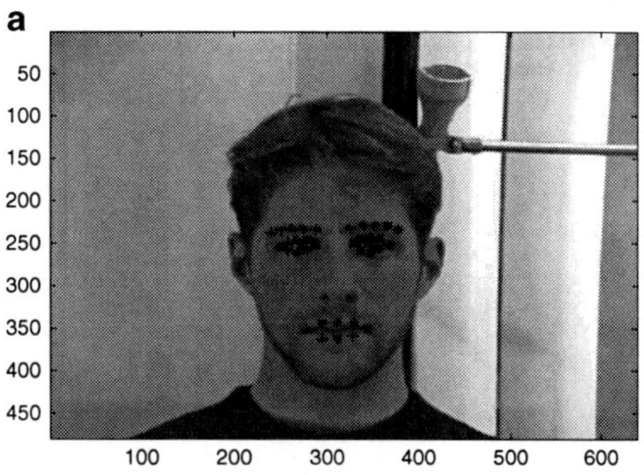

Fig. 1 Typical example of landmark configuration

As regards the data, we consider the FG-NET Database with Facial Expressions and Emotions from the Technical University Munich. This is an image database containing face images from 18 subjects performing the six basic expressions described in Ekman and Friesen (1971). For modelling purposes we consider here four expressions, namely: *neutral*, *happiness*, *sadness* and *surprise*.

On each image, representing the expression of interest, we have manually placed a set of 30 landmarks and Fig. 1 shows a typical example of a landmark configuration. In each of the $G = 4$ groups (i.e. expressions), there is thus complete information on $n_j = 18$ subjects and 30 landmarks, each with coordinates $\mathbf{s}_i = \{x_i, y_i\} \in \mathscr{R}^2$, $i = 1, \ldots, 30$.

Since the shape of an object determines its coordinates only up to a similarity transformation, it is necessary to reduce the data to just the shape information. One possible solution is that of using Procrustes tangent coordinates about a centered and scaled "mean" configuration estimated through Generalised Procrustes Analysis (Dryden and Mardia 1998).

An alternative approach is to work with Euclidean distances calculated among landmarks. The notion of distance is much more intuitive than procrustes tangent coordinates and hence, we wish to examine here the extent to which the use of distances among landmarks can be helpful for classifying the expressions.

In principle, all the distances among the 30 landmarks can be considered to construct the default feature vector, $\mathbf{z}$. However, there are additional distances to choose from to form a feature vector and these, for example, could be calculated from an additional landmark which is a function of some of the anatomical landmarks mentioned above. The construction of this landmark (e.g. points A or B) is shown in Fig. 2. The choice of working with the complete set of distances computed among the 30 landmarks, or with a reduced set of distances computed between each landmark and a reference one (e.g. points A or B), strongly affects the length, p, of the feature vector $\mathbf{z}$. In contrast with the first approach, the latter, henceforth called Fixed Point—FP—is a parsimonious one and, for example, avoids the problem of performing LDA under singularity conditions.

Note that distance data are invariant with respect to the transformations of translation and rotation. Hence, to transform the data into shape, the size effect

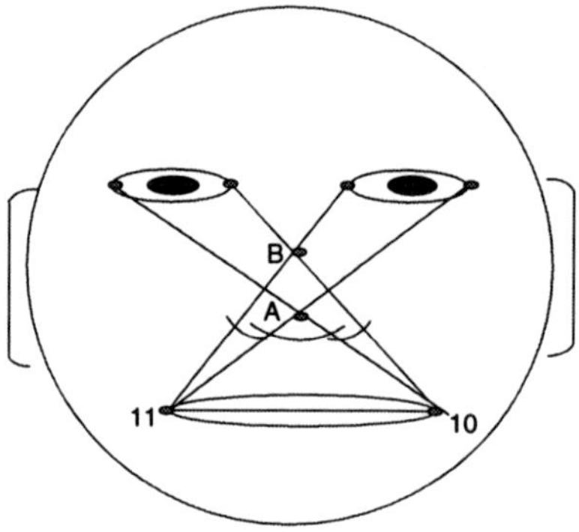

Fig. 2 Construction of the reference landmarks A and B

(scale) must be removed from the distance measurements. We remove the scale using a logarithmic transformation thus obtaining a shape space for distances. The shape mean for distances is the sample mean of the shape variables.

3 Background Theory for Linear Discriminant Analysis and the High Dimension/Small Sample Size Problem

In this section, we present the general theory of linear discriminant analysis (LDA), which is an efficient and simple multivariate technique useful to classify our expressions into the four considered categories. There exists a large amount of literature on the subject and we refer the interested reader to Krzanowsky and Marriott (1995), Mardia et al. (1979) and McLachlan (2004) for more details.

LDA examines the relationship between membership of one of several groups or populations and a set of interrelated variables. The classification of observations in G groups is based on the calculation of *canonical variates* obtained as $\mathbf{Y} = \mathbf{ZA}$, where $\mathbf{Z}$ is the $(n \times p)$ data matrix and $\mathbf{A}$ is a transformation matrix. LDA is thus a linear combination of the original p features which projects the data into a subspace of dimension, $r \leq min(p, G-1)$, in order to maximize the between-group to within-group variation, subject to the canonical variables being uncorrelated within groups and between groups.

Formally, it can be shown (Mardia et al. 1979) that the transformation matrix $\mathbf{A}$ is represented by the eigenvectors corresponding to the decreasing-order ranked eigenvalues of $\mathbf{W}^{-1}\mathbf{B}$. In practice, only r transformation vectors $\mathbf{a}_i$ (i.e. columns of $\mathbf{A}$) are useful so that the simplest and most frequent way to classify is by assigning the unit to the j-th group of the transformed group-mean vector, $(\mathbf{a}_1, \ldots, \mathbf{a}_r)^T \bar{\mathbf{z}}_j$, which is closest in the L_2-norm.

However, the choice of working with the complete set of distances obtained from the 30 landmarks generates a large number of features that raises the so called high dimension/small sample size problem. In this case, since it follows that

$p >> n$, $\mathbf{W}$ becomes singular and we thus need to replace its classical empirical estimate by alternative methods. As discussed below, one way to circumvent the singularity problem of $\mathbf{W}$ is to perform a dimensionality reduction of the data prior to the application of LDA on the scores values. Alternative methods come from the partitioned spectral decomposition of $\mathbf{W}$, see for example Tebbens and Schlesinger (2007).

3.1 *Dimension Reduction Methods*

One way to circumvent the singularity problem of $\mathbf{W}$ is to project the data in a reduced space by principal component analysis—PCA—Mardia et al. (1979), Jolliffe (2002), and apply LDA on the PCA-scores ordered according to the explained variability. However, by exploiting the available class information, the PCA-scores could be obtained from the eigenvectors of the within-group covariance matrix, $\mathbf{W}$. Hence, the principal components can be ranked according to the following criterion, see for example Devijver and Kittler (1982)

$$O_j = (\mathbf{a}_j^T \mathbf{B} \mathbf{a}_j)/\lambda_j,$$

where $\mathbf{a}_j$ and λ_j are the j-th eigenvector and eigenvalue of $\mathbf{W}$, respectively. In this case the first few ranked components provide the orthogonal projections of the original data that best highlight the between-group differences. Results from this procedure will be denoted throughout as PCA2.

3.2 *Rank Decomposition Methods*

Alternative methods come from the partitioned spectral decomposition of $\mathbf{W}$. In fact, if $rank(\mathbf{W}) = r < p$, then $\mathbf{W}$ can be decomposed as follows

$$\mathbf{W} = \begin{bmatrix} \mathbf{U}_1 & \mathbf{U}_2 \end{bmatrix} \begin{bmatrix} \mathbf{L}_1 & \mathbf{0} \\ \mathbf{0} & \mathbf{0} \end{bmatrix} \begin{bmatrix} \mathbf{U}_1^T \\ \mathbf{U}_2^T \end{bmatrix},$$

where $\mathbf{U}_1$ contains the first r columns of the $(p \times p)$ orthonormal matrix of eigenvectors $\mathbf{U}$, and $\mathbf{L}_1$ is the $(r \times r)$ diagonal matrix containing the non-zero eigenvalues, λ_i, only. $\mathbf{U}_2$ contains the first $p - r$ columns which are mutually orthogonal to each other and to those of $\mathbf{U}_1$, and $\mathbf{0}$ is a proper defined matrix of zeroes.

Hence, a possible approach to overcome the singularity problem is to consider a truncated SVD of $\mathbf{W}$ so that its Moore-Penrose pseudo-inverse, defined as $\mathbf{U}_1 \mathbf{L}_1^{-1} \mathbf{U}_1^T$, can be used. Among others, this approach was also proposed in

Krzanowski et al. (1995) and henceforth, it will be denoted as the Moore-Penrose discrimination method—MPD.

Another possibility is to concentrate on the *null space* of $\mathbf{W}$ and obtain $\mathbf{W}^{-1} = \mathbf{U}_2\mathbf{U}_2^T$, see Krzanowski et al. (1995). This approach corresponds to the zero-variance discrimination method—ZVD—and is motivated by the fact that in this null space the within-group variance is minimal.

Following the idea of minimal within-group variance, a further strategy is to exclude "a priori" the within-group variability from the analysis and perform LDA as standard PCA applied to the between-group matrix $\mathbf{B}$. $\mathbf{W}$ is thus assumed to be an identity matrix. This approach has been suggested in Dhillon et al. (2002) and is particularly useful when the data set is huge. We shall denote this procedure as the between-group based discrimination method (BBD).

4 Classification Results

In the following we discuss discrimination results from LDA applied on distance variables obtained as described in Sect. 2. By using the FP distance approach, a data matrix $\mathbf{Z}_1$, of dimension (72×30), is available for classification. In this case, for each individual, the feature vector is obtained as the set of the log-transformed distances computed between each anatomical landmark and the fixed point A (see Fig. 1). On the other hand, by using the full set of distances computed among the 30 anatomical landmarks, a new data matrix $\mathbf{Z}_2$, of dimension (72×435), is also available.

Performing a classification on $\mathbf{Z}_1$ results in a standard LDA procedure; in contrast, $\mathbf{Z}_2$ poses singularity problems of $\mathbf{W}$ and hence, dimension reduction or reduced rank decomposition methods must be considered for the analysis.

For the two data sets, the comparison of the results is based on the calculation of the misclassification error rate obtained for each classifier and by using a cross validation procedure. Since the same subject is involved 4 times, the procedure essentially consists in leaving out 4 observations (corresponding to 4 expressions provided by the same subject) from the original sample to use for validation, while the remaining $n - 4$ observations constitute the training data from which to classify the subject left out. The procedure is repeated until each subject in the sample is used once as a validation set. For each subject, this procedure allows for the construction of a subject-specific misclassification error rate which, considered through the different classifiers, provides a knowledge of difficulty in classifying each single subject.

By applying standard LDA on $\mathbf{Z}_1$ we note that the classification error is around 19.50%, which corresponds to 14 misclassified observations. The separate error rates for each group and the overall error rate for the whole data set are shown in Table 1.

Considering the data set $\mathbf{Z}_2$, we have first performed standard PCA to achieve dimension reduction and then applied LDA on the score variables for classification

Table 1 Data set $\mathbf{Z}_1$. Separate error rates for each group and the overall error rate for the whole data set

Classifier	Neutral	Happiness	Sadness	Surprise	Overall
LDA	11.11%	5.56%	33.33%	27.78%	19.44%

Table 2 Data set $\mathbf{Z}_2$. Separate error rates for each group and the overall error rate for the whole data set

Classifier	Neutral	Happiness	Sadness	Surprise	Overall
PCA	10.00%	5.00%	12.50%	17.90%	13.90%
PCA2	5.56%	0.00%	22.22%	16.67%	11.11%
MPD	16.67%	22.20%	50.00%	16.67%	27.80%
ZVD	22.20%	11.10%	33.30%	24.50%	20.80%
BBD	27.50%	5.60%	17.50%	19.91%	19.50%

purposes. The number of components is chosen through cross validation and the minimum is achieved for 13 components giving a classification error rate of 13.90%. By following the PCA2 method, the classification error rate decreases to 11.11% with 10 chosen components.

The classification results from the rank decomposition methods are not as good as those obtained through PCA, with the MPD method to be considered as the worst. A summary of the cross validation procedure for all the methods is reported in Table 2, where we show both the separate error rates for each group and the overall error rate for the whole data set.

5 Conclusion

In this paper we have discussed the problem of classifying facial expressions by using Euclidean distances computed among a set of landmarks. The prediction of group membership is performed by applying a set of LDA-based classifiers on a large set of features. Taking care of the singularity of the within-group covariance matrix, the classifiers have been compared through the error rate computed in a cross validation procedure, with PCA2 providing the best performance. Note that the misclassification error rate achieved here slightly improves the one obtained by applying the same procedure on procrustes tangent coordinates (12.50%).

The FP procedure is relatively simple and the classification can be performed using standard LDA. The error rate is larger than those obtained through PCA and PCA2 on $\mathbf{Z}_2$ although it is comparable with the results provided by the other classifiers.

The comparison of subject-specific misclassification error rates through different classifiers provides an estimate of the difficulties of classifying the single units. This comparison shows that some subjects are equally overrated or underrated by the classifiers so that their performance in simulating the expressions may be in question.

Finally, we acknowledge that there exist further proposals to extend LDA to the high-dimensional setting which have not been considered here. For example, some of these proposals involve sparse classifiers using *lasso* (Tibshirani 1996) or *elastic net* (Clemmensen et al. 2011) penalties. The study of the performance of these classifiers will be a topic for future works.

References

Clemmensen, L., Hastie, T., Witten, D., & Ersbøll, B. (2011). Sparse discriminant analysis. *Technometrics*, *53*, 406–413.

Devijver, P., & Kittler, J. (1982). *Pattern recognition: a statistical approach* (1st edn). London: Prentice Hall.

Dhillon, I. S., Modha, D. S., & Spangler, W. S. (2002). Class visualization of high-dimensional data with applications. *Computational Statistics and Data Analysis*, *41*, 59–90.

Dryden, I. L., & Mardia, K. V. (1998). *Statistical shape analysis*. Chichester: Wiley.

Ekman, P., & Friesen, W. (1971). Constants across cultures in the face and emotions. *Journal of Personality and Social Psychology*, *17*, 124–129.

Jolliffe, I. (2002). *Principal component analysis* (2nd edn). New York: Springer.

Krzanowsky, W. J., & Marriott, F. H. C. (1995). *Multivariate analysis: classification, covariance structures and repeated measurements* (vol. 2). London: Arnold, Holder Headline Group.

Krzanowski, W. J., Jonathan, P., McCarthy, W. V., & Thomas, M. R. (1995). Discriminant analysis with singular covariance matrices: methods and applications to spectroscopic data. *Journal of the Royal Statistical Society, Series C*, *44*, 101–115.

Mardia, K. V., Kent, J. T., & Bibby, J. M. (1979). *Multivariate analysis*. London: Academic Press.

Mase, K., & Pentland, A. (1991). Recognition of facial expression from optical flow. *IEICE Transaction E*, 3474–3483.

McLachlan, G. J. (2004). *Discriminant analysis and statistical pattern recognition*. Hoboken, New Jersey: Wiley.

Pantic, M., Cowie, R., Drrico, F., Heylen, D., Mehu, M., & Pelachaud, P. (2011). Social signal processing: the research agenda. In *Visual analysis of humans*, (Part 4, pp. 511–538). London: Springer.

Tebbens, J. D., & Schlesinger, P. (2007). Improving implementation of linear discriminant analysis for the high dimension/small sample size problem. *Computational Statistics and Data Analysis*, *52*, 423–437.

Tibshirani, R. (1996). Regression shrinkage and selection via the lasso. *Journal of Royal Statistical Society: Series B*, *58*, 267–288.

Grouping Around Different Dimensional Affine Subspaces

L.A. García-Escudero, A. Gordaliza, C. Matrán, and A. Mayo-Iscar

Abstract Grouping around affine subspaces and other types of manifolds is receiving a lot of attention in the literature due to its interest in several fields of application. Allowing for different dimensions is needed in many applications. This work extends the TCLUST methodology to deal with the problem of grouping data around different dimensional linear subspaces in the presence of noise. Two ways of considering error terms in the orthogonal of the linear subspaces are considered.

1 Introduction

Many non-hierarchical clustering methods are based on searching for groups around underlying features. For instance, the well-known k-means method creates groups around k point-centers. However, clusters found in a given data set are sometimes due to the existence of certain relationships among the measured variables.

On the other hand, the Principal Component Analysis method serves to find global correlation structures. However, some interesting correlations are non-global since they may be different in different subgroups of the data set or even be the distinctive characteristic of such groups. This idea has also been proposed with the aim at overcoming the "curse of dimensionality" trouble in high-dimensional problems by considering that the data do not uniformly fill the sample space and that data points are instead concentrated around low dimensional manifolds.

There exist many references about clustering around affine subspaces with equal dimensions within the statistical literature (see, e.g., Van et al. 2006 and Garcia-Escudero et al. 2008b and the references therein). We can distinguish between two

L.A. García-Escudero (✉) · A. Gordaliza · C. Matrán · A. Mayo-Iscar
Universidad de Valladolid, Facultad de Ciencias, Paseo Belén 7, 47011, Valladolid, Spain

Dpto. Estadística e I.O, Universidad de Valladolid and IMUVA
e-mail: lagarcia@eio.uva.es; alfonsog@eio.uva.es; matran@eio.uva.es; agustinm@eio.uva.es

P. Giudici et al. (eds.), *Statistical Models for Data Analysis*, Studies in Classification, Data Analysis, and Knowledge Organization, DOI 10.1007/978-3-319-00032-9_16,

different approaches: "clusterwise regression" and "orthogonal residuals methods". In clusterwise regression techniques, it is assumed the existence of a privileged response or outcome variable that we want to explain in terms of the explicative ones. Throughout this work, we will be assuming that no privileged outcome variables exist. Other model-based approaches have already been proposed based on fitting mixtures of multivariate normals assuming that the smallest groups' covariances eigenvalues are small (see, e.g, Dasgupta and Raftery 1998) but they are not directly aimed at finding clusters around linear subspaces (see Van et al. 2006).

It is not difficult to find problems where different dimensionalities appear. In fact, this paradigm has already been addressed by the Machine Learning community. For instance, we can find approaches like "projected clustering" (PROCLUS, ORCLUS, DOC, k-means projective clustering), "correlation connected objects" (4C method), "intrinsic dimensions", "Generalized PCA", "mixture probabilistic PCA", etc.

In Sect. 2, we will propose suitable statistical models for clustering around affine subspaces with different dimensions. These come from extending the TCLUST modeling in Garcia-Escudero et al. (2008a). The possible presence of a fraction α of outlying data is also taken into account. Section 3 provides a feasible algorithm for fitting these outliers. Finally, Sect. 4 shows some simulations and a real data example.

2 Data Models

Clustering around affine subspaces: We assume the existence of k feature affine subspaces in $\mathbb{R}^p$ denoted by H_j with possible different dimensions d_j satisfying $0 \le d_j \le p-1$ (a single point if $d_j = 0$). Each subspace H_j is thus determined from $d_j + 1$ independent vectors. Namely, a group "center" m_j where the subspace is assumed to pass through and d_j unitary and orthogonal vectors $u_j^l, l = 1, \dots, d_j$, spanning the subspace. We can construct a $p \times d_j$ orthogonal matrix U_j from these u_j^l vectors such that each subspace H_j may be finally parameterized as $H_j \equiv \{m_j, U_j\}$.

We assume that an observation x belonging to the j-th group satisfies $x = \mathrm{Pr}_{H_j}(x) + \varepsilon_j^*$, with Pr_{H_j} denoting the orthogonal projection of x onto the subspace H_j given by $\mathrm{Pr}_{H_j}(x) = m_j + U_j U_j'(x - \mu_j)$ and ε_j^* being a random error term chosen in the orthogonal of the linear subspace spanned by the columns of U_j. If ε_j is a random distribution in $\mathbb{R}^{p-d_j}$, we can chose $\varepsilon_j^* = U_j^\perp \varepsilon_j$ with $U_j^\perp$ being a $p \times (p - d_j)$ orthogonal matrix whose columns are orthogonal to the columns of U_j (the Gram-Schmidt procedure may be applied to obtain the matrix $U_j^\perp$). We will further assume that ε_j has a $(p - d_j)$-elliptical distribution with density $|\Sigma_j|^{-1/2} g(x' \Sigma_j^{-1} x)$.

Given a data set $\{x_1, \dots, x_n\}$, we define the clustering problem through the maximization of the "classification log-likelihood":

$$\sum_{j=1}^{k}\sum_{i\in R_j}\log(p_j f(x_i;H_j,\Sigma_j)), \tag{1}$$

with $\cup_{j=1}^{k} R_j = \{1,\ldots,n\}$, $R_j \cap R_l = \emptyset$ for $j \neq l$ and

$$f(x_i;H_j,\Sigma_j) = |\Sigma_j|^{-1/2} g\big((x_i - \Pr_{H_j}(x_i))' U_j^{\perp} \Sigma_j^{-1} (U_j^{\perp})' (x_i - \Pr_{H_j}(x_i))\big). \tag{2}$$

Furthermore, we assume the existence of some underlying unknown weights p_j's which satisfy $\sum_{j=1}^{k} p_j = 1$ in (1). These weights lead to more logical assignments to groups when they overlap.

Robustness: The term "robustness" may be used in a twofold sense. First, in Machine Learning, this term is often employed to refer to procedures which are able to handle a certain degree of internal within-cluster variability due, for instance, to measurement errors. This meaning obviously has to do with the consideration of data models as those previously presented. Another meaning for the term "robustness" (as referred to in the statistical literature) is related to the ability of the procedure to resist to the effect of a certain fraction of "gross errors". The presence of gross errors is unfortunately the rule in many real data sets.

To take into account gross errors, we can modify the "spurious-outliers model" in Gallegos and Ritter (2005) to define a unified suitable framework when considering these two possible meanings for the term "robustness". Starting from this "spurious-outliers model", it makes sense to search for linear affine subspaces H_j, group scatter matrices Σ_j and a partition of the sample $\cup_{j=0}^{k} R_j = \{1,2,\ldots,n\}$ with $R_j \cap R_l = \emptyset$ for $j \neq l$ and $\#R_0 = n - [n\alpha]$ maximizing the "trimmed classification log-likelihood":

$$\sum_{j=1}^{k}\sum_{i\in R_j}\log(p_j f(x_i;H_j,\Sigma_j)). \tag{3}$$

Note that the fraction α of observations in R_0 is no longer taken into account in (3).

Visual and normal errors: Although several error terms may be chosen under the previous general framework, we focus on two reasonable and parsimonious distributions that they follow from considering $\Sigma_j = \sigma_j I_{p-d_j}$ and the following g functions in (2):

(a) *Visual errors model* (VE-model): We assume that the mechanism generating the errors follows two steps. First, we randomly choose a vector v in the sphere $S_{p-d_j} = \{x \in \mathbb{R}^{p-d_j} : \|x\| = 1\}$. Afterwards, we obtain the error term ε_j as $\varepsilon_j = v \cdot |z|$ with z following a $N_1(0,\sigma_j^2)$ distribution. We call these "visual" errors because we "see" (when $p \leq 3$) the groups equally scattered when the σ_j's are equal independently of the dimensions. The VE-model leads to use:

$$f(x; H_j, \sigma_j) = \tag{4}$$
$$= \frac{\Gamma((p-d_j)/2)}{\pi^{(p-d_j)/2}\sqrt{2\pi\sigma_j^2}} \|x - \Pr_{H_j}(x)\|^{-(p-d_j-1)/2} \exp\left(-\|x - \Pr_{H_j}(x)\|^2/2\sigma_j^2\right).$$

To derive this expression, consider the stochastic decomposition of a spherical distribution X in $\mathbb{R}^{p-d_j}$ as $X = RU$ with R a "radius" variable and U a uniform distribution on S_{p-d_j}. If h denotes the p.d.f. of R and g the density generator of the spherical family then $h(r) = \frac{2\pi^{(p-d_j)/2}}{\Gamma((p-d_j)/2)} r^{(p-d_j)-1} g(r^2)$. Thus, if $R = |Z|$ with Z being a $N(0,1)$ random variable, we get $h(r) = 2/\sqrt{2\pi} \cdot \exp(-x^2/2)$. Expression (4) just follows from (2). Note that $g(x) = C_{p-d_j} x^{N-1} \exp(-rx^s)$ with $N-1 = -(p-d_j-1)/2, r = 1/2 > 0, s = 1 > 0$ (and satisfying the condition $2N + p > 2$). Therefore, this density is reduced to the univariate normal distribution whenever $p - d_j = 1$ and, in general, belongs to the symmetric Kotz type family (Kotz 1975).

(b) *"Normal" errors model* (NE-model): With this approach, the mechanism generating the error terms is based on adding a normal noise in the orthogonal of the feature space H_j. That is, we take ε_j following a $N_{p-d_j}(0, \sigma_j^2 I_{p-d_j})$ distribution:

$$f(x; H_j, \sigma_j) = (2\pi\sigma_j^2)^{-(p-d_j)/2} \exp\left(-\|x - \Pr_{H_j}(x)\|^2/2\sigma_j^2\right) \tag{5}$$

The use of "normal" errors has already been considered in Banfield and Raftery (1993) and "visual" errors in Standford and Raftery (2000) when working with two-dimensional data sets and grouping around (one-dimensional) smooth curves.

Figure 1 shows two data sets generated with VE- and NE-models. It also shows the boundaries of sets $\{x : d(x, H_j) \leq z_{0.025}/2\}$ with $z_{0.025}$ being the 97.5% percentile of the $N_1(0,1)$ and $d(x, H) = \inf_{y \in H} \|x - y\|$ when H_1 is a point (a ball) and when H_2 is a line (a "strip"). Note the great amount of observations that fall outside the ball in the normal errors case even though the same scatters were considered in both groups.

Constraints on the scatter parameters: Let us consider $d_j + 1$ observations and H_j the affine subspace determined by them. We can easily see that (3) (and (1) too) become unbounded when $|\Sigma_j| \to 0$. Thus, the proposed maximization problems would not be mathematically well-defined without posing any constraint on the Σ_j's.

When $\Sigma_j = \sigma_j I_{p-d_j}$, the constraints introduced in Garcia-Escudero et al. (2008a) are translated into

$$\max_j \sigma_j^2 / \min_j \sigma_j^2 \leq c \text{ for a given constant } c \geq 1. \tag{6}$$

The constant c avoids non interesting clustering solutions with clusters containing very few almost collinear observations. This type of restrictions goes back to Hathaway (1985).

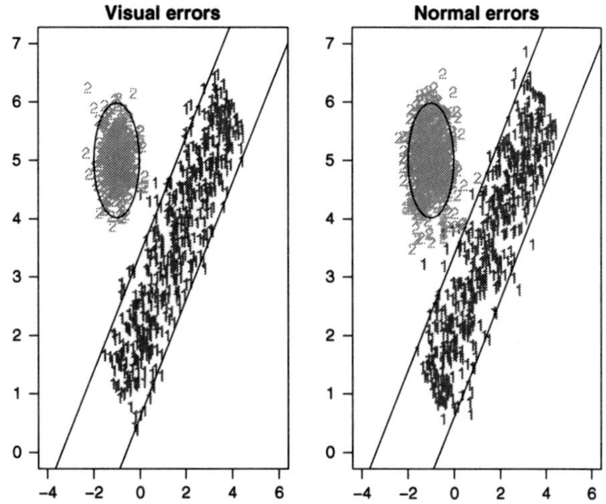

Fig. 1 Simulated data set from the VE- and NE- models

3 Algorithm

The maximization of (3) under the restriction (6) has high computational complexity. We propose here an algorithm based on the TCLUST one. Some ideas behind the classification EM algorithm (Celeux and Govaert 1992) and from the RLGA (Garcia-Escudero et al. 2008b) also underlie.

1. *Initialize the iterative procedure:* Set initial weights values $p_1^0 = \ldots = p_k^0 = 1/k$ and initial scatter values $\sigma_1^0 = \ldots = \sigma_k^0 = 1$. As starting k linear subspaces, randomly select k sets of $d_j + 1$ data points to obtain k initial centers m_j^0 and k initial matrices U_j^0 made up of orthogonal unitary vectors.
2. *Update the parameters in the l-th iteration as*:

2.1. Obtain

$$D_i = \max_{j=1,\ldots,k} \{p_j^l f(x_i; m_j^l, U_j^l, \sigma_j^l)\} \tag{7}$$

and keep the set R^l with the $n - [n\alpha]$ observations with largest D_i's. Split R^l into $R^l = \{R_1^l, \ldots, R_k^l\}$ with $R_j^l = \{x_i \in R^l : p_j^l f(x_i; m_j^l, U_j^l, \sigma_j^l) = D_i\}$.

2.2. Update parameters by using:

- $p_j^{l+1} \hookleftarrow$ "$n_j^l/[n(1-\alpha)]$ with n_j^l equal to the number of data points in R_j^l".
- $m_j^{l+1} \hookleftarrow$ "The sample mean of the observations in R_j^l".
- $U_j^{l+1} \hookleftarrow$ "A matrix whose columns are equal to the d_j unitary eigenvectors associated to the largest eigenvalues of the sample covariance matrices of observations in R_j^l".

Use the sum of squared orthogonal residuals to obtain initial scatters $s_j^2 = \frac{1}{n_j^l} \sum_{x_i \in R_j^l} \|x_i - P_{H_j^l}(x_i)\|^2$ with $H_j^l \equiv \{m_j^l, U_j^l\}$. To satisfy the constraints, they must be "truncated" as:

$$[s_j^2]_t = \begin{cases} s_j^2 \text{ if } s_j^2 \in [t, ct] \\ t \ \text{ if } s_j^2 < t \\ ct \text{ if } s_j^2 > ct \end{cases} . \tag{8}$$

Search for $t_{\text{opt}} = \arg\max_t \sum_{j=1}^k \sum_{x_i \in R_j^l} \log f(x_i; m_j^{l+1}, U_j^{l+1}, [s_j^2]_t)$ and take

- $\sigma_j^{l+1} \hookleftarrow \sqrt{[s_j^2]_{t_{\text{opt}}}}$.

3. *Compute the evaluation function:* Perform L iterations of the process described in step 2 and compute the final associated target function (3).
4. *Repeat several times:* Draw S random starting values and keep the solution leading to the maximal value of the target function.

Determining t_{opt} implies solving a one-dimensional optimization problem. This can be easily done by resorting to numerical methods. More details concerning the rationale of this algorithm can be found in Garcia-Escudero et al. (2008a). We denote the previous algorithm as VE-method when the density (4) is applied and as NE-method when using (5).

4 Examples

Simulation study: Let us consider a clustering problem where observations are generated around a point, a line and a plane in $\mathbb{R}^3$. We generate uniformly-distributed points on the sets $C_1 = \{(x_1, x_2, x_3) : x_1 = x_2 = x_3 = 3\}$ (no random choice), on $C_2 = \{(x_1, x_2, x_3) : 1 \le x_1 \le 6, x_2 = x_3 = 3\}$, and, on $C_3 = \{(x_1, x_2, x_3) : x_1 = -2, 1 \le x_2 \le 6, 1 \le x_3 \le 6\}$. Later, we add error terms in the orthogonal of the C_j's considering the models introduced in Sect. 2. Finally, points are randomly drawn on the cube $[-4, 6] \times [-4, 6] \times [-4, 6]$ as "gross errors". Figure 2 shows the result of the proposed clustering approach for a data set drawn from that scheme of simulation.

A comparative study based on the previous simulation scheme with VE- and NE-methods has been carried out. We have also considered an alternative (Euclidean distance) ED-method where the D_i's in (7) are replaced by the more simple expressions $D_i = \inf_{j=1,\ldots,k} \|x_i - P_{H_j^l}(x_i)\|$ and no updating of the scatter parameters is done. The ED-method is a straightforward extension of the RLGA in Garcia-Escudero et al. (2008b).

Hundred random samples of size $n = 400$ from the previously described simulation schemes with VE- and NE-models for the orthogonal errors are randomly drawn and the associated results for the three clustering VE-, NE- and ED-methods are monitored. Figure 3 shows the mean proportion of misclassified observations along these 100 random samples. The NE-model seems to have a higher complexity

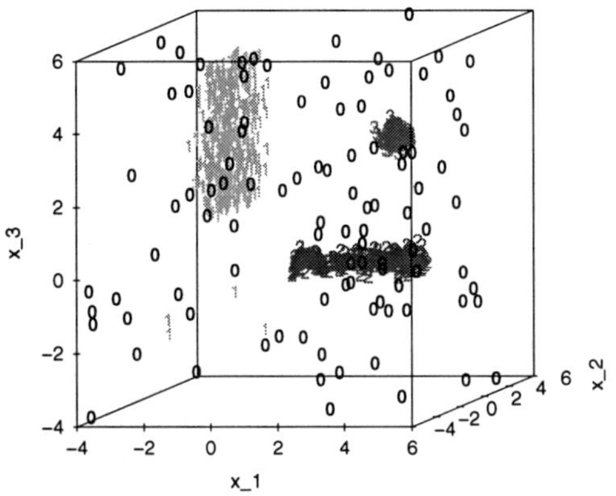

Fig. 2 Result of the NE-method with $k = 3$, $d_j{}'s = (0, 1, 2)$, $c = 2$ and $\alpha = .1$

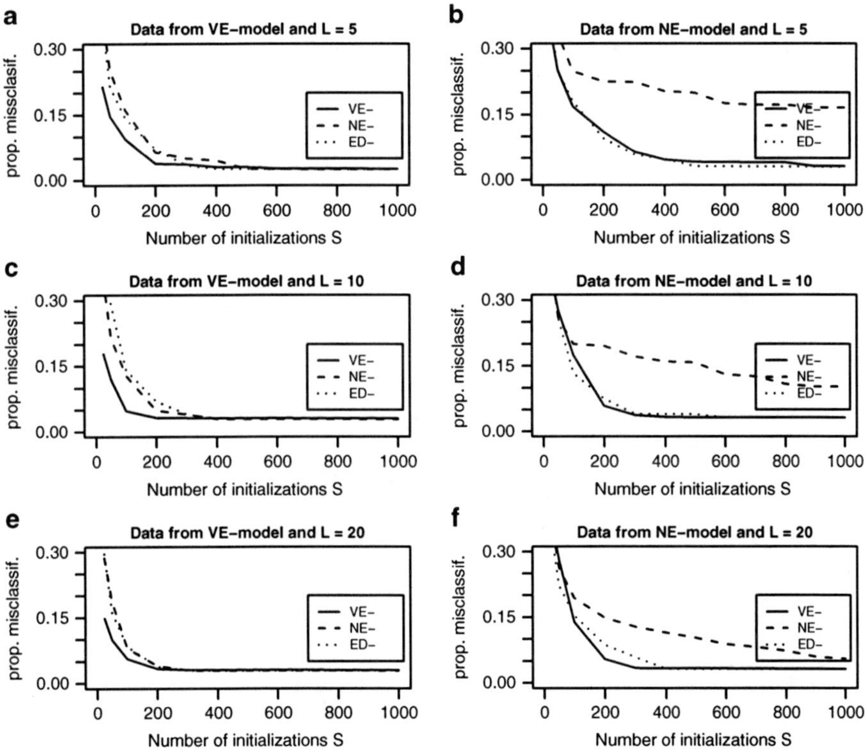

Fig. 3 Proportion of misclassified observations in the simulation study described in the text

since a higher number of random initializations is needed. Note that the results favor the VE-method even when the true model generating the data was indeed the NE-model. We can also see that parameter S is more critical than L.

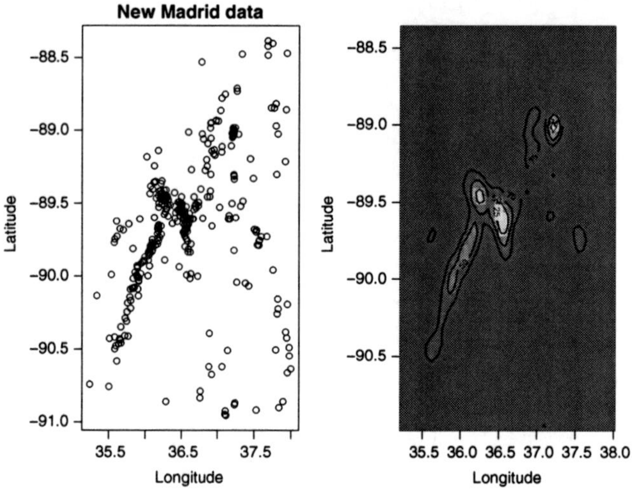

Fig. 4 Earthquake positions in the New Madrid seismic region

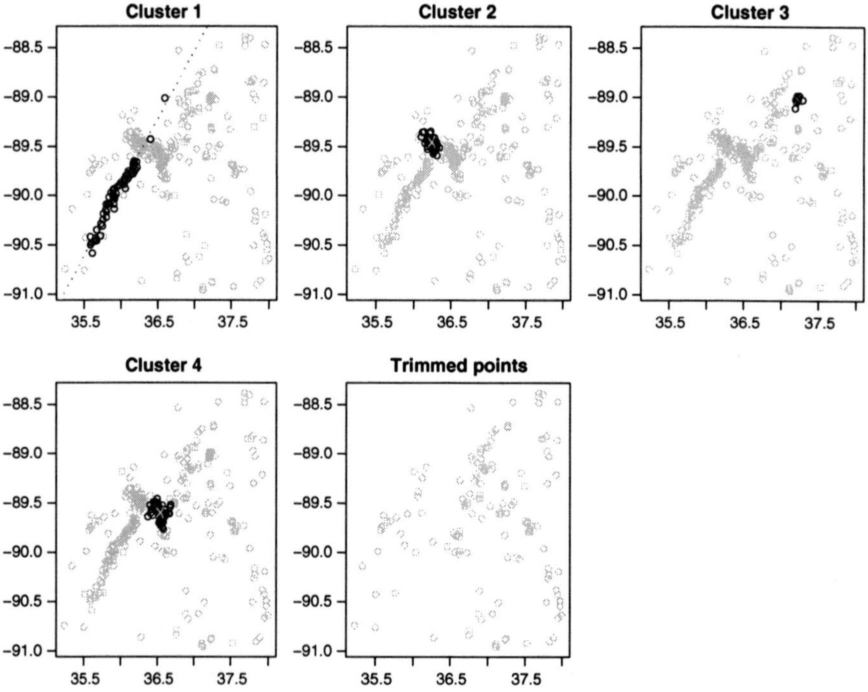

Fig. 5 Clustering results of the VE-method for $k = 4$, d_j's $= (1, 0, 0, 0)$, c =2 and α =.4

Real data example: As in Standford and Raftery (2000), we consider position data on some earthquakes in the New Madrid seismic region from the CERI. We include all earthquakes in that catalog from 1974 to 1992 with magnitudes 2.25 and above. Figure 4 shows a scatter plot of the position of earthquakes and a non-parametric kernel-based density estimation suggesting the existence of a linear tectonic fault and three main point foci.

Figure 5 shows the clustering results when $k = 4$ and dimensions $(1, 0, 0, 0)$. We have considered a high trimming level $\alpha = .4$ which allows for discarding earthquakes that take place in regions where they are not spatially concentrated.

5 Future Research Directions

The proposed methodology needs to fix parameters k, d_j's, α and c. Sometimes these are known in advance but other times they are completely unknown. "Split and merge", BIC and geometrical-AIC concepts could then be applied. Another important issue is how to deal with remote observations wrongly assigned to higher dimensional linear subspaces due to their "not-bounded" spatial extension. A further second trimming or nearest neighborhood cleaning could be tried.

References

Banfield, J. D., & Raftery, A. E. (1993). "Model-based Gaussian and non-Gaussian clustering". *Biometrics, 49*, 803–821.

Celeux, G., & Govaert, A. (1992). "Classification EM algorithm for clustering and two stochastic versions". *Computational Statistics & Data Analysis, 13*, 315–332.

Dasgupta, A., & Raftery, A. E. (1998). "Detecting features in spatial point processes with clutter via model-based clustering." *Journal of the American Statistical Association, 93*, 294–302.

Gallegos, M. T., & Ritter, G. (2005). "A robust method for cluster analysis." *Annals of Statistics, 33*, 347–380.

García-Escudero, L. A., Gordaliza, A., Matrán, C., & Mayo-Iscar, A. (2008a). "A general trimming approach to robust clustering". *Annals of Statistics, 36*, 1324–1345.

García-Escudero, L. A., Gordaliza, A., San Martín, R., Van Aelst, S., & Zamar, R. (2008b). "Robust linear grouping". *Journal of the Royal Statistical Society: Series B, 71*, 301–319.

Hathaway, R. J. (1985). "A constrained formulation of maximum likelihood estimation for normal mixture distributions." *Annals of Statistics, 13*, 795–800.

Kotz, S. (1975). "Multivariate distributions at a cross-road". In G. P. Patil, S. Kotz, J. K. Ord (Eds.) *Statistical distributions in scientific work* (vol. 1, pp. 247270).

Standford, D. C., & Raftery, A. E. (2000). "Finding curvilinear features in spatial point patterns: principal curve clustering with noise". *IEEE Transactions on Pattern Recognition, 22*, 601–609.

Van Aelst, S., Wang, X., Zamar, R. H., & Zhu, R. (2006). "Linear grouping using orthogonal regression". *Computational Statistics & Data Analysis, 50*, 1287–1312.

Hospital Clustering in the Treatment of Acute Myocardial Infarction Patients Via a Bayesian Semiparametric Approach

Alessandra Guglielmi, Francesca Ieva, Anna Maria Paganoni, and Fabrizio Ruggeri

Abstract In this work, we develop Bayes rules for several families of loss functions for hospital report cards under a Bayesian semiparametric hierarchical model. Moreover, we present some robustness analysis with respect to the choice of the loss function, focusing on the number of hospitals our procedure identifies as "unacceptably performing". The analysis is carried out on a case study dataset arising from MOMI^2 (Month MOnitoring Myocardial Infarction in MIlan) survey on patients admitted with ST-Elevation Myocardial Infarction to the hospitals of Milan Cardiological Network. The major aim of this work is the ranking of the health-care providers performances, together with the assessment of the role of patients' and providers' characteristics on survival outcome.

1 Introduction

Performance indicators have recently received increasing attention; they are mainly used with the aim of assessing quality in health-care research (Austin 2008; Austin and Lawrence 2008; Grieco et al. 2012; Normand et al. 1997; Normand and Shahian 2007; Ohlssen et al. 2007; Racz and Sedransk 2010). In this work, we suitably model the survival outcome of patients affected by a specific disease in different clinical structures; the aim is to point out similar behaviors among groups of hospitals and then classify them according to some acceptability criteria. In general, provider profiling of health-care structures is obtained producing report cards comparing

A. Guglielmi (✉) · F. Ieva · A.M. Paganoni
Politecnico di Milano, piazza Leonardo da Vinci 32 - 20133 Milano, Italy
e-mail: alessandra.guglielmi@polimi.it; francesca.ieva@mail.polimi.it; anna.paganoni@polimi.it

F. Ruggeri
CNR IMATI Milano, via Bassini 15 - 20133 Milano, Italy
e-mail: fabrizio@mi.imati.cnr.it

P. Giudici et al. (eds.), *Statistical Models for Data Analysis*, Studies in Classification, Data Analysis, and Knowledge Organization, DOI 10.1007/978-3-319-00032-9_17,

their global outcomes or performances of their doctors. These cards have mainly two goals:

- To provide information that can help individual consumers (i.e. patients) making a decision.
- To identify hospitals that require investments in quality improvement initiatives.

Here we are interested not only in point estimation of the mortality rate, but also to decide whether investing in quality improvement initiatives for each hospital with "unacceptable performances". The paper presents Bayes rules under several families of loss functions for hospital report cards. In particular, we adopt a Bayesian semiparametric hierarchical model in this case, since it is known that they are more flexible than "traditional" Bayesian parametric models. Moreover, we did some robustness analysis with respect to the choice of the loss function, focusing on the hospitals our procedure identifies as "unacceptably performing".

Our aim is to profile health-care providers in our regional district, i.e. Regione Lombardia. Indeed, the health governance of Regione Lombardia is very sensitive to cardiovascular issues, as proved by the huge amount of social and scientific projects concerning these syndromes, which were promoted and developed during the last years. Details on some of the most important clinical and scientific local projects can be found in Barbieri et al. (2010). The data we have analyzed in our application come from a survey called MOMI^2, which is a retrospective longitudinal clinical survey on a particular type of infarction called STEMI (STsegment Elevation Myocardial Infarction). STEMI has very high incidence all over the world and it causes approximately 700 events each month only in our district. These cases are mainly treated through the surgical practice of primary angioplasty (a collapsed balloon is inserted through a catheter in the obstructed vessel, and then inflated, so that the blood flow is restored). It is well known (Gersh et al. 2005; Giugliano and Braunwald 2003) that within this pathology, the more prompt the intervention is, the more effective the therapy is; for this reason the main process indicators used to evaluate hospitals performances are in-hospital treatment times. The MOMI^2 survey consists of six time periods data collection in the hospitals belonging to the cardiological network of the urban area of Milan. It contains 841 statistical units, and, for each patient, personal data, mode of admission, symptoms and process indicators, reperfusion therapy and outcomes have been collected. After each collection, all the hospital performances (in terms of patients' survival) were evaluated; moreover, a feedback was given to providers (especially those with "unacceptable performances") in order to let them improve their performances.

The article is organized as follows: in Sect. 2 we present the statistical method used to support decisions in this health-care context, while Sect. 3 shows how the proposed model and method have been applied to data coming from MOMI^2 survey. Finally, conclusions and open problems are discussed in Sect. 4.

2 Statistical Support to Decision-Making in Health-Care Policy

Since random errors can be present even in a perfect risk-adjustment framework, some mistakes could occur when classifying hospital performances as "acceptable" or not, so that some hospitals could be misclassified. Anyway, different players in the health-care context would pay different costs on misclassification errors. By *False Positive* we mean the hospital that truly had acceptable performances but was classified as "unacceptably performing", and by *False Negative* the hospital that truly had unacceptable performances but was classified as "acceptably performing". Then a health-care consumer would be presumably willing to pay a higher payoff for decisions that minimize false negatives, whereas hospitals might pay a higher cost for information that minimizes false positives. On the other hand, the same argument could be used to target hospitals for quality improvement: false positives would yield unneeded investments in quality improvement, but false negatives would lead to loose opportunities in improving the hospital quality. According to its plans, any health-care government could be interested in minimizing false positives and/or false negatives.

In order to provide support to decision-making in this context, we carry out the statistical analysis in the following way: first we estimate the in-hospital survival rates after fitting a Bayesian semiparametric generalized linear mixed-effects model, in particular modelling the random effect parameters via a Dirichlet process; then we develop Bayes decision rules in order to minimize the expected loss arising from misclassification errors, comparing four different loss functions for hospital report cards.

We fit a Bayesian generalized mixed-effects model for binary data. For unit (patient) $i = 1, \ldots, n_j$, in group (hospital) $j = 1, \ldots, J$, let Y_{ij} be a Bernoulli random variable with mean p_{ij}, i.e.,

$$Y_{ij} | p_{ij} \overset{ind}{\sim} Be(p_{ij}).$$

The p_{ij}s are modelled through a logit regression of the form

$$logit(p_{ij}) = \log \frac{p_{ij}}{1 - p_{ij}} = \gamma_0 + \sum_{h=1}^{p} \gamma_h x_{ijh} + \sum_{l=1}^{J} b_l z_{jl} \tag{1}$$

where $z_{jl} = 1$ if $j = l$ and 0 otherwise. In this model, $\boldsymbol{\gamma} = (\gamma_0, \ldots, \gamma_p)$ represents the $(p + 1)$-dimensional vector of the fixed effects, $\mathbf{x}_{ij}$ is the vector of patient covariates and $\mathbf{b} = (b_1, \ldots, b_J)$ is the vector of the additive random-effects parameters of the grouping factor. According to Kleinman and Ibrahim (1998), we assume a nonparametric prior for $b_1, \ldots, b_J$, namely the b_js will be i.i.d. according to a Dirichlet process (see Ferguson 1973), to include robustness with respect to miss-specification of the prior at this stage, since it is known that the regression parameters can be sensitive to the standard assumption of normality of the random effects; the prior for $\boldsymbol{\gamma}$ is parametric. Prior details will be given in Sect. 3. Model (1) is a generalized linear mixed model with $p + 1$ regression coefficients and one random effect. In Guglielmi et al. (2012b) the same model was fitted on a different dataset to classify hospitals taking advantage of the in-built clustering property of the Dirichlet process prior. Here we use Bayesian estimates to address a new decision problem concerning hospitals' performances.

Bayesian inferences are based on the posterior distribution, i.e., the conditional distribution of the parameters vector, given the data. Once the posterior distribution has been computed, suitable loss functions can be defined in order to *a posteriori* weigh the decision of wrongly classifying the hospital as having acceptable or unacceptable performances. The random intercepts of model (1), i.e., $\gamma_0 + b_1, \gamma_0 + b_2, \ldots, \gamma_0 + b_J$ represent the hospital performances quantifying the contribution to the model after patients' covariates adjustment. Let us denote by β_j the sum of γ_0 and b_j. The class of loss functions we are going to assume is then

$$L(\beta_j, d) = c_I \cdot f_1(\beta_j) \cdot d \cdot \mathbb{I}(\beta_j > \beta_t) + c_{II} \cdot f_2(\beta_j) \cdot (1 - d) \cdot \mathbb{I}(\beta_j < \beta_t), \quad (2)$$

where d is the decision to take ($d = 1$ means that the hospital has "unacceptable performances", $d = 0$ stand for "acceptable performances"), c_I is the weight assigned to the cost $f_1(\beta_j)$, occurring for a false positive, c_{II} is the weight assigned to cost $f_2(\beta_j)$, occurring for a false negative and β_t is defined as $\log(p_t/(1 - p_t))$, p_t being a reference value for survival probabilities.

Without loss of generality, we can assume a proportional penalization, i.e., $f_2(\beta_j) = k \cdot f_1(\beta_j)$, taking k as the ratio c_{II}/c_I. In this sense, the parameter k quantifies our beliefs on cost, being greater than 1 if we think that accepting a *false negative* should cost more than rejecting a true negative and less than 1 otherwise. An acceptable performance is then defined comparing the posterior expected losses associated with the decision that the hospital had "acceptable performances"

$$R(\mathbf{y}, d = 0) = E_\pi\left(L(\beta_j, d = 0)|\mathbf{y}\right) = \int f_2(\beta_j)\mathbb{I}(\beta_j < \beta_t)\Pi(\beta_j|\mathbf{y})d\beta_j$$

and the decision that the hospital had "unacceptable performances"

$$R(\mathbf{y}, d = 1) = E_\pi\left(L(\beta_j, d = 1)|\mathbf{y}\right) = \int f_1(\beta_j)\mathbb{I}(\beta_j > \beta_t)\Pi(\beta_j|\mathbf{y})d\beta_j.$$

In short, we classify an hospital as being "acceptable" (or with "acceptable performances") if the risk associated with the decision $d = 0$ is less than the risk associated with the decision $d = 1$, i.e., if $R(\mathbf{y}, d = 0) < R(\mathbf{y}, d = 1)$.

Within this setting, four different loss functions (2) will be considered in the next section, to address the decision problem, namely

0/1 Loss :

$$L(\beta_j, d) = d \cdot \mathbb{I}(\beta_j > \beta_t) + k \cdot (1 - d) \cdot \mathbb{I}(\beta_j < \beta_t),$$

Absolute Loss :

$$L(\beta_j, d) = |\beta_j - \beta_t| \cdot d \cdot \mathbb{I}(\beta_j > \beta_t) + k \cdot |\beta_j - \beta_t| \cdot (1 - d) \cdot \mathbb{I}(\beta_j < \beta_t),$$

Squared Loss :

$$L(\beta_j, d) = (\beta_j - \beta_t)^2 \cdot d \cdot \mathbb{I}(\beta_j > \beta_t) + k \cdot (\beta_j - \beta_t)^2 \cdot (1 - d) \cdot \mathbb{I}(\beta_j < \beta_t),$$

LINEX Loss :

$$L(\beta_j, d) = l(\beta_j - \beta_t) \cdot d \cdot \mathbb{I}(\beta_j > \beta_t) + k \cdot l(\beta_j - \beta_t) \cdot (1 - d) \cdot \mathbb{I}(\beta_j < \beta_t).$$

For instance, this means that, to recover the 0/1 loss function above, the functions $f_i(\beta_j)$, $i = 1, 2$ in (2) are both constant, $f_i(\beta_j) = |\beta_j - \beta_t|, i = 1, 2$ for the Absolute Loss case, $f_i(\beta_j) = (\beta_j - \beta_t)^2$, $i = 1, 2$ for the Squared Loss case and $f_i(\beta_j) = l(\beta_j - \beta_t) = \exp\{a \cdot (\beta_j - \beta_t)\} - a \cdot (\beta_j - \beta_t) - 1$, $i = 1, 2$ to obtain the LINEX Loss function. Note that all the loss functions, but the last one, are symmetric, and the parameter k is used to introduce an asymmetry in weighing the misclassification error costs.

3 Application to MOMI^2 Data

In this section we apply the model and the method proposed in Sect. 2 to 536 patients, from MOMI^2 data, who underwent PTCA treatment. For this sample, 17 hospitals of admission are involved, and in-hospital survival rate of 95% is observed. Among all possible covariates (mode of admission, clinical appearance, demographic features, time process indicators, hospital organization etc.) available in the survey, only age and Killip class (which quantifies the severity of infarction on a scale ranging from 1, the less severe case, to 4, the most severe one) have been selected as being statistically significant. The killip class is made dichotomic here, i.e., the killip covariate is equal to 1 for the two more severe classes (Killip 3 and 4) and equal to 0 otherwise (Killip 1 and 2). Moreover, we considered the total ischemic time (namely Onset to Balloon time or briefly OB) in the logarithmic scale too, because of clinical best practice and know-how. The choice of the covariates and the link function was suggested in Ieva and Paganoni (2010), according to frequentist selection procedures and clinical best-practice, and confirmed in Guglielmi et al. (2012a) using Bayesian tools.

Summing up, the model (1) we considered for our dataset is

$$logit\,(\mathbb{E}[Y_{ij}|b_j]) = logit(p_{ij}) = \gamma_0 + \gamma_1 \cdot age_i + \gamma_2 \cdot \log(OB)_i + \gamma_3 \cdot killip_i + b_j \quad (3)$$

Table 1 Providers labelled as "unacceptable", under (3)–(4), for different loss functions and different values of k and β_t

Loss	$k = 0.5$ $p_t = 0.96$ $\beta_t = 3.178$	$k = 1$ $p_t = 0.96$ $\beta_t = 3.178$	$k = 2$ $p_t = 0.96$ $\beta_t = 3.178$
0/1	None	None	None
Absolute	None	None	None
Squared	None	None	9
LINEX	None	None	9
Loss	$k = 0.5$ $p_t = 0.97$ $\beta_t = 3.476$	$k = 1$ $p_t = 0.97$ $\beta_t = 3.476$	$k = 2$ $p_t = 0.97$ $\beta_t = 3.476$
0/1	None	9	3,5,9,10
Absolute	None	9	3,5,9,10
Squared	9	9	3,5,9,10
LINEX	9	3,5,9,10	3,5,9,10
Loss	$k = 0.5$ $p_t = 0.98$ $\beta_t = 3.892$	$k = 1$ $p_t = 0.98$ $\beta_t = 3.892$	$k = 2$ $p_t = 0.98$ $\beta_t = 3.892$
0/1	3,5,9,10	All	All
Absolute	2,3,4,5,9,10, 13,15	1,2,3,4,5,6,7,8,9,10, 11,13,14,15,16,17	All
Squared	2,3,4,5,9,10, 13,15	1,2,3,4,5,6,7,8,9, 10,13,14,15,17	All
LINEX	2,3,4,5,6,7,8,9, 10,13,15,17	All	All

for patient i ($i = 1, \ldots, 536$) in hospital j ($j = 1, \ldots, 17$). As far as the prior is concerned, we assume

$$\begin{gathered} \boldsymbol{\gamma} \perp \mathbf{b} \qquad \boldsymbol{\gamma} \sim \mathcal{N}_4(\mathbf{0}, 100 \cdot \mathbb{I}_4) \\ b_1, \ldots b_J | G \overset{iid}{\sim} G \qquad G|\alpha, G_0 \sim Dir(\alpha G_0) \\ G_0|\sigma \sim \mathcal{N}(0, \sigma^2) \quad \sigma \sim Unif(0, 10) \quad \alpha \sim Unif(0, 30). \end{gathered} \tag{4}$$

See details in Guglielmi et al. (2012b). The estimated posterior expected number of distinct values among the b_js, computed on 5,000 iterations of the MCMC output, is close to 7. In Table 1 the performances of different loss functions for different values of k and different threshold β_t are reported. The different values of p_t we considered (that determine the β_t values) were fixed in a range of values close to the empirical survival probability, in order to stress the resolution power of different losses in detecting unacceptable performances. Of course, when increasing the threshold p_t (and therefore β_t), more hospitals will be labelled as unacceptable. The tuning depends on the sensitivity required by the analysis. The parameter a of the LINEX loss is set to be equal to -1. Some comments are due, observing the results of

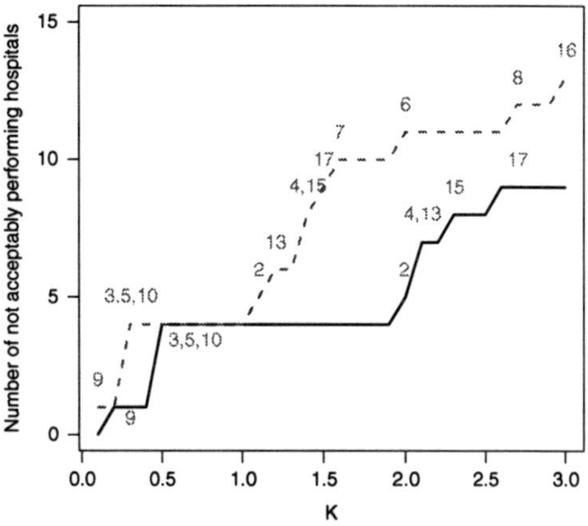

Fig. 1 Number of hospitals labelled as "unacceptable" as a function of k, under the Squared Loss function (*solid black*) and the LINEX Loss function (*dotted blue*). The threshold parameter β_t is 3.6635

Table 1. First, as mentioned before, k describes the different approach to evaluating misclassification errors. For example, people in charge with health-care government might be more interested in penalizing useless investments in quality improvements, choosing a value less than 1 for k. On the other hand, patients admitted to hospitals are more interested in minimizing the risk of wrongly declaring as "acceptably performing" providers that truly behave "worse" than the gold standards; therefore, they would probably choose a value greater than 1 for k. Moreover, when fixing the loss functions among the four proposed here, and k equal to 0.5, 1 or 2, as the threshold β_t increases, we obtain the same "implicit ranking" of providers:

$$9, 3, 5, 10, 2, 4, 13, 15, 6, 7, 8, 17, 1, 14, 16$$

(i.e., hospital 9 was classified as "unacceptable" even for small values of β_t, then, when increasing β_t, hospital 3 was classified as "unacceptable", etc.). This result is in agreement with the provider profiling pointed out also in Grieco et al. (2012). On the other hand, Fig. 1 shows the number of hospitals labelled as "unacceptable" as k increases, for a fixed value of the threshold β_t, under the Squared and the LINEX loss functions. Of course, the choice of the most suitable loss function is problem-driven: in our case, it seems reasonable to consider an asymmetric loss in order to penalize departures from threshold in different ways. For this reason we suggest the LINEX Loss with $k \neq 1$.

4 Conclusions and Further Developments

In this work we have considered data coming from a retrospective survey on STEMI to show an example of Operational Research applied to Regione Lombardia health-care policy. Using a logit model, we have represented the survival outcome by patient's covariates and process indicators, comparing results of different loss functions on decisions about provider's performances. In doing so, information coming from clinical registries was used to make the hospital network more effective, improving the overall health-care process and pointing out groups of hospitals with similar behavior, as it is required by the health-care decision makers of Regione Lombardia.

Currently, we are working on the extension of this paradigm to the whole Regional district, having designed and activated a new registry, called STEMI Archive (see Direzione Generale Sanità 2005), for all patients with STEMI diagnosis admitted to any hospital in Regione Lombardia. The analysis applied here to this sort of decision problems is relatively simple and effective. We believe that this approach could be considered by people in charge of the health-care governance in order to support decision-making in the clinical context.

References

Austin, P. C. (2008). Bayes rules for optimally using Bayesian hierarchical regression models in provider profiling to identify high-mortality hospitals. *BMC Medical Research Methodology*, 8–30.

Austin, P. C., & Lawrence, J. B. (2008). Optimal Bayesian probability levels for hospital report cards. *Health Service and Outcomes Research Method*, *8*, 80–97.

Barbieri, P., Grieco, N., Ieva, F., Paganoni, A. M., & Secchi, P. (2010). Exploitation, integration and statistical analysis of Public Health Database and STEMI archive in Lombardia Region. In P. Mantovan, & P. Secchi (Eds.) *Complex data modeling and computationally intensive statistical methods* (pp. 41–56). New York: Springer. "Contribution to Statistics".

Direzione Generale Sanità. (2005). Regione Lombardia: Patologie cardiocerebrovascolari: Interventi di Prevenzione, Diagnosi e Cura, Decreto N° 20592, 11/02/2005.

Ferguson, T. S. (1973). A Bayesian analysis of some nonparametric problems. *Annals of Statistics*, *1*, 209–230.

Gersh, B. J., Stone, G. W., White, H. D., & Holmes, D. R. (2005). Pharmacological facilitation of primary percutaneous coronary intervention for acute myocardial infarction: is the slope of the curve the shape of the future? *Journal of American Medical Association*, *293*(8), 979–986.

Giugliano, R. P., & Braunwald, E. (2003). Selecting the best reperfusion strategy in ST-Elevation Myocardial Infarction it's all a matter of time. *Circulation*, *108*, 2828–2830.

Grieco, N., Ieva, F., & Paganoni, A. M. (2012). Performance assessment using mixed effects models: a case study on coronary patient care. *IMA Journal of Management Mathematics*, *23*(2), 117–131.

Guglielmi, A., Ieva, F., Paganoni, A. M., & Ruggeri, F. (2012a). A Bayesian random-effects model for survival probabilities after acute myocardial infarction. *Chilean Journal of Statistics*, *3*(1), 1–15.

Guglielmi, A., Ieva, F., Paganoni, A. M., & Ruggeri, F. (2012b). Process indicators and outcome measures in the treatment of Acute Myocardial Infarction patients. In F. Faltin, R. Kenett, & F. Ruggeri (Eds.) *Statistical methods in healthcare*. Chichester: Wiley.

Ieva, F., & Paganoni, A. M. (2010). Multilevel models for clinical registers concerning STEMI patients in a complex urban reality: a statistical analysis of MOMI^2 survey. *Communications in Applied and Industrial Mathematics*, *1*(1), 128–147.

Kleinman, K. P., & Ibrahim, J. G. (1998). A semi-parametric Bayesian approach to generalized linear mixed models. *Statistics in Medicine*, *17*, 2579–2596.

Normand, S. T., Glickman, M. E., & Gatsonis, C. A. (1997). Statistical methods for profiling providers of medical care: issues and applications. *Journal of the American Statistical Association*, *92*, 803–814.

Normand, S. T., & Shahian, D. M. (2007). Statistical and clinical aspects of hospital outcomes profiling. *Statistical Science*, *22*(2), 206–226.

Ohlssen, D. I., Sharples, L. D., & Spiegelhalter, D. J. (2007). A hierarchical modelling framework for identifying unusual performance in health care providers. *Journal of the Royal Statistical Society: Series A (Statistics in Society)*, *170*(4), 865-890.

Racz, J., & Sedransk, J. (2010). Bayesian and frequentist methods for provider profiling using risk-adjusted assessments of medical outcomes. *Journal of the American Statistical Association*, *105*(489), 48–58.

A New Fuzzy Method to Classify Professional Profiles from Job Announcements

Domenica Fioredistella Iezzi, Mario Mastrangelo, and Scipione Sarlo

Abstract In the last years, Universities have created an office of placement to facilitate the employability of graduates. University placement offices select for companies, which offer a job and/or training position, a large number of graduates only based on degree and grades.

We adapt c-means algorithm to discover professional profiles from job announcements. We analyse 1,650 job announcements collected in DB SOUL since January 1st, 2010 to April 5th, 2011.

1 Introduction

In recent years, the number of Italian graduates has increased, although Italy is still far from targets set out in Europe 2020 (CEDEFOP 2010). In Italy, graduates have not find an adequate response from the employment point of view, especially if we compare our nation to other industrialized countries (OCDE 2010). Italian University reform designed new professional profiles, unlikely usable on the labour market (Aureli and Iezzi 2006; Iezzi 2005, 2008, 2009). The share of workers belonging to two large groups of ISCO88, which classifies occupations based on a master degree, is about 11 %. The European average is 15 % (OCDE 2010). Moreover graduates in Engineering, Pharmacy, Economics, Dentistry and Dental Implants quickly find a job consistent with their degree, but for many other graduates, e.g. in Communication Science, it is more difficult (ISTAT 2010a).

D.F. Iezzi (✉)
Tor Vergata University, Rome, Italy
e-mail: stella.iezzi@uniroma2.it

M. Mastrangelo • S. Sarlo
Sapienza University, Rome, Italy
e-mail: m.mastrangelo@uniroma1.it; scipione.sarlo@uniroma1.it

P. Giudici et al. (eds.), *Statistical Models for Data Analysis*, Studies in Classification, Data Analysis, and Knowledge Organization, DOI 10.1007/978-3-319-00032-9_18,

The contractual framework marks a further difference. Master's degrees in Biology and Architecture guarantees more consistent work with the university courses, their salary is really not as expected for highly qualified professions (ISTAT 2010b). In many cases, the job market doesn't recognize professional profiles designed by university reforms (Ministerial Degree [MD] 509/99, MD 270/04 and amendments to MD 270/04).

In order to facilitate the employability of graduates, all universities, often associated, have created an office of placement (Fabbris 2009; Iezzi 2011). Offices of placement collect job advertisements of companies accredited with Universities. The procedure to select a candidate is very simple: companies send to placement offices a job advertisement in which they describe features of candidates (degree, skills, knowledge, etc.). Then, Universities ask to companies to classify job profiles using their rules. Frequently companies classify incorrectly their announcements or simply they select the category "other."

The aim of this paper is to classify professional profiles of job announcements. We use a fuzzy approach, because job announcements for graduates frequently cannot be classified into mutually exclusive groups. Many different job announcements could require similar competences, e.g., a good knowledge of English or available to travel. A hard clustering could produce groups that are not good respect to compactness and separability. Moreover this approach helps to identify the shading off of each ad.

We adapt fuzzy *c*-means algorithm to have partial membership in each class (Dunn 1973; Pal et al. 1996). The paper is organized as follows: in Sect. 2, we present the method; in Sect. 3, we describe an application and the main results and, finally, in Sect. 4, we expose the conclusions and the future developments.

2 Data and Methods

We analyze 1,650 job announcements, published from January 1st, 2010 to April 5th, 2011 on DB SOUL (System University Orientation and Job) by 496 companies. SOUL[1] is a network of eight Universities (Sapienza, Roma Tre, Tor Vergata, Foro Italico, Accademia di belle Arti, Tuscia, Cassino, LUMSA—Libera Università degli Studi Maria SS. Assunta) in the area of Lazio. The main goal of SOUL is to make a bridge between the job market and the university, so that the university students and graduates can have their best chances to improve their employability.

Our method is composed of six steps:

1. Pre-processing.
2. Lexical analysis.

[1]Currently the DB SOUL collects 52,000 graduate CVs, of which about 27,000 come from "Sapienza," 7,500 from "Roma Tre," 2,500 from "Tor Vergata," and 15,000 from other universities not only of the Lazio region (LUMSA, LUISS, Tuscia, and Cassino), but also from other regions (e.g., Napoli Federico II, Salerno, Bari, Bologna, Chieti-Pescara, Lecce).

3. Selecting of weighing scheme.
4. Construction of proximity matrix.
5. Applying different clustering methods.
6. Comparison of final partitions.

In the first step, we work on alphabetic characters to reduce the variability of tokens due to grammatical or orthographical mistakes. For this reason, we corrected typing errors, e.g. "canditato," instead of "candidato" or "diresione" instead of "direzione." In the second step, we realized two different forms of normalization: the first one "soft" and the second one based on lists. In the first one we uniform numbers, dates, adding space after the apostrophe and we transform accents in apostrophes. Normalization based on lists recognizes multiple words, grammatical phrases and nominal groups to preserve their specificity inside the corpus. To realize this kind of normalization we used several lists, some of these were provided as resources by software Taltac2 (Bolasco 2010; Giuliano and La Rocca 2008) other lists were built by ourselves during the pre-processing. These lists contain multiple words, e.g. "ad esempio" (for example) "all'interno di" (inside of), "alla ricerca di" (looking for), "partita iva" (VAT), "orario di lavoro" (working hours). In particular, we focused our attention on the creation of two lists: the first one, concerning several types and levels of degrees, e.g. "laurea specialistica in scienze statistiche e attuariali" (master degree in statistics and actuarial science), the second one made of professions, e.g. "ingegnere elettronico" (electronic engineer) or "programmatore junior" (junior programmer). These steps involved text reformatting (e.g., whitespace removal) and stop word removal, using an other specific list that we created.

In the third step, we transform the corpus into vectors of weighted terms by a term-document matrix $\mathbf{T} = [w_{ij}]$, where w_{ij} is the weight *i*th word in a *j*th text (for $i = 1, \ldots, p$ and $j = 1, \ldots k$). We use two weighting schemes: (1) Term Frequency (TF), where w_{ij} is the frequency of word type *i* in a document *j* ($w_{ij} = n_{ij}$); (2) Term Frequency Inverse Document Frequency (TFIDF): $w_{ij} = \frac{n_{ij}}{\max n_{ij}} \log \frac{N}{n_i}$, where $\max n_{ij}$ is the maximum frequency of word *i* in a corpus, N is the total number of documents and n_i is the number of documents in which the word *i* appears.

TF scheme assumes that word order has no significance, and TFIDF put the emphasis on the fact that terms that occur in many documents are more general, but less discriminative terms, whereas a rare term will be a more precise document descriptor (Iezzi 2012a). Bolasco and Pavone (2008) specifies that TFIDF is used "as an indicator of the importance of terms as for example on the web to measure the relevance of the contents of a document in relation to a specific query, which in most cases consists in a simple list or combinations of words." The thresholds 5 and 10 allow to build a TFIDF reduced matrix, respectively, of 64 % and 78 %. This drastic cut makes the TFIDF matrix less dispersed, improving the performance of clustering algorithm, but it deletes also the most relevant information on the professional profiles. The TF matrix has the disadvantage to have a big size, incurring in "the curse of dimensionality" (Houle et al. 2010).

In the fourth step, we calculate cosine distance on both TF and TFIDF matrices to measure the similarity between the job announcements. The job announcements with many common terms will have vectors closer to each other, than document with fewer overlapping terms (Iezzi 2010).

In the fifth step, we apply the *c*-means algorithm (Bezdek 1981). The objective of fuzzy clustering is to partition a data set into *c* homogeneous fuzzy clusters. This algorithm is based on minimization of the following objective function:

$$Fcm_m = \sum_{i=1}^{N}\sum_{j=1}^{C} u^m{}_{ij}\left\|x_i - c_j\right\|^2,\ 1 \leq m < \infty$$

where m is any real number greater than 1, u_{ij} is the degree of membership of x_i in the cluster j, x_i is the ith of d-dimensional measured data, c_j is the d-dimension center of the cluster, and ||*|| is any norm expressing the similarity between any measured data and the center. This algorithm is a version soft of the popular *k*-means clustering (Iezzi 2012b). As well known, the *k*-means method begins with an initial set of randomly selected exemplars and iteratively refines this set so as to decrease the sum of squared errors. *k*-centers clustering is moderately sensitive to the initial selection of centers, so it is usually rerun many times with different initializations in an attempt to find a good solution. This algorithm requires the user to pre-define the number of clusters (c). We select the number of clusters (c) using silhouette index (Rousseeuw 1987).

Due to the fuzzy nature of job announcements, fuzzy methods performed better than the classical ones. According to the results of the preliminary data exploration and fuzzy clustering with different values of the input parameters for fuzzy *c*-means algorithm, the best parameter combination was chosen and applied to training data set. In Fuzzy *c*-means new data is compared to the cluster centers in order to assign clustering membership values to the test data. A common approach is to use data to learn a set of centers such that the sum of squared errors between data points and their nearest centers is small. We perform 12 strategies to select the best classification (Table 1).

To summarize, the realization of the best-adapted *c*-means clustering (M3C) is as follows:

1. Initialize: we calculate cosine distance on **T** matrix.
2. Randomly we select k of the n data points as the medoids.
3. We associate each data point to the closest medoid ("closest" here is defined using Euclidean distance).
4. We select the configuration with the lowest cost.
5. We repeat steps 3–4 until there is no change in the medoid.
6. We save the final medoids into a matrix $\mathbf{U} = [u_{ij}]$, that initializes centroids of *c*-means algorithm.

Table 1 Steps of the twelve adopted methods

Method	Input matrix	Intermediate algorithms (output: initial centroids)			Final algorithm
M1A	TFIDF				c-means
M1B	TFIDF	Cosine distance	k-means		c-means
M1C	TFIDF	Cosine distance	PAM[a]		c-means
M2A	TFIDF	Cosine distance	MDS		c-means
M2B	TFIDF	Cosine distance	MDS	k-means	c-means
M2C	TFIDF	Cosine distance	MDS	PAM	c-means
M3A	TF	Cosine distance			c-means
M3B	TF	Cosine distance	k-means		c-means
M3C	TF	Cosine distance	PAM		c-means
M4A	TF	Cosine distance	MDS		c-means
M4B	TF	Cosine distance	MDS	k-means	c-means
M4C	TF	Cosine distance	MDS	PAM	c-means

[a]The Partitioning Around Medoids (PAM) algorithm (Theodoridis & Koutroumbas, 2006) breaks the dataset up into groups, minimizing squared error and chooses data points as centers (medoids or exemplars), introducing a method that simultaneously considers, as initial centers. It is more robust to noise and outliers as compared to k-means because it minimizes a sum of pairwise dissimilarities instead of a sum of squared Euclidean distances

7. At k-step, we calculate the centers vectors $c_j = \frac{\sum_{i=1}^{N} u_{ij}^m x_i}{\sum_{i=1}^{N} u_{ij}^m}$, $\mathbf{C}^{(k)} = [c_j]$ with $\mathbf{U}^{(k)}$.
8. We update $\mathbf{U}^{(k)}$, $\mathbf{U}^{(k+1)}$, $u_{ij} = \frac{1}{\sum_{k=1}^{c} \left(\frac{\|x_i - c_j\|}{\|x_i - c_k\|} \right)^{\frac{2}{m-1}}}$. If $\left\| \mathbf{U}^{(k+1)} - \mathbf{U}^{(k)} \right\| < \varepsilon$ then stop; otherwise we return to step 7.

An R program for performing steps 1 through 8 has been developed by the authors.

3 Results

Before the pre-processing, the corpus of ads is composed of 2,50,368 tokens and 11,899 words. The ads are very short with a number average of tokens equals to 151, lexical measures underline a language very standardized (Table 2). The most frequent words are "languages," "English," "software," "engineer" and "development."

After the pre-processing, we involved text reformatting (e.g., whitespace removal) and stopword removal, using an ad hoc list we created (Table 3).

Overall the 1,650 announcements refer to 156 different job positions classified by the companies. The most frequent job announcements (454 corresponding to 27.2 % of total ads) belong to category "other."

Table 2 Lexical measures of the corpus before the pre-processing

Tokens (N)	2,50,368	Lexical richness (V/N) × 100	4.75 %
Types (V)	11,899	Hapax percentage	33.78 %
Hapax[a]	4,013	Guiraud index[b]	23.78

[a]A hapax is a word (token) which occurs only once within a corpus
[b]Guiraud index is a measure of lexical richness. It assumes the following formula: $G = \frac{V}{\sqrt{N}}$, where V is types, and N is tokens

Table 3 Lexical measures of the corpus after the pre-processing

Tokens (N)	1,94,930	Lexical richness (V/N) × 100	18.597 %
Types (V)	10,482	Hapax percentage	34.745 %
Hapax	3,642	Guiraud index	23.741

Table 4 The most popular job positions in DB SOUL

Rank	Job profile	Job ads no. (%)
1	Other category	454 (27.52)
2	Analyst programmer	109 (6.66)
3	Programmer	95 (5.75)
4	Account	48 (2.90)
5	Systems analyst	36 (2.18)
6	Engineer	30 (1.18)
7	Administrative assistant	26 (1.57)
8	Insurance agent	20 (1.12)
9	Business analyst	18 (1.09)

The job positions in DB SOUL are mainly connected to information technology, and secondly to financial bank and insurance sector (Table 4).

The maximum number of announcement published by one firm was 65, but 44 % of 496 firms published only one announcement, whereas three out of four of all firms published not more than three. We deleted 113 job ads, because they were written in English. At the end, we analyzed 1,537 job announcements. Our method detected ten big groups. System engineer, analyst programmer and insurance agent are the largest groups (Fig. 1). Candidates belong to big groups should have a degree in Engineering, Statistics or Mathematics. The category "other" belongs to different clusters.

We searched for the best number of clusters iterating the analyses from two to ten classes. The most powerful group is one of ten groups with the method M3C. Table 5 shows that the methods M3A, M3B, and M3C have all classes with a high number of job announcements.

The method M3C allows to clearly identify the macro groupings of ads, while other procedures make a confused reading of data.

The most requested job position in DB SOUL is connected with Information Technology (IT). The professional profiles detected are overlapped. Degree in

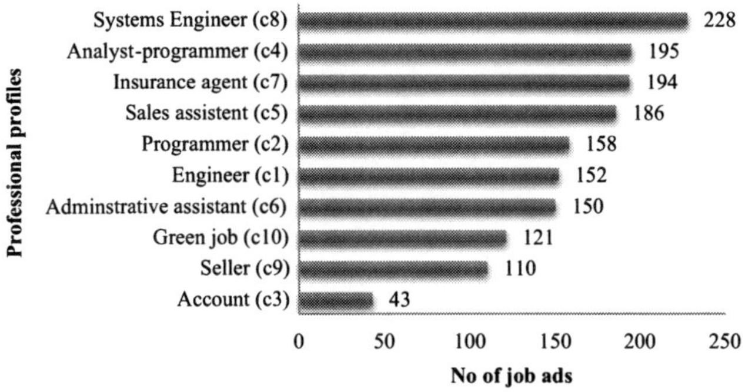

Fig. 1 Hard cluster profiles

Table 5 Size of the clusters composed of ten classes

	Size									
Methods	c1	c2	c3	c4	c5	c6	c7	c8	c9	c10
M1A	6	568	308	5	6	285	7	7	345	0
M1B	341	1	9	5	568	10	284	3	309	3
M1C	16	3	355	14	391	514	244	0	0	0
M2A	286	130	162	146	10	190	265	203	5	140
M2B	152	140	130	190	2	203	286	162	13	259
M2C	237	271	275	215	225	243	71	0	0	0
M3A	224	41	190	123	224	125	173	192	126	119
M3B	206	126	195	124	125	41	225	197	172	126
M3C	152	158	43	195	186	150	194	228	110	121
M4A	1	126	37	1,372	1	0	0	0	0	0
M4B	1	1,385	76	2	1	44	22	1	4	1
M4C	1,334	1	1	19	33	66	78	3	1	1

Engineering is demanded in interdisciplinary field. In particular, Systems Engineer deals with work-processes and tools to manage risks on such projects, and it overlaps with both technical and human-centered disciplines such as control engineering, industrial engineering, organizational studies, and project management. This procedure allows detecting even new professions, such as green job, not covered in the national classification system (Duda et al. 2000; Iezzi 2008). According to the Bureau of Labor Statistics of United Nations, green jobs are either: (1) Jobs in businesses that produce goods or provide services that benefit the environment or conserve natural resources. (2) Jobs in which workers' duties involve making their establishment's production processes more environmentally friendly or use fewer natural resources.

Table 6 shows that the centroids of clusters belong to hard cluster the maximum 57.6 % (c8—System Engineer); in general, the clusters are very overlapped each other. In particular, the seller ads require skills in common with many groups. System Engineer is a well-defined group; generally, the ads ask a degree in

Table 6 Fuzzy cluster profiles of the centroids

	fc1	fc2	fc3	fc4	fc5	fc6	fc7	fc8	fc9	fc10
c1	0.225	0.136	0.085	0.049	0.018	0.035	0.013	0.014	0.166	0.243
c2	0.109	0.294	0.048	0.141	0.039	0.092	0.028	0.023	0.216	0.083
c3	0.055	0.008	0.503	0.007	0.003	0.005	0.003	0.004	0.013	0.036
c4	0.053	0.071	0.031	0.203	0.137	0.250	0.093	0.049	0.103	0.039
c5	0.037	0.036	0.024	0.118	0.266	0.135	0.265	0.099	0.056	0.026
c6	0.050	0.064	0.030	0.192	0.157	0.244	0.107	0.053	0.094	0.037
c7	0.032	0.027	0.022	0.087	0.231	0.091	0.327	0.151	0.044	0.022
c8	0.023	0.017	0.017	0.046	0.100	0.043	0.131	0.576	0.028	0.016
c9	0.113	0.308	0.049	0.135	0.038	0.088	0.027	0.022	0.218	0.086
c10	0.304	0.040	0.192	0.023	0.010	0.016	0.007	0.009	0.063	0.411

fc final centroid, *c* cluster

Engineer, fluently English and propose full time contract. It is interesting to note that Administrative Assistant is overlap with competences required also by Analyst Programmer.

4 Conclusions

University offices of placement select for companies, which offer a job and/or training position, a large number of graduates only based on degree and grades. The classifier used by universities is frequently incomplete and, moreover, companies select often the category "other." This method allows detecting new professions, because it uses a bottom up procedure. Another advantage is that we do not lose valuable information on documents, because we didn't apply a procedure to reduce dimensions of TF matrix. We could apply Multidimensional Scaling (MDS) to reduce high dimensional data in a low dimensional space with preservation of the similarities between texts (Borg and Groenen 2006). In this case, this procedure could reduce dimensionality, but it may not reveal the genuine structure hidden in the data. In fact, the job announcements are very short and this method considers noisy information also hapax and, in this case, they could be a keyword of professional profiles. In this case, MDS would select a big one group and many small clusters. The ads would be classified on the basis of transversal skills, such as knowledge of English.

The future step is to use this method to detect new classifiers that we could be used for a supervised classification to measure distance between graduate CVs and job ads. We could analyze graduate CVs encoding data on university courses (NUP to three digits, declaratory of course) and information on experience of candidate, including the title and abstract of thesis. In this way, we could select for companies a large number of CVs not only based on degree and grades, but also taking into account characteristics and experience of graduates.

Acknowledgment The authors would like to particularly thank the SOUL organization for data collection, and signately Prof. Carlo Magni for his contribution to the discussion of this paper.

References

Aureli, E., & Iezzi, D. F. (2006). Recruitment via web and information technology: A model for ranking the competences in job market. *JADTm 1*, 79–88.

Bezdek, J. C. (1981). *Pattern recognition with fuzzy objective function algorithms*. New York: Plenum.

Bolasco, S. (2010). *TALTAC 2.10. Sviluppi, esperienze ed elementi essenziali di analisi automatica di testi.* ROMA: LED.

Bolasco, S., & Pavone, P. (2008). Multi-class categorization based on cluster analysis and TFIDF. In S. Heiden & B. Pincemin (Eds.), *JADT 2008* (Vol. 1, pp. 209–218). Lyon: Presses Universitaires de Lyon.

Borg, I., & Groenen, P. (2006). *Modern multidimensional scaling: Theory and applications (Springer series in statistics)*. New York: Springer.

CEDEFOP. (2010). *Skills supply and demand in Europe Medium-term forecast up to 2020.* Luxembourg: Publication Office of the European Union.

Duda, R. O., Hart, P. E., & Stork, D. G. (2000). *Pattern classification* (2nd ed.). New York: Wiley.

Dunn, J. C. (1973). A fuzzy relative of the ISODATA process and its use in detecting compact well-separated clusters. *Journal of Cybernetics, 3*, 32–57.

Fabbris, L. (2009). *I servizi a supporto degli studenti universitari*. Padova: CLEUP.

Giuliano, L., & la Rocca, G. (2008). *L'Analisi automatic e semi-automatica dei dati testuali. Software e istruzioni per l'uso*. Milano: LED.

Houle, M. E., Kriegel, H. P., Kröger, P., Schubert, E., Zimek, A. (2010). Can shared-neighbor distances defeat the curse of dimensionality? Scientific and statistical database management. *Lecture Notes in Computer Science, 6187*, 482. doi:10.1007/978-3-642-13818-8_34. ISBN 978-3-642-13817-1.

Iezzi, D. F. (2005). A new method to measure the quality on teaching evaluation of the university system: the Italian case. *Social Indicators Research, 73*(3), 459–477. ISSN: 0303-8300.

Iezzi, D. F. (2008). I lavoratori come informatori delle qualità delle professioni. In F. Luigi (Ed.), *Definire figure professionali tramite testimoni privilegiati*, vol. 1 (pp. 135–151). Padova: Cleup.

Iezzi, D. F. (2009). La valutazione dell'efficacia esterna dei corsi universitari: problemi e prospettive. In: M. Di Monte, M. Rotili (Eds), *Vincoli/constraints*, vol. Sensibilia 2 (pp. 229–246). Milano: Mimesis. ISBN: 978-88-5750-065-2.

Iezzi D. F. (2010). Topic connections and clustering in text mining: An analysis of the JADT network. *Statistical Analysis of Textual Data, Rome, Italy, 2*(29), 719–730.

Iezzi, D. F. (Ed.). (2011). *Indicatori e metodologie per la valutazione dell'efficacia del sistema universitario*. Padova: CLEUP.

Iezzi, D. F. (2012a). Centrality measures for text clustering. *Communications in Statistics – Theory and Methods, 41*(1), 3179–3197.

Iezzi, D. F. (2012b). A new method for adapting the *k*-means algorithm to text mining. *Italian Journal of Applied Statistics, 22*(1), 69–80.

ISTAT. (2010a). *I laureati e il mondo del lavoro*. Roma: ISTAT.

ISTAT. (2010b). *Forze di lavoro. Media 2009*. Roma: ISTAT.

OCDE. (2010). *Education at a glance 2010: OECD indicators*. Paris: OCDE.

Pal, N. R., Bezdek, J. C., & Hathaway, R. J. (1996). Sequential competitive learning and the fuzzy c-means clustering algorithms. *Neural Networks, 9*(5), 787–796.

Rousseeuw, P. J. (1987). Silhouettes: A graphical aid to the interpretation and validation of cluster analysis. *Computational and Applied Mathematics, 20*, 53–65.

Theodoridis, S., & Koutroumbas, K. (2006). *Pattern recognition* (3rd ed.). Amsterdam: Elsevier.

A Metric Based Approach for the Least Square Regression of Multivariate Modal Symbolic Data

Antonio Irpino and Rosanna Verde

Abstract In this paper we propose a linear regression model for multivariate modal symbolic data. The observed variables are probabilistic modal variables according to the definition given in (Bock and Diday (2000). *Analysis of symbolic data: exploratory methods for extracting statistical information from complex data.* Springer), i.e. variables whose realizations are frequency or probability distributions. The parameters are estimated through a Least Squares method based on a suitable squared distance between the predicted and the observed modal symbolic data: the squared ℓ_2 Wasserstein distance. Measures of goodness of fit are also presented and an application on real data corroborates the proposed method.

1 Introduction

Symbolic Data Analysis (SDA) is a recent approach for the statistical analysis of multivalued data, gathering contributes from different scientific communities: statistics, machine learning, data mining and knowledge discovery. Differently from the classical data where each observation is a "punctual" realization of an observed variable, symbolic data are characterized by sets of values observed for each variable (intervals, sets of categories, empirical frequency or probability distributions). Indeed, symbolic data are generally referred to the description of groups of individuals, typologies or concepts. Many data analysis techniques were extended to the study of symbolic data (for a wide overview of the SDA methods see Bock and Diday 2000 and Diday and Noirhomme-Fraiture 2008). In this paper we focus our proposals on the regression analysis for symbolic data. In SDA, linear regression models were proposed to study the structure of dependence of a response

A. Irpino (✉) · R. Verde
Dipartimento di Scienze Politiche "J. Monnet", Seconda Universit degli Studi di Napoli, Caserta, Italy
e-mail: antonio.irpino@unina2.it; rosanna.verde@unina2.it

P. Giudici et al. (eds.), *Statistical Models for Data Analysis*, Studies in Classification, Data Analysis, and Knowledge Organization, DOI 10.1007/978-3-319-00032-9_19,

"symbolic" variable from a set of independent or explicative variables of the same nature.

Some initial proposals were regression models for interval data as extensions of the classical linear model for numerical variables (see Diday and Noirhomme-Fraiture 2008 and the therein references for a full overview of regression models for interval data).

In Billard and Diday (2006) was presented a first regression model for histogram variables. However, that regression model allows to predict only punctual values and only indirectly a histogram response from a set of explicative histograms. Indeed, the authors left open the problem of how to predict directly symbolic data from a set of symbolic descriptions. To meet such requirement, in Verde and Irpino (2010) was proposed a simple linear regression model for modeling the relationship between a histogram response variable and an histogram explicative variable. The main novelty of the proposed model consists into using a suitable distance, the ℓ_p Wasserstein metric (also known as Mallow's distance, see Gibbs and Su 2002) for measuring the Sum of Squared Errors in the Ordinary Least Squares estimate procedure. The distance is computed stating from the quantile functions associated (in bijection) with histogram descriptions. So, the method allows to predict quantile functions, and then the corresponding histograms. According to the nature of the mathematical entities involved in the analysis, some constraints on the parameters space are required.

In order to solve such constrained problem, Dias and Brito (2011) proposed a linear regression model for histogram data based on the Wasserstein distance and on a constrained Least Square approach, but introducing supplementary variables related to the original ones.

In this paper, we extend the model proposed in Verde and Irpino (2010) to the multivariate case. We propose a matrix formulation of the LS problem based on a novel scalar product between vectors of quantile functions related to the decomposition of the Wasserstein metric proposed by Irpino and Romano (2007). We show that the LS problem in the space of quantile functions can be decomposed into two independent LS problems (one for the averages of the histograms and one for the centered histograms), and we suggest the solution of the LS problem trough a Non Negative LS formulation. Indeed, a linear combination of a set of quantile functions returns a quantile function too only if it is a conical combination (i.e. the quantile functions are multiplied by non negative scalars).

The paper is organized as follows: in section 2 symbolic data are presented according to the definition given in Bock and Diday (2000) and Diday and Noirhomme-Fraiture (2008); in section 3 the new multiple linear regression model for Numerical Probabilistic (Modal) Symbolic Variables is introduced; finally in section 4, we apply the model to a climatic data set and compare the obtained results with two concurrent models.

2 Numerical Probabilistic Modal Symbolic data

In a classic data table ($n \times p$ individuals per variables) each individual is described by a vector of values that are the realizations of a set of variables. Similarly, in a *symbolic data table* each individual is described by a vector of set-valued descriptions (like intervals, set of categories, sequences of values, associated with frequency or probability distributions), that, for extension, are the realizations of a set of symbolic variables. According to the taxonomy of symbolic variables presented in Bock and Diday (2000) we may consider as numeric symbolic variables all those symbolic variables having as realizations numeric set-valued data.

Given a set of n individuals (concepts, classes) $\Omega = \{\omega_1, \ldots, \omega_n\}$ a *symbolic variable* X, with domain D, is a map $X : \Omega \rightarrow D$ such that $X(\omega_i) \in D$. In this paper we refer only to *Modal Symbolic Variables*. According to Bock and Diday (2000) the domain of a *Modal Variable* is a set of mappings. Let us consider $D \subseteq M$ where $M_i \in M$ is a map $M_i : S_i \rightarrow W_i$, such that for each element of the support $s_i \in S_i$ it is associated $w_i = M_i(s_i) \in \Re^+$.

If $M_i(s_i)$ has the same properties of a random variable (i.e. $\int_{s \in S_i} w(s)ds = 1$, or $\sum_{s \in S_i} w_s = 1$), X can be defined as a *Numerical Probabilistic* (Modal) *Symbolic Variable* (NPSV) and $M_i(s_i)$ can described through a probability density function, a histogram, an empirical frequency distribution $f_i(x)$. In this paper we refer only to *Numerical Probabilistic (Modal) Symbolic Data* (NPSD), that are in the domain of *Numerical Probabilistic* (Modal) *Symbolic Variables*. Histogram data are the most common case of NPSD. Given the generic individual ω_i, $X(\omega_i)$ is its histogram description for the NPSV X. It consists in a set of disjoint K_i intervals (a.k.a. bins) $I_{ki} = [a_{ki}, b_{ki}]$ $(k = 1, \ldots, K_i)$ with associate a set of positive K_i weights w_{ki} such that $\sum_{k=1}^{K_i} w_{ki} = 1$, as follows:

$$X(\omega_i) = f_i(x) = \{(I_{1i}, w_{1i}), \ldots, (I_{ki}, w_{ki}), \ldots, (I_{K_i i}, w_{K_i i})\}.$$

3 Least Squares Linear Regression Analysis of NPSV's

Given p explicative NPSV's $X_1, \ldots, X_j, \ldots, X_p$ and a dependent NPSV Y observed on a set Ω, the aim is to predict the variable Y as a linear combination of the X_j's according to the nature of the symbolic variables.

Denoting with X and Y the matrix of the observed independent symbolic data Xj (j = 1,. . . ,p) and the vector of the observed response symbolic data Y, the linear regression model f(.) to be fitted according to a set of parameters teta collected in a vector Teta, is:

Denoting with $\mathbf{X}$ and $\mathbf{Y}$, respectively, the matrix of the observed independent symbolic data x_{ij} $(i = 1, \ldots, n$ and $j = 1, \ldots, p)$ and the vector of the observed response symbolic data y_i, the regression model $\phi(\cdot)$ is determined by the estimates of a vector $\boldsymbol{\theta}$ of parameters, as follows:

$$\mathbf{Y} = \phi(\mathbf{X}, \boldsymbol{\theta}) + \varepsilon \tag{1}$$

where ε is vector of errors. In the regression analysis of NPSV's (in particular for histogram-valued variables), two main approaches for the estimation of the parameters were proposed: the first one (Billard and Diday 2006) is an extension of the classic LS to histogram-valued data treated as weighted punctual data, while a second approach is based on the ℓ_2 Wasserstein distance, where the sum of square errors in the LS problem is defined as the integrated squared difference between two quantile functions (which are in bijection w.r.t. their corresponding NSPD's). The idea behind the last approach is to predict the quantile functions (denoted $y_i(t)$ for $i = 1, \ldots, n$) having observed a set of quantile functions (denoted $x_{ij}(t)$ for $i = 1, \ldots, n$) of the p predictors. A simply generalization leads to estimate a set of parameters of a linear combination of $x_{ij}(t)$'s (for $j = 1, \ldots, p$) which allow to predict the $y_i(t)$'s (for $i = 1, \ldots, n$) except for a error term $e_i(t)$. It is worth noting that $e_i(t)$ is a residual function, which is not necessarily a quantile function. The resulting regression model to fit is:

$$y_i(t) = \beta_0 + \sum_{j=1}^{p} \beta_j x_{ij}(t) + e_i(t). \tag{2}$$

The parameters β_j are estimated using a LS method, where the Sum of Squared Errors (*SSE*) to minimize is computed by using the (squared) ℓ_2 Wasserstein distance between distributions. We recall that this distance permits to explain and interpret in an easy way the proximity relations between the characteristics of two distributions (Verde and Irpino 2007). Consequently, the SSE function can be expressed as

$$SSE = \sum_{i=1}^{n} \int_0^1 [e_i(t)]^2 \, dt = \sum_{i=1}^{n} d_W^2 \left(y_i(t), \left[\beta_0 + \sum_{j=1}^{p} \beta_j x_{ij}(t) \right] \right). \tag{3}$$

However, a problem arises for the linear combination of quantile functions: only if $\beta_j \geq 0$ ($j = 1, \ldots, p$) it is assured that $y_i(t)$ is a quantile function (i.e. a not decreasing function). In order, to overcome this problem, in Verde and Irpino (2010) and Dias and Brito (2011) are proposed novel formulations of the regression model for histogram-valued data based on the Wasserstein distance.

3.1 The Proposed Model

Before introducing the new model we recall two particular decompositions of ℓ_2^2 Wasserstein distance. Given f and g, two NPSD and $x_f^c(t)$ and $x_g^c(t)$ the respective centred quantile functions (i.e., the quantile functions shifted by the means of the distributions) in Cuesta-Albertos et al. (1997) is showed that the ℓ_2^2 Wasserstein distance can be rewritten as:

$$d_W^2(f,g) = \int_0^1 \left[x_f(t) - x_g(t)\right]^2 dt = \left(\bar{x}_f - \bar{x}_g\right)^2 + \int_0^1 \left[x_f^c(t) - x_g^c(t)\right]^2 dt. \quad (4)$$

This property allows to consider the squared distance as the sum of two components, the first related to the location of the distributions and the second related to their variability structure. In Irpino and Romano (2007) was expanded such decomposition, showing that the d_W^2 can be finally decomposed into three quantities:

$$d_W^2(f,g) = \left(\bar{x}_f - \bar{x}_g\right)^2 + \left(s_f - s_g\right)^2 + 2s_f s_g \left(1 - \rho(x_f, x_g)\right) \quad (5)$$

where $\rho(x_f, x_g)$ is the correlation between the two quantile functions, i.e.:

$$\rho(x_f, x_g) = \frac{\int_0^1 x_f(t) \cdot x_g(t) dt - \bar{x}_f \cdot \bar{x}_g}{s_f \cdot s_g}. \quad (6)$$

Equation (6) allows to express the inner product between two quantile functions, as follows:

$$\left\langle x_f(t), x_g(t) \right\rangle = \int_0^1 x_f(t) \cdot x_g(t) dt = \rho(x_f, x_g) \cdot s_f \cdot s_g + \bar{x}_f \cdot \bar{x}_g. \quad (7)$$

Thus, given two vectors of quantile functions $\mathbf{x} = [x_i(t)]_{n\times 1}$and $\mathbf{y} = [y_i(t)]_{n\times 1}$, we can define the scalar product of two vectors of NPSD as:

$$\mathbf{x}^{\mathbf{T}}\mathbf{y} = \sum_{i=1}^{n} \left\langle x_f(t), x_g(t) \right\rangle = \sum_{i=1}^{n} \left[\rho(x_i, y_i) \cdot s_{x_i} \cdot s_{y_i} + \bar{x}_i \cdot \bar{y}_i\right]. \quad (8)$$

If we consider the centred quantile functions $y_i^c(t) = y_i(t) - \bar{y}_i$ and $x_{ij}^c(t) = x_{ij}(t) - \bar{x}_{ij}$, denoting with $\bar{\mathbf{Y}} = [\bar{y}_i]_{n\times 1}$ the vector of the means of the $f_i(y)$, with $\mathbf{Y}^c = \left[y_i^c(t)\right]_{n\times 1}$ the vector of the centred quantile functions, with $\bar{\mathbf{X}} = \left[\bar{x}_{ij}\right]_{n\times p}$ the matrix of the means, with $\mathbf{X^c} = \left[x_{ij}^c(t)\right]_{n\times p}$ the matrix of the centred quantile functions of $f_i(x_j)$'s, and with $\bar{\mathbf{X}}_+ = [\mathbf{1}|\bar{\mathbf{X}}]$, we express the model as follows:

$$\mathbf{Y} = \bar{\mathbf{X}}_+\mathbf{B} + \mathbf{X}^c\Gamma + \mathbf{e}. \quad (9)$$

In order to estimate the parameters, we define the Sum of Square Errors criterion (SSE) as follows:

$$SSE(\mathbf{B}, \Gamma) = \mathbf{e}^T e = \left[\mathbf{Y} - \bar{\mathbf{X}}_+\mathbf{B} - \mathbf{X}^c\Gamma\right]^T \left[\mathbf{Y} - \bar{\mathbf{X}}_+\mathbf{B} - \mathbf{X}^c\Gamma\right]. \quad (10)$$

Using Eqs. (7) and (8) it is possible to prove[1] that $\bar{\mathbf{X}}_+^T \mathbf{X}^\mathbf{c} = \mathbf{0}_{(p+1)\times p}$, $\bar{\mathbf{X}}_+^T \mathbf{Y} = \bar{\mathbf{X}}_+^T \bar{\mathbf{Y}}$ and $\mathbf{X}^{\mathbf{c}T} \mathbf{Y} = \mathbf{X}^{\mathbf{c}T} \mathbf{Y}^\mathbf{c}$. Thus, the $SSE(\mathbf{B}, \Gamma)$ can be expressed as follows:

$$SSE(\mathbf{B}, \Gamma) = SSE(\mathbf{B}) + SSE(\Gamma) = \bar{\mathbf{e}}^T \bar{\mathbf{e}} + (\mathbf{e}^\mathbf{c})^\mathbf{T} \mathbf{e}^\mathbf{c} \tag{11}$$

The global minimization problem is divided into the minimization of two independent terms: $SSE(\mathbf{B})$ related to the means of the predictor quantile functions $\bar{x}_{ij}$'s in $\bar{\mathbf{X}}_+$, and $SSE(\Gamma)$ related to the variability of the centered quantile distributions $x_{ij}^c(t)$'s in $\mathbf{X}^c$. Therefore, the two independent models are:

$$\bar{\mathbf{Y}} = \bar{\mathbf{X}}_+ \mathbf{B} + \bar{\mathbf{e}} \qquad \mathbf{Y}^\mathbf{c} = \mathbf{X}^c \Gamma + \mathbf{e}^c. \tag{12}$$

The vector $\mathbf{B}$ can be solved using a classical OLS model, while the vector of Γ's is estimated using the NNLS (Non Negative Least Squares) algorithm proposed by Lawson and Hanson (1974), integrated with the scalar product of vectors of NPSD's. This is necessary, because the Γ's cannot be negative, considering that they are multiplied by quantile functions.

Considering the nature of the data, the evaluation of the goodness of fit of the model is not straightforward. The extension of the classic R^2 index to the regression of NPSD produced the following two indices:

Ω **index** (Dias and Brito 2011) **Pseudo $-$ R^2** (Verde and Irpino 2010)

$$\Omega = \sum_{i=1}^{n} d_W^2\left(\hat{y}_i(t), \bar{y}\right) \Big/ \sum_{i=1}^{n} d_W^2\left(y_i(t), \bar{y}\right) \; ; \; PseudoR^2 = \min\left[\max\left[0; 1 - \tfrac{SSE}{SSY}\right]; 1\right].$$

Further, we propose to use the classic Root Mean Square Index ($RMSE_W$) in order to compare concurrent models based on the Wasserstein distance:

$$RMSE_W = \sqrt{\frac{\sum_{i=1}^{n} \int_0^1 \left(\hat{y}_i(t) - y_i(t)\right)^2 dt}{n}} = \sqrt{\frac{SSE}{n}}.$$

4 Application on Real Data

The Clean Air Status and Trends Network (CASTNET)[2] is an air quality monitoring network of United States which is designed to provide data to assess trends in air quality, atmospheric deposition, and ecological effects due to changes in air

[1]Note that if $\mathbf{x}$ is a vector of scalars Eq. (8) becomes $\mathbf{x}^\mathbf{T}\mathbf{y} = \sum_{i=1}^{n} x_i \cdot \bar{y}_i$.

[2]http://java.epa.gov/castnet/.

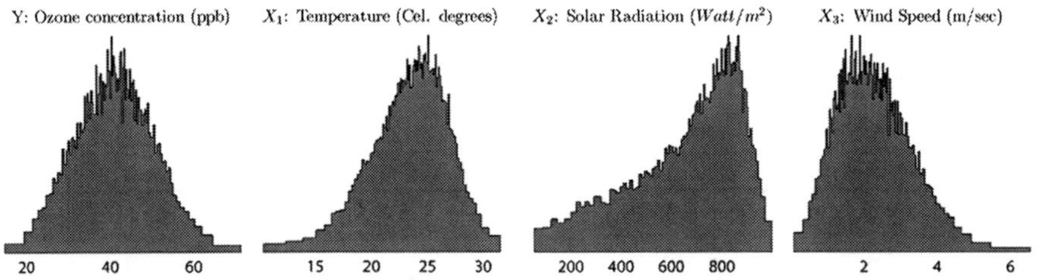

Fig. 1 Ozone dataset. Representation of the barycenters for each variable according to Verde and Irpino (2008)

Table 1 Ozone dataset: Standard deviation of the NPSV's according to Verde and Irpino (2008)

	Ozone Concentration (Y in Ppb)	Temperature (X_1 in Celsius deg.)	Solar Radiation (X_2 Watt/m^2)	Wind Speed (X_3 m/s)
Standard dev.	9.5295	3.8422	113.4308	1.1337

pollutant emissions. From the CASTNET repository, we have chosen to select data about the Ozone concentration repository, observed in 78 USA sites. Ozone is a gas that can cause respiratory diseases and, in the literature, there exist studies that relates the Ozone concentration level to the Temperature, the Wind speed and the Solar radiation (see for example Dueñas et al. 2002). CASTNET collects hourly data for each variable and, in this paper, we choose to study the summer season of 2010 and the central hours of the days (10 a.m.–5 p.m.). For each sites we have collected the NPSD's of the four variables in terms of histograms[3] and, in order to give a visual reference, in Fig. 1 we show the *average* station described by the *sample Wasserstein mean histogram* (in the sense of Verde and Irpino 2007) for each variable. We recall that the *Wasserstein mean histogram* is the histogram corresponding to the mean quantile function of all the quantile functions of a set of NPSD's. In Table 1, we reported the standard deviation of each variable according to the dispersion measures proposed in Verde and Irpino (2008).

For a site, given the distribution of Temperature (X_1) (Celsius degrees), the distribution of Solar Radiation (X_2) (Watts per square meter) and the distribution of Wind Speed (X_3) (meters per second), the main objective is to predict the distribution of Ozone Concentration (Y) (Particles per billion) using a linear model. We have considered the different regression models proposed by Billard and Diday (2006) and by Dias and Brito (2011), and we compared them with the here proposed model. The estimates of the parameter and the goodness of fit indices are reported in Table 2.

Considering the goodness of fit indices, we observe that the Wasserstein distance based methods perform better than the Billard–Diday one, while, the new model is more accurate than the Dias–Brito one. However, the main differences among

[3]We supply the full table of histogram data, the Matlab™ code and workspace upon request.

Table 2 Ozone dataset: estimates of the three regression models and goodness of fit indices. Asterisks indicate that the indices have been computed after a Montecarlo experiment on the Billard–Diday model

		Goodness of fit		
Model	Estimates	Ω	Ps-R^2	*RMSE*
Billard–Diday	$\hat{y}_i = 18.28 + 0.357\,x_{i1} + 0.017\,x_{i2} + 1.550\,x_{i3}$	0.203*	0.024*	9.41*
Dias–Brito	$\hat{y}_i(t) = 13.32 + 0\,x_{i1}(t) + 0.037\,x_{i2}(t) +$ $1.691\,x_{i3}(t) + 0\,\tilde{x}_{i1}(t) + 0\,\tilde{x}_{i2}(t) + 0\,\tilde{x}_{i3}(t)$	0.670	0.371	7.56
Irpino–Verde	$\hat{y}_i(t) = \hat{\bar{y}}_i + \hat{y}_i^c(t)$ $\hat{\bar{y}}_i = 2.93 - 0.346\,\bar{x}_{i1} + 0.07\,\bar{x}_{i2} + 0.395\,\bar{x}_{i3}$ $\hat{y}_i^c(t) = 0.915\,x_{i1}^c(t) + 0.018\,x_{i2}^c(t) + 1.887\,x_{i3}^c(t)$	0.742	0.460	7.00

the models are related to the interpretation of the parameters. While the Billard–Diday model is interpretable like a classic regression model, it does not suggest relationships between the internal variability of the NPSD's. The Dias and Brito model, introducing new entities (the quantile functions of the symmetric distributions $\tilde{x}_{ij}(t)$), solves the non negativity constraint of the Wasserstein distance based model parameters, but, in the opinion of the authors, it seems quite artful in introducing new variables which are also strongly correlated with the original ones.

The here proposed method represent a new and a reasonable trade-off between the goodness of fit and interpretation issue: a first part of the model allows to predict the location (the mean) of a NPSD (linearly) depending from the location (a set of means) of p independent NPSD's (similar to the Billard and Diday); the second part of the model is devised for predicting the internal variability of a NPSD (conically, considering the non negativeness constraint) depending from the internal variabilities of p explicative NPSD's: i.e., being the Γ's related to centered quantile functions, and if the NPSD's have not very different shapes, their estimates are measures of how much the variability of the response variable is inflated (when the parameter is grater than one) or deflated (if the parameter is lower than one) when an increase of one of the internal variability of a predictor variable occurs (considering the other constants).

References

Billard, L., & Diday, E. (2006). *Symbolic data analysis: conceptual statistics and data mining*. New York: Wiley.

Bock, H., & Diday, E. (2000). *Analysis of symbolic data: exploratory methods for extracting statistical information from complex data*. New York: Springer.

Cuesta-Albertos, J. A., Matrán, C., & Tuero-Díaz, A. (1997). Optimal transportation plans and convergence in distribution. *Journal of Multivariate Analysis, 60*, 72–83.

Dias, S., & Brito, P. (2011). A new linear regression model for histogram-valued variables. In *58th ISI World Statistics Congress*. Dublin, Ireland.

Diday, E., & Noirhomme-Fraiture, M. (2008). *Symbolic data analysis and the SODAS software*. New York: Wiley.

Dueñas, C., Fernández, M., Cañete, S., Carretero, J., & Liger, E. (2002). Assessment of ozone variations and meteorological effects in an urban area in the mediterranean coast. *Science of The Total Environment, 299*(1–3), 97–113.

Gibbs, A., & Su, F. (2002). On choosing and bounding probability metrics. *International Statistical Review, 70*(3), 419–435.

Irpino, A., & Romano, E. (2007). Optimal histogram representation of large data sets: Fisher vs piecewise linear approximation. *Revue des Nouvelles Technologies de l'Information, RNTI-E-9*, 99–110.

Lawson, C. L., & Hanson, R. J. (1974). *Solving least square problems*. Edgeworth Cliff, NJ: Prentice Hall.

Verde, R., & Irpino, A. (2007). Dynamic clustering of histogram data: Using the right metric. In P. e. a. Brito (Ed.) *Selected contributions in data analysis and classification* (pp. 123–134). New York: Springer.

Verde, R., & Irpino, A. (2007). Dynamic clustering of histogram data: Using the right metric. In P. Brito, G. Cucumel, P. Bertrand, & F. De Carvalho (Eds.) *Selected contributions in data analysis and classification* (Chap. 12, pp. 123–134). Berlin, Heidelberg: Springer.

Verde, R., & Irpino, A. (2008). Comparing histogram data using a mahalanobis-wasserstein distance. In P. Brito (Ed.) *COMPSTAT 2008* (Chap. 7, pp. 77–89). Heidelberg: Physica-Verlag HD.

Verde, R., & Irpino, A. (2010). Ordinary least squares for histogram data based on wasserstein distance. In Y. Lechevallier, & G. Saporta (Eds.) *Proceedings of COMPSTAT'2010*. (Chap. 60, pp. 581–588). Heidelberg: Physica-Verlag HD.

A Gaussian–Von Mises Hidden Markov Model for Clustering Multivariate Linear-Circular Data

Francesco Lagona and Marco Picone

Abstract A multivariate hidden Markov model is proposed for clustering mixed linear and circular time-series data with missing values. The model integrates von Mises and normal densities to describe the distribution that the data take under different latent regimes, with parameters that depend on the evolution of an unobserved Markov chain. Estimation is facilitated by an EM algorithm that treats the states of the latent chain and missing values as different sources of incomplete information. The model is exploited to identify sea regimes from multivariate marine data.

1 Introduction

In multivariate analysis, mixture-based classification studies are typically carried out by assuming that the data are in the form of independent multivariate samples. This assumption is a shortcoming when classification is based on observations that are collected in the form of multivariate time series, because a clustering procedure should account for the potential redundancy of the data, due to temporal autocorrelation. Multivariate hidden Markov models (MHMMs; Zhang et al. 2010 and the references therein) are mixture models that allow for temporal correlation, because observations are modeled by a mixture of multivariate distributions, whose parameters depend on the states of latent Markov chain. As a result, classification

F. Lagona (✉)
University Roma Tre, Rome, Italy
e-mail: lagona@uniroma3.it

M. Picone
University Roma Tre, Rome, Italy

Marine Service ISPRA
e-mail: marco.picone@uniroma3.it

P. Giudici et al. (eds.), *Statistical Models for Data Analysis*, Studies in Classification, Data Analysis, and Knowledge Organization, DOI 10.1007/978-3-319-00032-9_20,

is not only based on similarities in the variables space, but also on similarities that occur in a temporal neighborhood.

The literature on MHMM-based classification studies is dominated by Gaussian MHMMs for multivariate continuous data. MHMMs for non-normal data are less developed and traditionally specified by approximating the joint distribution of the data by a mixture whose components are products of univariate densities, therefore assuming that measurements in a multivariate profile are conditionally independent (CI), given the states of the hidden Markov chain. Although this assumption facilitates both the specification and the estimation of a non-normal MHMM, CI-based MHMMs often require an unnecessary large number of states to obtain a reasonable goodness of fit, complicating the interpretation of the results. This has motivated a number of efforts in order to relax the CI assumption in nonnormal MHMMs, for example in the analysis of categorical data (Zhang et al. 2010). Motivated by classification issues that arise in marine studies, we extend this strand of the literature in the context of mixed linear and circular time series data. We focus in particular on quadrivariate time series of wave and wind directions, wind speed and wave height, typically collected by a buoy to describe sea conditions. Previous work on the classification of linear-circular data includes either mixtures whose components are specified as products of univariate or bivariate densities (Lagona and Picone 2011, 2012), which ignore temporal correlation, or hidden Markov models where single measurements are assumed as conditionally independent given the latent states of a Markov chain, which ignore correlation within latent classes (Holzmann et al. 2006). We integrate temporal autocorrelation and within-classes correlation in a Gaussian–von Mises MHMM, by approximating the joint distribution of the data by a mixture with components that are specified as the product of a bivariate von Mises and a bivariate normal density and with parameters that depend on the evolution of a latent Markov chain. In this way, circular measurements (e.g., wind and wave directions) are clustered according to a number of toroidal clusters, while linear measurements (e.g., wind speed and wave heights) are clustered within standard elliptical clusters. Toroidal and elliptical clusters are then paired according to the states of the latent Markov chain, which can be hence interpreted as latent regimes of the observation process.

2 A Gaussian–von Mises Multivariate Hidden Markov Model

Our data are in the form of a quadrivariate time series, say $\mathbf{z} = (\mathbf{z}_t, t = 0, \ldots, T)$, where each profile $\mathbf{z}_t = (\mathbf{x}_t, \mathbf{y}_t)$ includes two circular and two linear components, $\mathbf{x}_t \in (-\pi, \pi)^2$ and $\mathbf{y}_t \in \mathbb{R}^2$. In MHMM-based classification studies, the temporal evolution of class membership is driven by a latent Markov chain, which can be conveniently described as a vector $\boldsymbol{\xi} = (\boldsymbol{\xi}_t, t = 0, \ldots, T)$ of multinomial variables $\boldsymbol{\xi}_t = (\xi_{t1}, \ldots, \xi_{tK})$ with one trial and K classes, whose binary components represent class membership at time t. The joint distribution of the chain, say $p(\boldsymbol{\xi};\mathbf{p},\mathbf{P})$, is fully known up an initial probabilities vector $\mathbf{p}=(p_1,\ldots,p_K)$, $p_k=P(\xi_{0k}=1)$, and a transition probabilities matrix $\mathbf{P}=(p_{hk},h,k=1,\ldots K)$, $p_{hk}=P(\xi_{tk}=1|\xi_{t-1,h}=1)$.

The specification of the proposed Gaussian–von Mises MHMM is completed by assuming that circular and linear data profiles are conditionally independent, given the states of the Markov chain, say

$$f(\mathbf{z}|\boldsymbol{\xi};\mathbf{a},\mathbf{b})=\prod_{t=0}^{T}\prod_{k=1}^{K}\left(f(\mathbf{x}_t;\mathbf{a}_k)f(\mathbf{y}_t;\mathbf{b}_k)\right)^{\xi_{tk}}, \tag{1}$$

where

$$\begin{aligned} f(\mathbf{x};\mathbf{a}) \propto \exp\,&(a_{11}\cos(x_1-a_1)+a_{22}\cos(x_2-a_2)\\ &+a_{12}\sin(x_1-a_1)\sin(x_2-a_2)) \end{aligned} \tag{2}$$

is a bivariate von Mises density (Singh et al. 2002) on the torus $(-\pi,\pi)^2$, indexed by a multivariate parameter $\mathbf{a}=(a_1,a_2,a_{11},a_{22},a_{12})$, and

$$f(\mathbf{y};\mathbf{b})\propto\exp\left(b_{11}(y_1-b_1)^2+b_{22}(y_2-b_2)^2+b_{12}(y_1-b_1)(y_2-b_2)\right) \tag{3}$$

is a bivariate normal density, indexed by a parameter $\mathbf{b}=(b_1,b_2,b_{11},b_{22},b_{12})$.

We assume that the observed time series is a sample drawn from the K-states Gaussian–von Mises MHMM distribution

$$f(\mathbf{z};\boldsymbol{\theta})=\sum_{\boldsymbol{\xi}}p(\boldsymbol{\xi};\mathbf{p},\mathbf{P})f(\mathbf{z}|\boldsymbol{\xi};\mathbf{a},\mathbf{b}), \tag{4}$$

known up the parameter $\boldsymbol{\theta}=(\mathbf{p},\mathbf{P},\mathbf{a},\mathbf{b})$, $\mathbf{a}=(\mathbf{a}_1,\ldots\mathbf{a}_K)$, $\mathbf{b}=(\mathbf{b}_1,\ldots\mathbf{b}_K)$. Under this setting, the data can be clustered by assigning each profile $\mathbf{z}_t$ to the class k with the highest posterior probability $\hat{\pi}_{tk}=P(\xi_{tk}=1|\mathbf{z}_T;\hat{\boldsymbol{\theta}})$, where $\hat{\boldsymbol{\theta}}$ is the maximum likelihood estimate of $\boldsymbol{\theta}$.

3 Parameter Estimation

Let $\mathbf{x}_{\text{mis}}=(\mathbf{x}_{t,\text{mis}},t=1,\ldots,T)$ and $\mathbf{x}_{\text{obs}}=(\mathbf{x}_{t,\text{obs}},t=0,\ldots T)$ be the missing and observed circular observations, and let $\mathbf{y}_{\text{mis}}=(\mathbf{y}_{t,\text{mis}},t=1,\ldots T)$ and $\mathbf{y}_{\text{obs}}=(\mathbf{y}_{t,\text{obs}},t=0,\ldots T)$ be the missing and observed linear observations. Accordingly, $\mathbf{z}_{\text{obs}}=(\mathbf{x}_{\text{obs}},\mathbf{y}_{\text{obs}})$ and $\mathbf{z}_{\text{mis}}=(\mathbf{x}_{\text{mis}},\mathbf{y}_{\text{mis}})$ indicate the vectors of the observed and the missing values, respectively. If the data are missing at random, the MLE of $\boldsymbol{\theta}$ is the maximum point of the marginal log-likelihood function

$$\log L(\boldsymbol{\theta}) = \log \sum_{\boldsymbol{\xi}} p(\boldsymbol{\xi}; \mathbf{p}, \mathbf{P}) \prod_{t=0}^{T} \prod_{k=1}^{K} \left(\int f(\mathbf{x}_t; \mathbf{a}_k) d\mathbf{x}_{t,\text{mis}} \int f(\mathbf{y}_t; \mathbf{b}_k) d\mathbf{y}_{t,\text{mis}} \right)^{\xi_{tk}}. \tag{5}$$

Because direct maximization of (5) can be computationally problematic, we describe an EM algorithm that is based on the complete-data log-likelihood function,

$$\begin{aligned} \log L_{\text{comp}}(\boldsymbol{\theta}) &= \sum_{k=1}^{K} \xi_{0k} \log p_k + \sum_{t=1}^{T} \sum_{h=1}^{K} \sum_{k=1}^{K} \xi_{t-1,h} \xi_{t,k} \log p_{h,k} \\ &+ \sum_{t=0}^{T} \sum_{k=1}^{K} \xi_{tk} \log f(\mathbf{x}_{t,\text{mis}}, \mathbf{x}_{t,\text{obs}}; \mathbf{a}_k) \\ &+ \sum_{t=0}^{T} \sum_{k=1}^{K} \xi_{tk} \log f(\mathbf{y}_{t,\text{mis}}, \mathbf{y}_{t,\text{obs}}; \mathbf{b}_k), \end{aligned} \tag{6}$$

and generates a sequence of points toward a maximum point of the likelihood, by alternating a E step and a M step.

The E step reduces to the evaluation of the expected value of the complete data log-likelihood with respect to the conditional distribution $p(\boldsymbol{\xi}, \mathbf{x}_{\text{mis}}, \mathbf{y}_{\text{mis}} | \mathbf{z}_{\text{obs}}, \hat{\boldsymbol{\theta}})$ of the unobserved quantities given the available data, say

$$Q(\boldsymbol{\theta} | \hat{\boldsymbol{\theta}}) = \mathbb{E}\left(\log L_{\text{comp}}(\boldsymbol{\theta}) | \mathbf{z}_{\text{obs}} \right). \tag{7}$$

The expectation (7) can be evaluated in terms of iterated expectations, by observing that the conditional distribution of the unobserved quantities can be factorized as follows

$$f(\boldsymbol{\xi}, \mathbf{x}_{\text{mis}}, \mathbf{y}_{\text{mis}} | \mathbf{z}_{\text{obs}}; \hat{\boldsymbol{\theta}}) = p(\boldsymbol{\xi} | \mathbf{z}_{\text{obs}}; \hat{\boldsymbol{\theta}}) f(\mathbf{x}_{\text{mis}} | \boldsymbol{\xi}, \mathbf{x}_{\text{obs}}; \hat{\mathbf{a}}) f(\mathbf{y}_{\text{mis}} | \boldsymbol{\xi}, \mathbf{y}_{\text{obs}}; \hat{\mathbf{b}}) \tag{8}$$

where

$$\begin{aligned} f(\mathbf{x}_{\text{mis}} | \boldsymbol{\xi}, \mathbf{x}_{\text{obs}}; \hat{\mathbf{a}}) &= \prod_{t=0}^{T} \prod_{k=1}^{K} \left(\frac{f(\mathbf{x}_t; \hat{\mathbf{a}}_k)}{\int_{\mathbf{x}_{t,\text{mis}}} f(\mathbf{x}_t; \hat{\mathbf{a}}_k) d\mathbf{x}_{t,\text{mis}}} \right)^{\xi_{tk}} \\ f(\mathbf{y}_{\text{mis}} | \boldsymbol{\xi}, \mathbf{y}_{\text{obs}}; \hat{\mathbf{b}}) &= \prod_{t=0}^{T} \prod_{k=1}^{K} \left(\frac{f(\mathbf{y}_t; \hat{\mathbf{b}}_k)}{\int_{\mathbf{y}_{t,\text{mis}}} f(\mathbf{y}_t; \hat{\mathbf{b}}_k) d\mathbf{y}_{t,\text{mis}}} \right)^{\xi_{tk}}. \end{aligned} \tag{9}$$

As a result, the expected complete data log-likelihood can be computed as follows,

$$Q(\boldsymbol{\theta}|\hat{\boldsymbol{\theta}}) = \sum_{k=1}^{K} \hat{\pi}_{0k} \log p_k + \sum_{t=1}^{T}\sum_{h=1}^{K}\sum_{k=1}^{K} \hat{\pi}_{t-1,t,hk} \log p_{h,k}$$
$$+ \sum_{t=0}^{T}\sum_{k=1}^{K} \hat{\pi}_{tk} \mathbb{E}\left(\log f(\mathbf{x}_{t,\text{mis}}, \mathbf{x}_{t,\text{obs}}; \mathbf{a}_k)|\mathbf{x}_{t,\text{obs}}, \xi_{tk} = 1; \hat{\mathbf{a}}\right)$$
$$+ \sum_{t=0}^{T}\sum_{k=1}^{K} \hat{\pi}_{tk} \mathbb{E}\left(\log f(\mathbf{y}_{t,\text{mis}}, \mathbf{y}_{t,\text{obs}}; \mathbf{b}_k)|\mathbf{y}_{t,\text{obs}}, \xi_{tk} = 1; \hat{\mathbf{b}}\right), \quad (10)$$

The expected values of $\log f(\boldsymbol{x}_t; \mathbf{a}_k)$ and $\log f(\boldsymbol{y}_t; \mathbf{b}_k)$ with respect to the conditional distributions of the circular and linear missing values, within each latent class, can be computed by replacing the sufficient statistics with their expected values; these expected sufficient statistics are available in closed form in both the case of normal data (Shafer 1997) and in the case of circular data (Lagona and Picone 2012). The expected values $\hat{\pi}_{tk} = \mathbb{E}(\xi_{tk}|\mathbf{z}, \hat{\boldsymbol{\theta}})$ and $\hat{\pi}_{t-1,t,hk} = \mathbb{E}(\xi_{t-1,h}\xi_{t,k}|\mathbf{z}, \hat{\boldsymbol{\theta}})$ denote the posterior state probabilities of first and second order and can be evaluated by a standard backward–forward iteration procedure, well known in the literature of MHMMs (Cappe et al. 2005).

The M step is carried out by maximizing the expected complete data log-likelihood, with respect to $\boldsymbol{\theta}$. We observe that function $Q(\boldsymbol{\theta}|\hat{\boldsymbol{\theta}})$ depends on three functions, say $Q(\mathbf{p}, \mathbf{P}|\hat{\boldsymbol{\theta}})$, $Q(\mathbf{a}|\hat{\boldsymbol{\theta}}) = \sum_{k=1}^{K} Q_k(\mathbf{a}_k|\hat{\boldsymbol{\theta}})$ and $Q(\mathbf{b}|\hat{\boldsymbol{\theta}}) = \sum_{k=1}^{K} Q_k(\mathbf{b}_k|\hat{\boldsymbol{\theta}})$, that are known up to independent parameters and can be therefore maximized separately. However, while closed formulas are available for the maximum point of $Q(\mathbf{p}, \mathbf{P}|\hat{\boldsymbol{\theta}})$ and $Q(\mathbf{b}_k|\hat{\boldsymbol{\theta}})$ (Cappe et al. 2005), the maximum point of $Q(\mathbf{a}_k|\hat{\boldsymbol{\theta}})$ is the solution of a system of five trigonometric equations, which has to be solved iteratively (Lagona and Picone 2012).

Standard errors can be conveniently computed by a parametric-bootstrap procedure, by re-fitting the model to the bootstrap data, simulated from the estimated model, and computing the standard deviation of the distribution of the bootstrap estimates. Simulation of a Gaussian–von Mises MHMM is straightforward. We first simulate a sequence of states from the Markov chain. Given a sequence of states, bivariate circular and linear observations are at each time t drawn according to the appropriate bivariate von Mises or Gaussian density, respectively evaluated at $\mathbf{a} = \mathbf{a}_{k_t}$ and $\mathbf{b} = \mathbf{b}_{k_t}$, where k_t is the state that has been drawn at time t. To obtain a bivariate circular sample, we follow an acceptance-rejection algorithm, as suggested in Mardia et al. (2007).

4 Application

Sea conditions can be described in terms of representative wave regimes in specific areas, characterized by the probability of occurrence and corresponding to dominant environmental conditions (e.g., wind conditions), acting in the area and during a

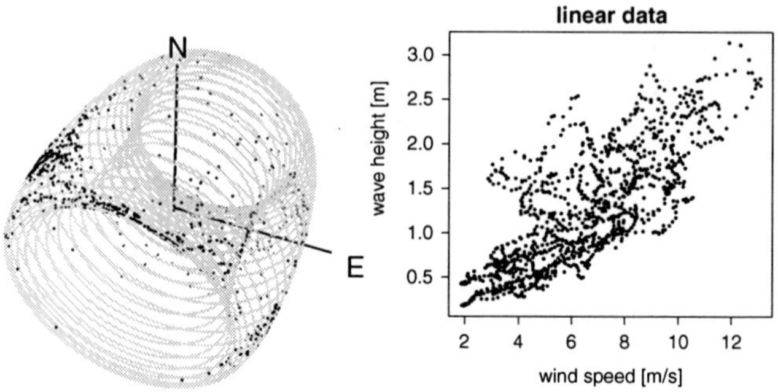

Fig. 1 Marine data, observed at the buoy of Ancona in the period 12/12/2009–12/1/2010; *left*: semi-hourly wind and wave directions (in radians), projected on a toroidal surface; *right*: semi-hourly wind speed (meter/sec) and wave height (meters)

period of interest (Lagona and Picone 2011). The data that motivated this paper are semi-hourly, quadrivariate profiles with two linear and two circular components: wind speed and wave height, wind direction and wave direction, taken in the period 12/12/2009–12/1/2010 by the buoy of Ancona, which is located in the Adriatic sea at about 30 Km from the coast. Of the resulting 1,501 profiles of wind and wave observations, about 20% include at least one missing value. In this paper we assume that the data are missing at random (MAR). In marine studies, missing values occur because devices transmission errors or malfunctioning. Because buoys are normally equipped in a way that they are able to transmit data even in the case of severe environmental conditions, marine observations are often missing completely at random (MCAR), i.e. the missingness probability does not depend on observed and unobserved data. The MCAR assumption is a particular case of the MAR hypotesis and is often likely for marine data that are obtained in semi-enclosed seas, such as the Adriatic sea, where severe environmental conditions seldom occur. Figure 1 displays the circular and the linear observations, after discarding the incomplete profiles. In particular, points in the left-hand side plot of the figure indicate hourly directions *from* which the wind blows and the wave travels, respectively, projected on the torus $(-\pi, \pi)^2$.

We clustered these data by fitting Gaussian–von Mises HMMs with $K = 2\ldots5$ states. To select the number of components, we computed both the Bayesian Information Criterion (BIC) and the Integrated Complete Likelihood (ICL) statistic (Table 1). The BIC statistic suggests a model with $K = 4$ components. However, a model with 4 components distinguishes the same three clusters provided by a model with 3 components, using two overlapping components to approximate the distribution of the data under a single latent state. This behavior of BIC has been extensively discussed in Baudry et al. (2010) in the context of mixture models. In our case study, however, overlapping components lack of physical interpretation, and cluster separation is more important than goodness of fit. We therefore use the ICL criterion, which approximates the integrated complete log-likelihood (Biernacki

Table 1 Model selection results

Number of parameters	Number of parameters	BIC	ICL
2	23	1545.2	1577.3
3	38	1494.4	1551.8
4	55	1483.1	1585.2
5	74	1501.1	1674.0

Table 2 Estimated parameters and standard errors (within brackets) of a 3-states multivariate Hidden Markov model with mixed linear and circular components

		Components					
		1	(s.e.)	2	(s.e.)	3	(s.e.)
Circular Parameters	a_1	1.710	(0.020)	6.165	(0.011)	2.938	(0.036)
	a_2	3.885	(0.072)	6.074	(0.011)	3.314	(0.046)
	a_{11}	3.158	(0.187)	10.750	(0.187)	4.819	(0.662)
	a_{22}	0.482	(0.069)	9.420	(0.384)	3.465	(0.480)
	a_{12}	−0.889	(0.137)	12.450	(0.485)	2.007	(0.491)
Gaussian Parameters	b_1	0.967	(0.032)	1.170	(0.029)	0.555	(0.036)
	b_2	3.985	(0.115)	6.527	(0.117)	5.175	(0.245)
	b_{11}	0.643	(0.040)	0.649	(0.036)	0.148	(0.017)
	b_{22}	8.179	(0.478)	9.937	(0.553)	6.287	(0.738)
	b_{12}	2.040	(0.130)	2.355	(0.137)	0.936	(0.111)
		Destination state					
Origin State	1	0.975	(0.007)	0.015	(0.006)	0.010	(0.005)
	2	0.011	(0.005)	0.988	(0.005)	0.001	(0.002)
	3	0.020	(0.026)	0.009	(0.022)	0.972	(0.031)

et al. 2000) and reduces to a BIC statistic, penalized by substracting the estimated mean entropy $\sum_{i=1}^{n}\sum_{k=1}^{K}\hat{\pi}_{tk}\log\hat{\pi}_{tk}$. Because ICL includes cluster separation as an additional criterion for model choice, the minimum ICL is attained by a model with three components, which is the model we consider to analyze the data.

Table 2 displays the maximum likelihood estimates and the standard errors of the model. The dependence parameters a_{12} and b_{12} are significant under each component, making a CI assumption difficult to motivate. Figure 2 shows the contour plots of the toroidal and planar densities, paired into three latent regimes, and the data points, allocated to the most probable regime. For simplicity, points in the top row are displayed in a plane. The model detects three regimes of straightforward interpretation. Regime 1 is associated with periods of calm sea: weak winds are associated with waves of modest size, traveling from the Italian coast, along the minor axis of the Adriatic sea. Under this regime, wind and wave directions are poorly synchronized and, simultaneously, wind speed and wave height are weakly correlated, because when wind episodes are modest, then waves are essentially influenced by marine currents. Wind and wave data are instead strongly correlated under regimes 2 and 3. Regime 2 is associated with bora episodes,

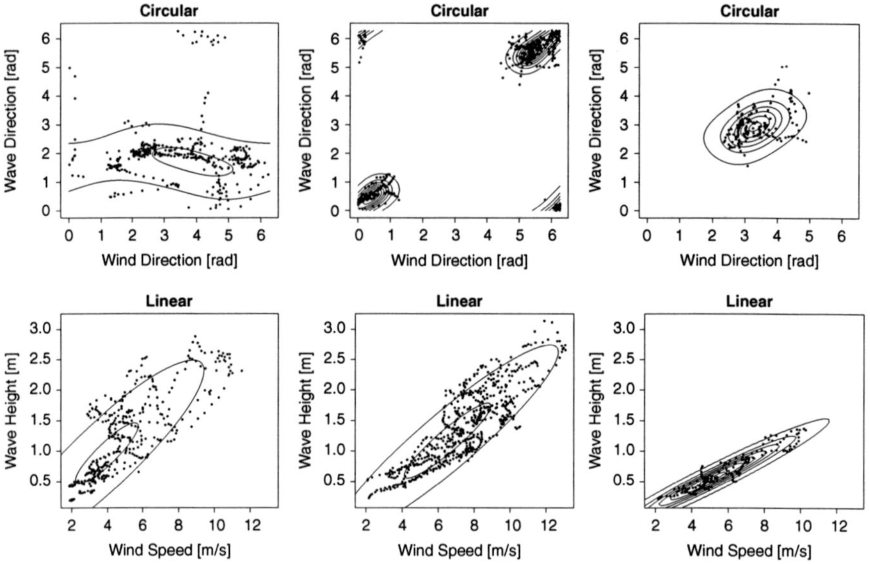

Fig. 2 Contour plots of the three conditional circular and linear bivariate densities, as estimated by fitting a multivariate hidden Markov model with three states, and points allocated to the most probable component

when fine structured wind jets blow within the Dinaric Alps toward the eastern Adriatic coast. Regime 3 is instead associated with sirocco episodes, when wind blows southeasterly, along the major axis of the Adriatic sea. In summary, regime switching does not only change directional and linear averages but also, and more interestingly, the correlation structure of the data, which is ignored by CI-based HMMs (Holzmann et al. 2006). The model hence indicates that weather conditions should not be used to predict wave direction and height, without accounting for the latent heterogeneity of the data.

5 Discussion

Motivated by classification issues that arise in marine studies, we introduce a new MHMM for segmenting environmental data according to latent classes or regimes, associated with toroidal and elliptical clusters. The combination of bivariate von Mises and normal distributions allows for a simple specification of the dependence structure between variables and for the computational feasibility of a mixture-based classification strategy where missing values can be efficiently handled within a likelihood framework.

References

Biernacki, C., Celeux, G., & Govaert, G. (2000). Assessing a mixture model for clustering with the integrated completed likelihood. *IEEE Transactions on Pattern Analysis and Machine Intelligence, 22*, 719–725.

Baudry, J. -P., Raftery, A. E., Celeux, G., Lo, K., & Gottardo, R. (2010). Combining mixture components for clustering. *Journal of Computational and Graphical Statistics, 19*, 332–353.

Holzmann, H., Munk, A., Suster, M., & Zucchini, W. (2006). Hidden Markov models for circular and linear-circular time series. *Environmental and Ecological Statistics, 13*, 325–347.

Cappe, O., Moulines, E., & Ryden, T. (2005). *Inference in hidden Markov models*. New York: Springer.

Lagona, F., & Picone, M. (2011). A latent-class model for clustering incomplete linear and circular data in marine studies. *Journal of Data Science, 9*, 585–605.

Lagona, F., & Picone, M. (2012). Model-based clustering of multivariate skew data with circular components and missing values. *Journal of Applied Statistics, 39*, 927–945.

Mardia, K., Taylor, C., & Subramaniam, G. (2007) Protein bioinformatics and mixtures of bivariate von mises distributions for angular data. *Biometrics, 63*, 505-512.

Shafer, J. L. (1997). *Analysis of incomplete multivariate data*. Boca Raton, FL: Chapman and Hall.

Singh, H., Hnizdo, V., & Demchuk, E. (2002). Probabilistic model for two dependent circular variables. *Biometrika, 89*, 719-723.

Zhang, Q., Snow Jones, A., Rijmen, F., & Ip, E. H. (2010). Multivariate discrete hidden Markov models for domain-based measurements and assessment of risk factors in child development. *Journal of Computational and Graphical Statistics, 19*, 746–765.

A Comparison of Objective Bayes Factors for Variable Selection in Linear Regression Models

Luca La Rocca

Abstract This paper deals with the variable selection problem in linear regression models and its solution by means of Bayes factors. If substantive prior information is lacking or impractical to elicit, which is often the case in applications, objective Bayes factors come into play. These can be obtained by means of different methods, featuring Zellner–Siow priors, fractional Bayes factors and intrinsic priors. The paper reviews such methods and investigates their finite-sample ability to identify the simplest model supported by the data, introducing the notion of full discrimination power. The results obtained are relevant to structural learning of Gaussian DAG models, where large spaces of sets of recursive linear regressions are to be explored.

1 Introduction

In a variety of applications, a numerical response vector $\mathbf{y} = [y_1 \ldots y_n]'$ has to be predicted, and there is an interest in determining a reduced set of predictors among the columns of a numerical matrix $\mathbf{X} = [\mathbf{X}_1 \ldots \mathbf{X}_p]$. This is known as the variable selection problem, and it is often dealt with by assuming that $\mathbf{y}$ follows a multivariate normal distribution with mean vector μ and covariance matrix $\phi^{-1}\mathbf{I}_n$, where ϕ is a precision (inverse variance) parameter and $\mathbf{I}_n$ is the n-dimensional identity matrix. Each set of predictors will be represented by a p-dimensional zero-one vector γ, with $\gamma_j = 1$ ($\gamma_j = 0$) meaning that $\mathbf{X}_j$ is included in (excluded from) the set, and γ will be associated with a linear regression model $\mathscr{M}_\gamma$ by letting

$$\mu = \alpha \mathbf{1}_n + \mathbf{X}_\gamma \beta_\gamma , \tag{1}$$

L. La Rocca (✉)
Dipartimento di Scienze Fisiche, Informatiche e Matematiche, University of Modena and Reggio Emilia, Edificio Matematica, Via Campi 213/b, 41125 Modena, Italy
e-mail: luca.larocca@unimore.it

P. Giudici et al. (eds.), *Statistical Models for Data Analysis*, Studies in Classification, Data Analysis, and Knowledge Organization, DOI 10.1007/978-3-319-00032-9_21,

where $\mathbf{1}_n$ is the n-dimensional vector of all ones, α is a scalar intercept parameter, $\mathbf{X}_\gamma$ is the matrix consisting of the columns of $\mathbf{X}$ identified by $\gamma_j = 1$, and β_γ is a vector of regression parameters. An "effective" method for variable selection will identify a model $\mathcal{M}_{\hat{\gamma}}$ with "good" predictive performance, but "small" dimension $|\hat{\gamma}| + 2$, where $|\hat{\gamma}| = \mathbf{1}'_p \hat{\gamma}$. More on variable selection can be found in George (2000).

Bayesian comparison of any two models $\mathcal{M}_0$ and $\mathcal{M}_1$ for data $\mathbf{y}$ can be performed through the Bayes Factor (BF): denoting by $f_k(\mathbf{y}|\theta_k)$ the sampling density of $\mathbf{y}$ under $\mathcal{M}_k$, $k = 0, 1$, and by $p_k(\theta_k)$ the corresponding parameter prior, the BF for $\mathcal{M}_1$ against $\mathcal{M}_0$ is defined as $\mathrm{BF}_{10}(\mathbf{y}) = f_1(\mathbf{y})/f_0(\mathbf{y})$, where $f_k(\mathbf{y}) = \int f_k(\mathbf{y}|\theta_k) p(\theta_k) \mathrm{d}\theta_k$ is the marginal likelihood of $\mathcal{M}_k$ given $\mathbf{y}$. If several models are to be compared, each of them will be assigned a prior probability $p(\mathcal{M}_k)$ and will receive the posterior probability $p(\mathcal{M}_k|\mathbf{y}) = p(\mathcal{M}_k)\mathrm{BF}_{k0}(\mathbf{y})/\{p(\mathcal{M}_0) + \sum_{h \neq 0} p(\mathcal{M}_h)\mathrm{BF}_{h0}(\mathbf{y})\}$ once the BF for each model against a baseline model $\mathcal{M}_0$ has been computed. Posterior model probabilities can be used for model selection or to average predictions across models; see Kass and Raftery (1995) for a classic review.

BFs are especially suited to variable selection, because they automatically implement *Ockham's razor*: they penalize unnecessarily complex models, as discussed by Jefferys and Berger (1992) among others. However, BFs critically depend on parameter priors, so that using them for objective analyses (analyses in lack of substantive prior information) is a challenging task; see Pericchi (2005) and the references therein. This paper reviews, in Sect. 2, different methods available in the literature to specify objective parameter priors for the comparison of linear regression models and compares, in Sect. 3, their performance as Ockham's razors. A brief discussion of the link to structural learning of Gaussian DAG models concludes the paper.

2 Miscellany of Bayes Factors

Let $\mathcal{M}_\gamma$ and $\mathcal{M}_{\tilde{\gamma}}$ be two distinct linear regression models for the response vector $\mathbf{y}$ conditional on the predictor matrix $\mathbf{X}$. A first possibility to obtain an objective BF for $\mathcal{M}_{\tilde{\gamma}}$ against $\mathcal{M}_\gamma$ is simply to neglect the effect of parameter priors by using *Schwarz criterion* to approximate the log-ratio of marginal likelihoods; see Kass and Raftery (1995). This gives

$$\mathrm{SCBF}_{\tilde{\gamma}\gamma}(B_{\gamma\tilde{\gamma}}, n) = n^{-(|\tilde{\gamma}|-|\gamma|)/2} B_{\gamma\tilde{\gamma}}^{-n/2}, \tag{2}$$

where $B_{\gamma\tilde{\gamma}} = \mathrm{RSS}_{\tilde{\gamma}}/\mathrm{RSS}_\gamma$ is the ratio of residual sum of squares for the two models fitted using ordinary least squares, n is the available sample size, and $|\tilde{\gamma}| - |\gamma|$ is the difference between the two model dimensions. As discussed by Kass and Raftery (1995), the SCBF in (2) is a rough tool mostly appealing for reference purposes.

More elaborate objective BFs for comparing linear regression models typically start by considering the non-informative limiting-conjugate parameter prior

$$p_\gamma^{\mathrm{N}}(\alpha, \beta_\gamma, \phi) \propto \phi^{-1} \tag{3}$$

under model $\mathcal{M}_\gamma$. If priors $p_\gamma^{\mathrm{N}}(\cdot)$ and $p_{\tilde{\gamma}}^{\mathrm{N}}(\cdot)$ are directly used to compare $\mathcal{M}_\gamma$ and $\mathcal{M}_{\tilde{\gamma}}$, the resulting BF will only be defined up to a constant factor. Spiegelhalter and Smith (1982) specified the constant factor so as to obtain a BF equal to one on special data from a minimal experiment, but their proposal is somehow arbitrary; see O'Hagan (1995). Alternatively, if $p_\gamma^{\mathrm{N}}(\cdot)$ in (3) is replaced by Jeffreys's prior $p_\gamma^{\mathrm{J}}(\alpha, \beta_\gamma, \phi) \propto \phi^{-1+|\gamma|/2}$, the constant factor could be uniquely determined using Jeffreys's measure without dropping its leading constant factor, as suggested by Dawid (1999); this would give a BF proportional to SCBF in (2). In order to make progress with parameter priors, more elaborate methods are needed, which transform $p_\gamma^{\mathrm{N}}(\cdot)$ so that BFs are uniquely defined. Three such methods are presented below.

Zellner–Siow Priors

Let $\mathcal{M}_\gamma$ be nested in $\mathcal{M}_{\tilde{\gamma}}$ and split the vector of regression parameters for $\mathcal{M}_{\tilde{\gamma}}$ as $\beta_{\tilde{\gamma}} = [\beta_\gamma' \, \beta_\delta']'$, where $\delta = \tilde{\gamma} - \gamma$ identifies the additional predictors in $\mathcal{M}_{\tilde{\gamma}}$ with respect to $\mathcal{M}_\gamma$. The BF for $\mathcal{M}_{\tilde{\gamma}}$ against $\mathcal{M}_\gamma$ will be uniquely defined if we replace $p_{\tilde{\gamma}}^{\mathrm{N}}(\cdot)$ with $p_{\tilde{\gamma}}^{g}(\alpha, \beta_\gamma, \beta_\delta, \phi) \propto \phi^{-1}\mathrm{dmn}(\beta_\delta | \mathbf{0}, g\phi^{-1}(\mathbf{X}_\delta'\mathbf{X}_\delta)^{-1})$, where $\mathrm{dmn}(\cdot|\mathbf{0}, \Sigma)$ denotes the centred multivariate normal density with covariance matrix Σ and g is a strictly positive scalar quantity, whose choice is discussed later. This amounts to giving the common parameters a common non-informative prior under the two models, regarding them as nuisance parameters, while conventionally adopting Zellner's g-prior for the additional parameters. The interpretation of α and the elements of β_γ as common parameters, following Zellner and Siow (1980), is based on a reparameterization leading to $\mathbf{1}_n'\mathbf{X}_\delta = \mathbf{0}$ and $\mathbf{X}_\gamma'\mathbf{X}_\delta = \mathbf{0}$, which is also instrumental in finding the following expression for the BF:

$$ZSBF_{\tilde{\gamma}\gamma}(B_{\gamma\tilde{\gamma}}, n|g) = \{1+g\}^{(n-|\tilde{\gamma}|-1)/2}\{1+gB_{\gamma\tilde{\gamma}}\}^{-(n-|\gamma|-1)/2}, \tag{4}$$

where $B_{\gamma\tilde{\gamma}} \leq 1$ as a consequence of $\mathcal{M}_\gamma$ being nested in $\mathcal{M}_{\tilde{\gamma}}$; see Liang et al. (2008).

Two non-nested models can be compared through a third model nested in both of them; using the null model $\mathcal{M}_\mathbf{0}$ for all pairs, the null-based approach will compare $\mathcal{M}_\gamma$ and $\mathcal{M}_{\tilde{\gamma}}$ using $ZSBF_{\tilde{\gamma}\gamma}^{\mathbf{0}}(B_{\mathbf{0}\tilde{\gamma}}, B_{\mathbf{0}\gamma}, n|g) = ZSBF_{\tilde{\gamma}\mathbf{0}}(B_{\mathbf{0}\tilde{\gamma}}, n|g)\{ZSBF_{\gamma\mathbf{0}}(B_{\mathbf{0}\gamma}, n|g)\}^{-1}$. In practice, this is equivalent to conventionally using g-priors under all models. Alternatively, the full-based approach would compare all pairs through the full model $\mathcal{M}_\mathbf{1}$, but this would introduce incoherences, because the prior under $\mathcal{M}_\mathbf{1}$ would change in each pairwise comparison. Finally, concerning the choice of g, a wide range of possibilities is discussed by Liang et al. (2008) and implemented in the R package BAS (Clyde 2010). If g has to be fixed, the choice $g = \max\{n, p^2\}$ is recommended as a benchmark on the basis of the work of Fernández et al. (2001). If g can be given a hyper-prior, then an Inverse-Gamma distribution depending on sample size, namely, $p_n(g) = (n/2)^{1/2}\Gamma(1/2)^{-1}g^{-3/2}\exp\{-n/(2g)\}$, results in the heavy-tailed multivariate Cauchy priors of Zellner and Siow (1980). These two choices guarantee consistency of the BF as a model selector (under a technical assumption and using strictly positive prior model probabilities). Other mixtures of g-priors, as well as empirical Bayes estimates of g, are also considered by Liang et al. (2008).

Fractional Bayes Factors

Let $p_k^{\mathrm{N}}(\theta_k)$ be the non-informative prior available under model $\mathcal{M}_k$, $k = 0, 1$, and choose a fraction b of the data for training. The fractional BF of O'Hagan (1995) is the BF obtained using the data-dependent *fractional prior* $p_k^{\mathrm{F}}(\theta_k) \propto f(\mathbf{y}|\theta_k)^b p_k^{\mathrm{N}}(\theta_k)$ and the *fractional likelihood* $f(\mathbf{y}|\theta_k)^{1-b}$ under model $\mathcal{M}_k$. A generally recommended fraction choice is $b = n_\bullet/n$, where $n_\bullet$ is the minimal (integer) sample size that makes both $p_0^{\mathrm{F}}(\theta_k)$ and $p_1^{\mathrm{F}}(\theta_k)$ proper; an argument in favour of this choice is given by Moreno (1997). The fractional BF is especially suited to the context of exponential families and conjugate priors, where the fractional likelihood is an ordinary likelihood with sample size $n - n_\bullet$ and summary statistics as in the whole data; see Consonni and La Rocca (2012) for details on this aspect. In particular, for any two linear regression models $\mathcal{M}_\gamma$ and $\mathcal{M}_{\tilde{\gamma}}$, the fractional BF for $\mathcal{M}_{\tilde{\gamma}}$ against $\mathcal{M}_\gamma$ admits the following closed-form expression:

$$\mathrm{FBF}_{\tilde{\gamma}\gamma}(B_{\gamma\tilde{\gamma}}, n|n_\bullet) = \frac{\Gamma((n_\bullet - |\gamma| - 1)/2)\Gamma((n - |\tilde{\gamma}| - 1)/2)}{\Gamma((n_\bullet - |\tilde{\gamma}| - 1)/2)\Gamma((n - |\gamma| - 1)/2)} B_{\gamma\tilde{\gamma}}^{-(n-n_\bullet)/2}, \quad (5)$$

where $\Gamma(\cdot)$ denotes the gamma function and $n_\bullet = \max\{|\gamma|, |\tilde{\gamma}|\} + 2$. Notice that $n_\bullet = p + 2$ is necessary to explore the whole model space without incoherences.

Intrinsic Priors

Let $\mathcal{M}_0$ be nested in $\mathcal{M}_1$ and $p_k^{\mathrm{N}}(\theta_k)$ be the non-informative prior available under $\mathcal{M}_k$, $k = 0, 1$. Choose a training sample size t. The intrinsic prior for θ_1 under $\mathcal{M}_1$, with respect to $p_0^{\mathrm{N}}(\theta_0)$ under $\mathcal{M}_0$, is $p_1^{\mathrm{I}}(\theta_1|t) = p_1^{\mathrm{N}}(\theta_1)\int \mathrm{BF}_{01}^{\mathrm{N}}(\mathbf{u}) f_1(\mathbf{u}|\theta_1)d\mathbf{u}$, where $\mathrm{BF}_{01}^{\mathrm{N}}(\mathbf{u})$ is the BF for $\mathcal{M}_0$ against $\mathcal{M}_1$ using the non-informative priors and observing an *auxiliary* training sample $\mathbf{u} = [u_1 \ldots u_t]'$; see Berger and Pericchi (1996). Notice that $\mathbf{u}$ is not part of the data, and that it is averaged out in the definition of $p_1^{\mathrm{I}}(\theta_1|t)$, so that the latter only depends on t. The larger is t, the more peaked $p_1^{\mathrm{I}}(\theta_1|t)$ will be on the subspace of $\mathcal{M}_1$ corresponding to $\mathcal{M}_0$, because values of θ_1 originating samples $\mathbf{u}$ supporting $\mathcal{M}_0$ are emphasized in the definition of $p_1^{\mathrm{I}}(\theta_1|t)$. This feature can also be grasped from the intrinsic prior representation as an expected-posterior prior of Perez and Berger (2002): $p_1^{\mathrm{I}}(\theta_1|t) = \int p_1^{\mathrm{N}}(\theta_1|\mathbf{u}) f_0^{\mathrm{N}}(\mathbf{u})d\mathbf{u}$. A typical choice of t is the minimal (integer) sample size that makes $p_1^{\mathrm{N}}(\theta_1|\mathbf{u})$ proper for all $\mathbf{u}$. With this choice, in the context of variable selection, starting from $p_\gamma^{\mathrm{N}}(\cdot)$ in (3), and pairing it with $p_{\tilde{\gamma}}^{\mathrm{I}}(\cdot)$, the following BF for $\mathscr{M}_{\tilde{\gamma}}$ against $\mathscr{M}_\gamma$ is obtained:

$$IPBF_{\tilde{\gamma}\gamma}(B_{\gamma\tilde{\gamma}},n) = \frac{2}{\pi}\int_0^{\frac{\pi}{2}} \frac{\{\|\tilde{\gamma}\|\sin^2\psi\}^{(|\tilde{\gamma}|-|\gamma|)/2}\{n+\|\tilde{\gamma}\|\sin^2\psi\}^{(n-|\tilde{\gamma}|-1)/2}}{\{nB_{\gamma\tilde{\gamma}}+\|\tilde{\gamma}\|\sin^2\psi\}^{(n-|\gamma|-1)/2}}\mathrm{d}\psi\ , \tag{6}$$

where $\|\tilde{\gamma}\| = |\tilde{\gamma}|+2$ is the dimension of $\mathscr{M}_{\tilde{\gamma}}$ and $B_{\gamma\tilde{\gamma}} \leq 1$, because $\mathscr{M}_\gamma$ is nested in $\mathscr{M}_{\tilde{\gamma}}$; see Casella et al. (2009) and the references therein. Note that implementing (6) amounts to a unidimensional integral over a bounded interval, and that non-nested models can be compared exactly as with Zellner–Siow priors.

3 Sharpness of Ockham's Razor

A general argument by Dawid (1999) suggests that the typical learning rate of $\mathrm{BF}_{10}(\mathbf{y})$, as $n \to \infty$, when two nested models $\mathscr{M}_0 \subset \mathscr{M}_1$ with respective dimensions $d_0 < d_1$ are compared, is given by $BF_{10}(\mathbf{y}) = O_{\mathbf{p}}(n^{-(d_1-d_0)/2})$, when the sampling distribution $\mathbf{p}$ belongs to $\mathscr{M}_0$, and by $BF_{01}(\mathbf{y}) = \exp(-Kn + O_{\mathbf{p}}(n^{1/2}))$, for some $K > 0$, when the sampling distribution $\mathbf{p}$ belongs to $\mathscr{M}_1 \setminus \mathscr{M}_0$. In this way, asymptotically, both wrong models and unnecessarily complex models are dropped, and the BF automatically implements Ockham's razor. The learning speed is clearly unbalanced, as pointed out by Johnson and Rossell (2010), but this aspect will not be pursued any further here. It will be enough to say that the above *asymptotic learning rate* can be verified for the objective BFs of Sect. 2. For variable selection in linear regression models, a different perspective on the ability of different methods to implement Ockham's razor is also of interest. The rest of this section is written from this perspective, which can be named as the study of *finite-sample discrimination power*.

All objective BFs for variable selection presented in Sect. 2 can be written as

$$\mathrm{BF}^{\mathbf{0}}_{\tilde{\gamma}\gamma}(B_{\mathbf{0}\tilde{\gamma}}, B_{\mathbf{0}\gamma}, n) = \mathrm{BF}_{\tilde{\gamma}\mathbf{0}}(B_{\mathbf{0}\tilde{\gamma}}, n)\{\mathrm{BF}_{\gamma\mathbf{0}}(B_{\mathbf{0}\gamma}, n)\}^{-1}\ , \tag{7}$$

using the null-based approach, although this is pleonastic when the pairwise BF is also defined for non-nested models and $\mathrm{BF}^{\mathbf{0}}_{\tilde{\gamma}\gamma}(B_{\mathbf{0}\tilde{\gamma}}, B_{\mathbf{0}\gamma}, n) = \mathrm{BF}_{\tilde{\gamma}\gamma}(B_{\gamma\tilde{\gamma}}, n)$. The limit of (7) as $B_{\mathbf{0}\tilde{\gamma}} \to 0$ when $\tilde{\gamma} \neq \mathbf{0}$ determines, for fixed n, how much the posterior probability mass will be able to concentrate on non-null models with good fit (relative to the null model). Specifically, the following definition identifies two kinds of BFs.

Definition 1. The BF in (7) has *full discrimination power with respect to wrong models* if $\mathrm{BF}^{\mathbf{0}}_{\tilde{\gamma}\gamma}(B_{\mathbf{0}\tilde{\gamma}}, B_{\mathbf{0}\gamma}, n) \to \infty$, as $B_{\mathbf{0}\tilde{\gamma}} \to 0$, when $|\tilde{\gamma}| > 0$.

Table 1 Discrimination power of objective BFs for variable selection in linear regression models. The first column identifies the BF. The second and third column report whether the BF satisfies or not Definitions 1 and 2. The following columns report two-digit posterior model probabilities for Hald data (identifying models by the set of selected predictors); models receiving zero two-digit posterior probability from all BFs are not shown. Not all rows sum to one due to rounding errors

BF	Def. 1	Def. 2	$\{3,4\}$	$\{1,4\}$	$\{1,2\}$	$\{2,3,4\}$	$\{1,3,4\}$	$\{1,2,4\}$	$\{1,2,3\}$	$\{1,2,3,4\}$	Total
BPBF[a]	No	No	0.03	0.22	0.34	0.06	0.10	0.11	0.11	0.03	1.00
SCBF	Yes	No	0.00	0.05	0.25	0.01	0.16	0.23	0.23	0.07	1.00
CFBF[b]	Yes	No	0.00	0.10	0.24	0.04	0.16	0.20	0.20	0.06	1.00
ZSBF	Yes	Yes	0.00	0.16	0.52	0.02	0.08	0.10	0.10	0.01	0.99
IFBF[c]	Yes	Yes	0.00	0.17	0.54	0.02	0.07	0.09	0.09	0.01	0.99
IPBF	Yes	Yes	0.00	0.18	0.55	0.02	0.07	0.09	0.09	0.01	1.01

[a]BPBF (Benchmark Prior BF) is ZSBF given $g = \max\{n, p^2\}$
[b]CFBF (Coherent Fractional BF) is FBF with $n_\bullet = p + 2$ in all pairwise comparisons
[c]IFBF (Incoherent Fractional BF) is FBF with $n_\bullet = |\gamma| + 2$ in $\mathrm{FBF}_{\gamma\mathbf{0}}$

A BF with full discrimination power with respect to wrong models, which is a BF resolving the *information paradox* in the terminology of Liang et al. (2008), will drop all non-null models with poor fit, as well as the null model, whenever a non-null model $\mathcal{M}_{\tilde{\gamma}}$ with good fit is present. However, $\mathcal{M}_{\tilde{\gamma}}$ will be nested in other non-null models with good fit, unless it is the full model, and there is no guarantee that the posterior probability mass will not spread on all these models, instead of concentrating on the simplest of them. Indeed, for Ockham's razor to be sharp when n is small, a further condition has to be verified by the BF.

Definition 2. The BF in (7) has *full discrimination power with respect to unnecessarily complex models* if $\mathrm{BF}^{\mathbf{0}}_{\tilde{\gamma}\gamma}(B_{\mathbf{0}\tilde{\gamma}}, B_{\mathbf{0}\gamma}, n) \to 0$, as $B_{\mathbf{0}\tilde{\gamma}} \to 0$ and $B_{\mathbf{0}\gamma} \to 0$, with $B_{\gamma\tilde{\gamma}} = B_{\mathbf{0}\tilde{\gamma}}/B_{\mathbf{0}\gamma} \to 1$, when $|\tilde{\gamma}| > |\gamma| > 0$.

A BF with full discrimination power with respect to unnecessarily complex models will concentrate the posterior probability mass on the simplest non-null model with good fit whenever one such model is present.

The discrimination power of six objective BFs, covering the whole spectrum of methods reviewed in Sect. 2, is illustrated in Table 1. Each BF was classified according to Definitions 1 and 2, then it was applied to *Hald data*: this very well-known example, where up to $p = 4$ ingredients can be used to predict the heat evolved during the hardening of a cement mix, based on $n = 13$ observations, provides a simple but effective demonstration of the different degrees of finite-sample discrimination power; the small sample size prevents the asymptotic Ockham's razor from coming into play, while the small number of predictors makes exhaustive model search possible and thus avoids confounding the features of BFs with model search issues. Computations were carried out in R (R Development Core Team 2011), using package BAS (Clyde 2010) for Zellner–Siow priors and Schwarz criterion, together with user-defined functions for fractional BFs and intrinsic priors; package BAS also provided a built-in version of Hald data (complete with documentation). Uniform prior model probabilities were used, as a default

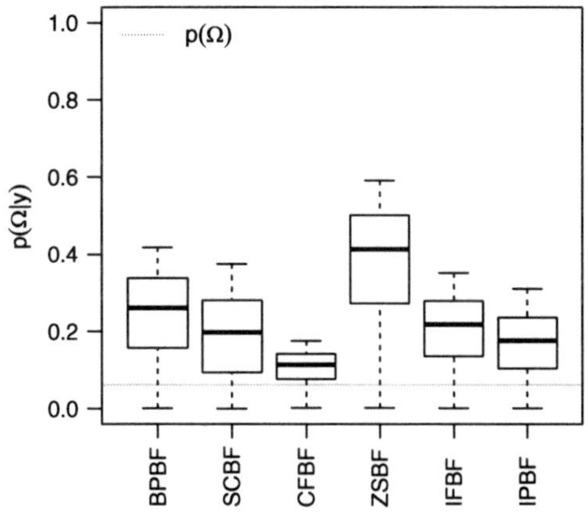

Fig. 1 Posterior probability of the null model $\Omega = \mathscr{M}_0$ on i.i.d. normal data: 2,500 response vectors of length $n = 13$ (Hald data sample size) were randomly generated and then analysed with the six objective BFs of Table 1, using Hald data predictors and uniform prior model probabilities. Results do not depend on the expected value and variance of the simulated data, because the distribution of $B_{0\gamma}$ under $\mathscr{M}_0$ is free of these parameters

choice for comparing BFs, although this assignment should not be given for granted in objective analyses; see in particular Scott and Berger (2010).

Intrinsic priors and Zellner–Siow priors (g-priors with random g) have full discrimination power both with respect to wrong models and to unnecessarily complex models. These two BFs essentially give the same results on Hald data, with more than half of the posterior probability mass concentrated on the model indexed by $\hat{\gamma} = [1100]'$, which clearly stands out for selection. Schwarz criterion and fractional BFs (implemented coherently) only achieve full discrimination power with respect to wrong models, and thus their posterior probability mass is more spread towards complex models. Interestingly, an incoherent usage of fractional BFs performs equivalently to intrinsic priors and Zellner–Siow priors. Finally, benchmark priors (g-priors with fixed g) lack discrimination power both with respect to wrong models and to unnecessarily complex models, and their posterior probability mass is spread towards both simple and complex models. Notice that, although all methods select the same model $\mathscr{M}_{\hat{\gamma}}$ under the highest posterior probability criterion, they give different results if the median posterior probability criterion of Barbieri and Berger (2004) is used: BPBF, SCBF and CFBF select $\{1, 2, 4\}$ in place of $\{1, 2\}$, because the inclusion probability of predictor 4 is also above 50%.

In order to complete the study of finite-sample discrimination power, the case of no non-null model fitting the data significantly better than $\mathscr{M}_0$ has to be considered. In this case, the ratio $B_{0\gamma}$ will be close to one for all γ, and the posterior probability of $\mathscr{M}_0$ will be close to the following upper bound:

$$p(\mathscr{M}_0|\mathbf{y}) \leq \frac{p(\mathscr{M}_0)}{p(\mathscr{M}_0) + \sum_{\gamma \neq \mathbf{0}} p(\mathscr{M}_\gamma)\mathrm{BF}_{\gamma \mathbf{0}}(1, n)} . \tag{8}$$

Clearly, due to consistency, the right hand side of (8) goes to one, as $n \to \infty$. The distribution of $p(\mathscr{M}_0|\mathbf{y})$ on simulated data from the null model is illustrated in Fig. 1.

4 Discussion

Variable selection in linear regression models is a factor in structural learning of Gaussian DAG models, because these are sets of recursive linear regressions. In this setting, Consonni and La Rocca (2011) and Altomare et al. (2013) dealt with the imbalance in the asymptotic learning rate of the fractional BF along the lines suggested by Johnson and Rossell (2010), and Consonni and La Rocca (2012) developed fractional BFs invariant with respect to Markov equivalence; work is in progress to apply intrinsic priors. Since large spaces of DAGs are to be explored, even with few variables, discrimination power is important to obtain a short list of models with good fit.

Acknowledgements This work was conceived while Luca La Rocca and Guido Consonni were visiting Elías Moreno in Granada: warm hospitality and interesting discussions were not forgotten. Partial financial support by the University of Granada and the University of Pavia is gratefully acknowledged. The author would also like to thank David Rossell for a beneficial conversation, and an anonymous referee for helpful comments.

References

Altomare, D., Consonni, G., & La Rocca, L. (2013). Objective Bayesian search of Gaussian directed acyclic graphical models for ordered variables with non-local priors. Biometrics, Early View Article.

Barbieri, M. M., & Berger, J. O. (2004). Optimal predictive model selection. *Annals of Statistics*, *32*(3), 870–897.

Berger, J. O., & Pericchi, L. R. (1996). The intrinsic Bayes factor for model selection and prediction. *Journal of the American Statistical Association*, *91*(433), 109–122.

Casella, G., Girón, F. J., Martínez, M. L., & Moreno, E. (2009). Consistency of Bayesian procedures for variable selection. *Annals of Statistics*, *37*(3), 1207–1228.

Clyde, M. A. (2010). BAS: Bayesian Adaptive Sampling for Bayesian Model Averaging. URL http://cran.r-project.org/web/packages/BAS/index.html. R package version 0.92.

Consonni, G., & La Rocca, L. (2011). Moment priors for Bayesian model choice with applications to directed acyclic graphs. In J. M. Bernardo, M. J. Bayarri, J. O. Berger, A. P. Dawid, D. Heckerman, A. F. M. Smith, & M. West (Eds.) *Bayesian statistics 9—Proceedings of the Ninth Valencia International Meeting* (pp. 119–133). Oxford: Oxford University Press. With discussion.

Consonni, G., & La Rocca, L. (2012). Objective Bayes factors for Gaussian directed acyclic graphical models. *Scandinavian Journal of Statistics*, *39*(4), 743–756.

Dawid, A. (1999). The trouble with Bayes factors. Tech. Rep. 202, Department of Statistical Science, University College London.

Fernández, C., Ley, E., & Steel, M. F. J. (2001). Benchmark priors for Bayesian model averaging. *Journal of Econometrics*, *100*(2), 381–427.

George, E. I. (2000). The variable selection problem. *Journal of the American Statistical Association*, *95*(452), 1304–1308.

Jefferys, W. H., & Berger, J. O. (1992). Ockham's razor and Bayesian analysis. *American Scientist*, *80*(1), 64–72.

Johnson, V. E., & Rossell, D. (2010). On the use of non-local prior densities in Bayesian hypothesis tests. *Journal of the Royal Statistical Society Series B Statistical Methodology*, *72*(2), 143–170.

Kass, R. E., & Raftery, A. E. (1995). Bayes factors. *Journal of the American Statistical Association*, *90*(430), 773–795.

Liang, F., Paulo, R., Molina, G., Clyde, M. A., & Berger, J. O. (2008). Mixtures of g priors for Bayesian variable selection. *Journal of the American Statistical Association*, *103*(481), 410–423.

Moreno, E. (1997). Bayes factors for intrinsic and fractional priors in nested models. Bayesian robustness. In Y. Dodge (Ed.) *L_1-Statistical procedures and related topics* (pp. 257–270). Hayward: Institute of Mathematical Statistics.

O'Hagan, A. (1995). Fractional Bayes factors for model comparison. *Journal of the Royal Statistical Society Series B Methodology*, *57*(1), 99–118. With discussion.

Perez, J. M., & Berger, J. O. (2002). Expected-posterior prior distributions for model selection. *Biometrika*, *89*(3), 491–511.

Pericchi, L. R. (2005). Model selection and hypothesis testing based on objective probabilities and Bayes factors. In D. Dey, & C. R. Rao (Eds.) *Bayesian thinking: modeling and computation. Handbook of statistics* (vol. 25, pp. 115–149). Amsterdam: Elsevier/North-Holland.

R Development Core Team. (2011). R: A Language and Environment for Statistical Computing. R Foundation for Statistical Computing, Vienna, Austria. URL http://www.R-project.org/

Scott, J. G., & Berger, J. O. (2010). Bayes and empirical-Bayes multiplicity adjustment in the variable-selection problem. *Annals of Statistics*, *38*(5), 2587–2619.

Spiegelhalter, D. J., & Smith, A. F. M. (1982). Bayes factors for linear and loglinear models with vague prior information. *Journal of the Royal Statistical Society Series B Methodology*, *44*(3), 377–387.

Zellner, A., & Siow, A. (1980). Posterior odds ratio for selected regression hypotheses. In J. M. Bernardo, M. H. DeGroot, D. V. Lindely, & A. F. M. Smith (Eds.) *Bayesian Statistics—Proceedings of the First Valencia International Meeting Held in Valencia* (pp. 585–603). University of Valencia Press, Valencia. With discussion.

Evolutionary Customer Evaluation: A Dynamic Approach to a Banking Case

Caterina Liberati and Paolo Mariani

Abstract Today, the most important asset for a bank is its customer and therefore, the main targets to achieve by management are: knowledge of his needs, anticipation of his concerns and to distinguish itself in his eyes. The awareness that a satisfied customer is a highly profitable asset effort to provide a satisfactory service to the customer by diversifying its services. This paper aims to analyze the customer evaluation evolution of the main attributes of banking services to catch differences among the clusters and time lags through a dynamic factorial model. We propose a new system of weights by which assessing the dynamic factor reduction that is not optimal for all the instances considered across different waves. An empirical study will be illustrated: it is based on customer satisfaction data coming from a national bank with a spread network throughout Italy which wanted to analyze its reduced competitiveness in retail services, probably due to low customer satisfaction.

1 Scenario

Customer centric vision has been successfully applied in the last years in banking sector. Such a concept has customer satisfaction as the most important asset of a company. According with this idea banks have increased their products differentiation in order to match with potential clients requests or expectations. This resulted in offer homologation making necessary a focus on service attributes

---

C. Liberati (✉)
Department of Economics, Statistics and Management, University of Milano-Bicocca, P.zza Ateneo Nuovo n.1, 20126 Milan, Italy
e-mail: caterina.liberati@unimib.it

P. Mariani
Department of Economics, Statistics and Management, University of Milano-Bicocca, via Bicocca degli Arcimboldi, n.8, 20126 Milan, Italy
e-mail: paolo.mariani@unimib.it

P. Giudici et al. (eds.), *Statistical Models for Data Analysis*, Studies in Classification, Data Analysis, and Knowledge Organization, DOI 10.1007/978-3-319-00032-9_22,

differentiation. Such scenario leads banks to employ intelligent systems to monitor their own clients in order to build and preserve a robust relationships with them. That generated an operative improvement in terms of efficiency and economic returns. However, as it is well known, the bank-consumer is an ever-changing relationship due to both environment and actors evolution. Therefore the core of our contribution focuses on analyzing the evolution of customer satisfaction and track patterns of customer evaluations related to bank service features. The tested hypothesis is the loss of bank retail services competitiveness, probably due to a drop of customer satisfaction.

2 Methodological Framework

A growing number of banks are directing their strategies towards customer satisfaction. In fact, researches have demonstrated, also by longitudinal examinations, that customer satisfaction serves as a link to critical consumer behaviors, such as cross-buying of financial services, positive word-of-mouth, willingness to pay a premium-price, and tendency to the relationship (Winstanley 1997; Bernhardt et al. 2000; Winkler and Schwaiger 2004).

Customer satisfaction data sets can be multidimensional and have a complex structure: especially when they are collected as sets (tables) of objects and variables obtained under different sampling periods (as in our case of study). Dynamic multivariate techniques allow the analysis of complex data structures in order to study a given instance phenomenon in both a structural (fixing base relationships among interesting objects (variables)) and a dynamic way (to identify change and development of in accordance to the occasions referred to). When a sufficiently long term series is not available a Multiway Factor Analysis (MFA) turns to be a suitable tool for the study of variable dynamics over various time periods (Tucker 1966; Kroonenberg 1993; Kiers 1989).

MFA main idea is to compare different data tables (matrices) obtained under various experimental conditions, but containing the same number of rows and/or columns. By analogy to N-way methods, the three-way data set is denoted by X with dimensions n, p and k, corresponding to the number of rows (individuals), columns (variables) and tables (occasions), respectively (Rizzi and Vichi 1995). Thus, an element of X is x_{ijh}, where $i = 1, \ldots, n$, $j = 1, \ldots, p$ and $h = 1, \ldots, k$. Following Escofier and Pagès (1994) we built a common factorial space, the "compromise space", in which the elements are represented according to their average configuration relative to data volume as a whole. This space is obtained by means of a Principal Component Analysis of the unfolded matrix X, solving the following the minimum equation:

$$\min ||X - A\Lambda B'||^2 \tag{1}$$

where Λ is a matrix $(m \times m)$ of eigenvalues, A is a matrix $(m \times m)$ of eigenvectors of the quadratic form $X'X$ and B is a matrix $(n \times n)$ of eigenvectors of the quadratic form XX'. The compromise plan (composed by the first ad the second factor axes) is the space spanned by a linear combination of three factor analysis. Such plan is the mean of the covariance matrices as in two-way PCA of unfolded matrix. Distribution of the subjects belonging to different occasions can be visualized in the space spanned by the principal components where the center is located according with the volume of the n objects observed in $k = 3$ occasions. Such distribution could be also explored via graphical analysis of subjects trajectories which consists in drawing instances route paths composing adjacent vectors in order to highlight the evolution, across the three waves, of the subjects position respect to the compromise plane (Carlier 1986; D'Urso 2000; D'Urso and Vichi 1998; Coppi and D'Urso 2001).

3 Weighted Factor Analysis

As we underlined in Sect. 2 Multiway Factor Analysis is a suitable technique for summarizing the variability of a complex phenomenon by highlighting both similarities/dissimilarities among the occasions considered and the main components of the average behavior in the time interval chosen. Visual inspection of the objects plotted onto the principal space derived via PCA on the unfolded X matrix (compromise plan) allows to drawn subjects routes covered in the three waves. Such solution has to be reviewed in the light of some limitations and geometrical properties of the orthogonal projection. As it is well known mapping data from a high dimensional space to a lower dimensional space (compromise plane) might cause new scatter point configuration. Thus, in such new plot, we obtain a representation where points do not have the same distance from the factor plan itself. According with this remark, in this work we propose, as new system of weights, the usage of the quality of representation of each point (Fig. 1) defined as follows:

$$QR(i) = \cos^2_\theta(x_i, u_\alpha) = \frac{c^2_\alpha(i)}{\sum_{\alpha=1}^{p} c^2_\alpha} \tag{2}$$

which is a measure of closeness of a point to the axe itself.

Thus, in order to adjust the multiway solution according with relevance of the point projection we re-weight each coordinates of the compromise plane with a linear combination in a fashion as follows:

$$c_{QR\alpha}(i) = QR_\alpha(i) \cdot c_\alpha(i) \tag{3}$$

Such transformation produces a rescaling in terms of value but not in terms of sign and it enhances further the representation of the points, with respect to the other rotations present in literature (VARIMAX) (Bolasco 1999).

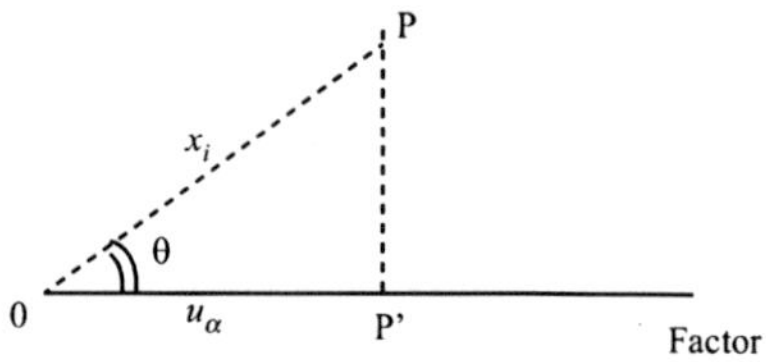

Fig. 1 Quality of representation

4 An Italian Bank Case

The managing board of an Italian bank, with a distribution network throughout the country, wanted to analyze its competitive positioning in retail services (ABI 2009). The starting point was a loss in the market share in some regions and an average customer lifetime shorter than before especially in some clusters: the loss of competitiveness was due to service level. Therefore a survey has been conducted, sampling 27.000 retail customers who effected at least 5 retail requests, conjoint with other contact points of the bank (call center, e-banking,..) within a year. The questionnaire was framed according to SERVQUAL model (Berry et al. 1985, 1993; Oliver 1977), therefore, with five dimensions to analyze perceived quality and expectation of the banking service. A variable number of items catches the quality dimensions (Tangibles, reliability, responsiveness, assurance, empathy). All the scores are on a Likert scale 1 to 10. There are 16 questions (type A), measuring expectations/importance of items, and 16 questions (type B), catching evaluations on perceived quality of a particular item. One final question aims summarizing the entire banking service satisfaction. The same questionnaire has been applied to the same sample for three waves: this is a perfect case for constructing a dynamic model to quantify variable changes across the three different occasions. A descriptive analysis of the sample shows a homogeneous distribution across different ages, sex, instruction levels and profession segments. This reflects the Italian banking population: more than 60% is between 26 and 55 years old; the sample is equally distributed between the two sexes and shows a medium low level of instruction. The sample has been analyzed across 9 different professional clusters: entrepreneurs, managers, employees, workers, farmers, pensioners, housewives, student, others. The sample is well distributed across the different professional segments employees 24%, pensioners 22%, housewives 14%. The customer satisfaction was analyzed according to three different indicators to avoid dependency on the metrics used. For the three indicators, satisfaction scores are high (above 7/10 in the three cases) with the same trend across the waves. There is an increase in satisfaction from the first to the second wave and a decrease in the third wave. A gap analysis between questions A and B shows that expectations/item importance are always higher than the perception of that item. A dynamic analysis will show evidence of a gap decrease between the first and third wave.

5 Results: Professional Clusters Trajectories and Satisfaction Decomposition

In this section, we evaluate the effectiveness of our Weighted Factor Analysis and we present the results on professional clusters trajectories and distances.

The first step of the study consists in estimating, the compromise matrix in order to represent the average position of the professional clusters with respect to the selected variables (regardless of the different occasions). The PCA results are very sturdy: the first two components explain about 84% of the total variance and in particular the first one explains 54% alone. The compromise matrix based on the first eigenvalue is robust and can provide a realistic view of the evolution of variables and individual positioning in the time horizon considered. Also KMO index (0,8) and Bartlett test (1.312,7; sig. .000) show the quality of the factorial model created. Representing the compromise matrix and the occasion-points of the k matrices on the factorial plan created by the first two components, a clear polarization of the variables on the two axis showed. The first component is characterized by the positive pole B variables (valuations on the different aspects of the service) and from the negative pole by A variables; the second component shows the contrary. In accordance to these results, the first axis is called service evaluation and the second service expectations. By projecting the unities on the factorial axis of the compromise phase, movements with respect to the different waves can be analyzed (Fig. 2).

We chose to monitor opposite professional clusters: Entrepreneurs, Workers, Students and Pensioners in order to compare and to contrast if and how satisfaction effects them in the 3 waves. Figure 2 illustrates evolutions of such professional categories computed according with the simple multiway coordinates (red trajectory) and the weighted multiway coordinates (blue trajectory). Visual inspection of the plots highlights a drop of satisfaction in the third wave using both coordinates system. In particular, the first line-segment (passage from 1st to 2nd wave) shows an increasing slope (Fig. 2 panel (a) and (c)) or constant one (Fig. 2 panel (b) and (d)) then the second passage (from the 2nd to the 3rd wave) evidences a decreasing trend for all the categories. Moreover all the category paths computed with the weighted coordinates remain similar to the original one except for the Students cluster which shows a new point configuration due to the rescale operates by the Quality of Representation: the compromise plane of the MFA, in fact, does not take into account the points inertia which are better represented by the other factor axes (Table 1). That produces a remodeling and a compression of such route.

Thus, it highlights the effectiveness of such transformation which retains the original value of the factor coordinates only if the quality of representation of a point reaches its maximum value ($QR(i) = 1$) otherwise it reduces it according with its closeness to the factor itself. Obviously the weights system might cause a rotation of the factor axes that could not be orthogonal any more, but if the percentage of

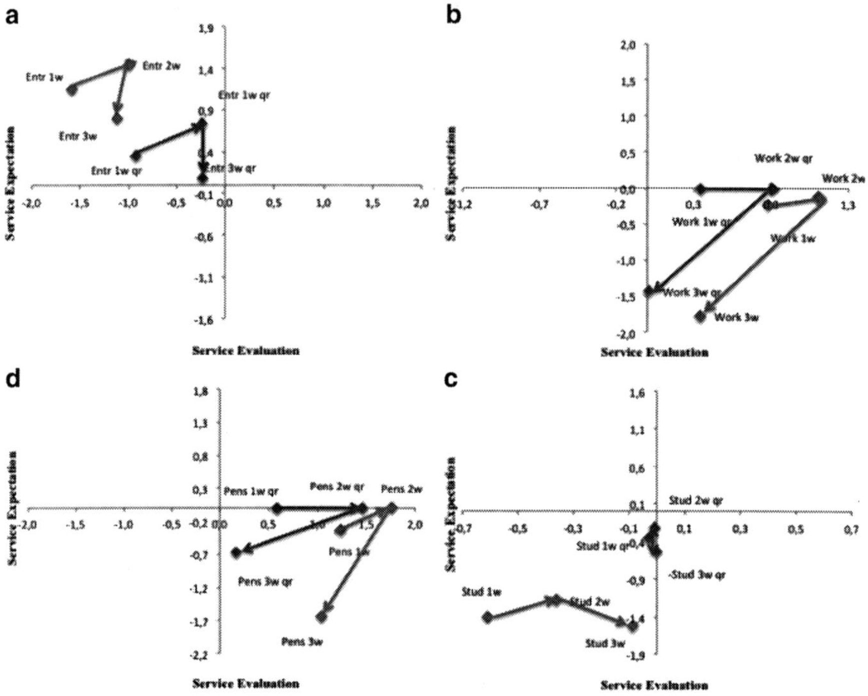

Fig. 2 Professional Cluster Trajectories: panel (**a**) Entrepreneurs, panel (**b**) Workers, panel (**c**) Students, panel (**d**) Pensioners

the inertia explained by the compromise plane is high we are confident that such rescaling does not effect the orthogonality.[1]

Graphical analysis is a suitable tool to describe and estimate the evolutions of the clusters: following our approach we interpret trajectories as a proxi of the customer sensibility to the bank marketing stimuli, therefore more route is covered more strong is the reaction of a professional category. Such approach can drive the management to recognize which cluster has to be monitored more closely.

In order to deepen if the overall satisfaction (or dissatisfaction) is mainly composed by positive (negative) perception of service performances or by the positive (negative) considerations about service expectation, we decompose the customer paths analyzing the shifts on each single axe in terms of value and in terms of sign. Tables 2 and 3 collect category movements which have been measured via the following equation:

$$shift(i) = c_{f_\alpha k}(i) - c_{f_\alpha k-1}(i) \tag{4}$$

[1]The significance of the correlation value between Weighted Multiway Factor Axes WMFA f1 and WMFAf2 has been computed via Pearson test ($\rho = -0.0163$ p-value $= 0.9356$) and Spearman test ($\rho = 0.0537$ p-value $= 0.7898$). It turned to be not zero but its value was not significative.

Table 1 Professional clusters quality of representation and multiway coordinates

Category per wave	QR_1	QR_2	MFA f1	MFA f2	WMFA f1	WMFA f2
Entrep 1w	0.5809	0.3055	−1.5842	1.1488	−0.9203	0.3510
Entrep 2w	0.2356	0.5065	−0.9914	1.4535	−0.2336	0.7362
Entrep 3w	0.2048	0.1061	−1.1128	0.8011	−0.2279	0.0850
Work 1w	0.4421	0.0412	0.7777	−0.2375	0.3438	−0.0098
Work 2w	0.7313	0.0064	1.1016	−0.1027	0.8056	−0.0007
Work 3w	0.0302	0.8083	0.3422	−1.7701	0.0103	−1.4307
Pens 1w	0.4763	0.0329	1.2343	−0.3247	0.5879	−0.0107
Pens 2w	0.8265	0.0000	1.7650	0.0054	1.4588	0.0000
Pens 3w	0.1629	0.4072	1.0384	−1.6418	0.1691	−0.6686
Stud 1w	0.0476	0.2533	−0.6101	−1.4068	−0.0291	−0.3563
Stud 2w	0.0176	0.1833	−0.3630	−1.1699	−0.0064	−0.2144
Stud 3w	0.0012	0.3542	−0.0883	−1.5184	−0.0001	−0.5379

Table 2 Net routes on factor plan

Professional	Routes on MFA f1			Routes on MFA f2		
Categories	1st w-2nd w	2nd w-3rd w	Net Route	1st w-2nd w	2nd w-3rd w	Net Route
Entrepreneurs	0.593	−0.122	0.471	0.305	−0.652	−0.348
Workers	0.324	−0.759	−0.435	0.135	−1.667	−1.533
Students	0.247	0.275	0.522	0.237	−0.349	−0.112
Pensioners	0.531	−0.727	−0.196	0.330	−1.648	−1.317

Table 3 Net routes on weighted factor plan

Professional	Routes on WMFA f1			Routes on WMFA f2		
Categories	1st w-2nd w	2nd w-3rd w	Net Route	1st w-2nd w	2nd w-3rd w	Net Route
Entrepreneurs	0.687	0.006	0.692	0.385	−0.651	−0.266
Workers	0.462	−0.795	−0.334	0.009	−1.430	−1.421
Students	0.023	0.006	0.029	0.142	−0.324	−0.182
Pensioners	0.871	−1.290	−0.419	0.011	−0.669	−0.658

where i is the index set of the professional category, c_{f_α} is the coordinate of the factor axe α relatives to the kth wave.

Again shifts have been computed employing the two system of coordinates. According with the original factors (Table 2) Entrepreneurs (0,471) and Students (0.522) are the only categories which show positive consideration about service performance, then Workers (−0.435) and Pensioners (−0.196) present a negative perception of the same service. If we use the new system of coordinates (Table 3) we obtain again a positive service evaluation for Entrepreneurs (0,692) and Students (0.029) but with a new ranking of the values which highlights a decrease of satisfaction for the "young people". As we underlined previously such path compression is due to quality of representation which is very poor for the Students in the compromise plan. For what concern service expectation we notice that all the categories considered show a negative net route paths using both coordinates system, also values ranking is invariant. Such decomposition highlights that the drop of satisfaction mainly intense for Pensioner and Workers, occurs in the third wave and weights dramatically on the final clusters expectation.

6 Conclusions

In this paper we presented a novel approach to perform a dynamic customer satisfaction analysis based on a three way factors analysis. We also introduced a clever system of weights which exploits the quality of representation of each point to adjust the original factor solution and to obtain more reliable trajectories. The empirical study presented aims to offer some ideas on the main trends in customer evaluation and expectation evolution relative to a set of bank service attributes. A systematic analysis of management action-customers reaction provides the effectiveness of every decision to increase satisfaction, loyalty and consequently bank profitability. By such continual analysis management can check the correct direction of action, and change it according to customer desiderata evolution. Therefore satisfaction decomposition performed is able to measure the effects of satisfaction/dissatisfaction in terms of customer instability. A valuable extension of this work would be to derive a synthetic index which takes into account both paths and shifting direction for operative usage.

References

ABI. (2009). *Dimensione cliente 2009*. Roma: Bancaria Editrice.

Bernhardt, K. L., Donthu, N., & Kennett, P. A. (2000). A longitudinal analysis of satisfaction and profitability. *Journal of Business Research*, *47*, 161–171.

Parasuram, A., Zeithaml, V. A., & Berry, L. L. (1985). A conceptual model of service quality and its implications for future research. *The Journal of Marketing*, *49*.

Parasuraman, A., Berry, L. L. & Zeithaml, V. A. (1993). Research Note: More on Improving Service Quality Measurement. *Journal of Retailing*, *69*.

Bolasco, S. (1999). *Analisi multidimensionale dei dati*. Roma: Carocci.

Escofier, B., & Pagès, J. (1994). Multiple factor analysis (AFMULT package). *Computational Statistics & Data Analysis*, *18*, 121–140.

Carlier, A. (1986). *Factor analysis of evolution and cluster methods on trajectories*. COMPSTAT'86. Heidelberg Wien: Physica-Verlag.

Coppi, R., & D'Urso, P. (2001). The geometric approach to the comparison of multivariate time trajectories. In S. Borra, R. Rocci, M. Vichi, & M. Schader (Eds.) *Advances in data science and classification*. Heidelberg: Springer.

D'Urso, P. (2000). Dissimilarity measures for time trajectories. *Journal of the Italian Statistical Society*, *9*, 53–83.

D'Urso, P., & Vichi, M. (1998). Dissimilarities between trajectories of a three-way longitudinal data set. In A. Rizzi, M. Vichi, & H. H. Bock (Eds.) *Advances in data science and classification*. New York: Springer.

Kiers, H. (1989). *Three way method for the analysis of qualitative and quantitive two way data*. Leiden: DSWO Press.

Kroonenberg, P. M. (1993). *Principal component analysis of three-mode data*. Leiden: DSWO Press.

Oliver, R. L. (1977). Effect of expectation and disconfirmation on post-exposure product evaluations: an alternative interpretation. *Journal of Applied Psychology, 4*, 480–486.

Rizzi, A., & Vichi, M. (1995). Representation, synthesis, variability and data preprocessing of three-way data set. *Computational Statistics & Data Analysis, 19*, 203–222.

Tucker, R. L. (1966). Some mathematical notes on three-mode factor analysis. *Psychometrika, 31*, 279–311.

Winkler, G., & Schwaiger, M. S. (2004). Is customer satisfaction driving revenue - a longitudinal analysis with evidence from the banking industry. *Journal of Business & Economics Research, 2*, 11–22.

Winstanley, M. (1997). What drives customer satisfaction in commercial banking. *Commercial Lending Review, 12*, 36–42.

Measuring the Success Factors of a Website: Statistical Methods and an Application to a "Web District"

Eleonora Lorenzini and Paola Cerchiello

Abstract In this paper we propose a statistical methodology to address the issue of measuring success factors of an ecommerce application, and in particular of a regional e-marketplace, using as a measurement framework based on the customers' satisfaction. In the first part of the paper, two different ranking methods have been compared in order to identify the more appropriate tool to analyse the opinions expressed by the visitors: a novel non parametric index, named the Stochastic dominance index, built on the basis of the cumulative distribution function alone, and a qualitative ranking based on the median and on the Leti Index. SDI has resulted to be more convenient for comparison purposes and, according to this measurement tool, the higher satisfaction has been expressed for the quality of the products. Then, a logistic regression has been performed to understand the impact of the different satisfaction factors on the overall satisfaction. The empirical evidence confirms the literature on the importance of the different success factors, showing that Website user friendliness and Information about purchase mechanisms have the major impact on the overall satisfaction.

1 Introduction and Literature Review

Directly measuring the success of an e-commerce (EC) application has been found to be impractical and perhaps impossible (Galletta and Lederer 1989). However, EC application success can be measured using several frameworks, one being the customers' perception of utility and satisfaction (Garrity and Sanders 1998; Cho 1999; Lu 2001). An increased awareness of the importance of customer satisfaction

E. Lorenzini (✉) • P. Cerchiello
Department of Economics and Management, University of Pavia, Lombardy, Italy
e-mail: eleonora.lorenzini@unipv.it; paola.cerchiello@unipv.it

P. Giudici et al. (eds.), *Statistical Models for Data Analysis*, Studies in Classification, Data Analysis, and Knowledge Organization, DOI 10.1007/978-3-319-00032-9_23,

issue has prompted the research community to explore how to measure and model customers satisfaction and their preferences (Kurniawan 2000).

Among the success factors identified in the literature, Trepper (2000) found that convenient site design and financial security had a significant effect on customer assessment for EC applications.

Lu (2003) found that customers would not pay for products or services over the web if financial information was not clear, or could not be transmitted securely. This is why businesses and web developers should actively seek ways to improve information and service quality provided through websites, and focus on the way the customer uses a website. Liu and Arnett (2000) surveyed Web-masters from Fortune 1,000 companies and found four factors that are critical to success: information and service quality, system use, playfulness, and system design quality.

In this paper the results of a web survey are used in order to understand which factors affect the overall satisfaction of the users of a particular EC application: the regional e-marketplace Store Valtellina. Valtellina is a mountain area in Lombardy Region, known both for its eno-gastronomic typical products but also for winter tourism and spas. In 2010 some local producers, supported by a University, a bank, a logistic company and other territorial agents, promoted a project of e-commerce of the territorial products.

The peculiarity of this portal is that it can be considered as a "web district." In fact, it gathers about 40 producers belonging to the Valtellina area that became partners in this e-commerce experience with the aim of proposing in an integrated manner the differentiated supply of quality products and services of the area, ranging from eno-gastronomic products and handcrafts, to tourism services.

2 Research Design and Methodology

The data for this study were gathered by means of a questionnaire emailed to all the registered users of the website after 3 months of activity, at the beginning of January 2011. The redemption rate was around 20 %, that is 65 visitors.

The interviewed expressed their opinions about the Store with reference to the quality of the products, the website-related attributes and the overall satisfaction (see Table 1 for a list of the factors).

A 5-point Likert scale was used, where A means high satisfaction and E represents low satisfaction.

In order to understand the relative importance of the factors, a double level of inquiry has been used. Firstly, two different ranking methods have been compared: the Stochastic Dominance Index (SDI) and a ranking method based on the median and the Normalised Leti Index. Secondly, logistic regression is applied on the data in order to analyse how the evaluation of the different attributes of the site impacts on the total satisfaction of the visitors.

Table 1 Ranking produced by ML index, based on the NLI, for the satisfaction factors of the store

Factors	Median	NLI	Ranking_ML
Info about products	B	0.27	B
Info about purchase mechanism	B	0.31	B
Website graphics	B	0.34	BB
Promotions	B	0.45	BB
Customer care	B	0.45	BB
Terms of payement	B	0.49	BB
Shipping time	B	0.58	BB
Website user friendliness	B	0.41	BB
Quality of the products	B	0.42	BB
Info about producers	B	0.34	BB
Overall satisfaction	B	0.32	B

3 Comparison of Different Ranking Methods

In the first part of the paper, two different ranking methods have been compared in order to identify the more appropriate tool to analyse the opinions expressed by the visitors to the Store with reference to the quality of the products, the website-related attributes and the overall satisfaction. We can apply to the data a novel non parametric index, built on the basis of the cumulative distribution function alone, which embodies the essential information for ordinal data (for more details see Cerchiello and Giudici 2012). We propose to consider the sum of the values of the cumulative distribution function according to the stochastic dominance approach to model selection, that we now describe. On the basis of the cumulative distribution function, a summary index, that we name SDI (Stochastic dominance index) can be calculated as follows:

$$SDI = \sum_{i=1}^{J} F_i$$

where F_i is the cumulative distribution function and J is the number of classes. SDI has been chosen due to the good performance of this index when used for comparative purpose. In addition, a qualitative ranking (henceforth ML) has been provided based on the median and on the Leti Index (Leti 1983; Cerchiello et al. 2010). Leti index is defined as:

$$L = 2\sum_{i=1}^{K-1} F_i(1 - F_i)$$

The median is useful to select the evaluation, while the Leti index is useful to express the level of consensus on the selected evaluation.

While the SDI index is a second order stochastic dominance measure, based on the arithmetic mean, ideal for comparison purposes, ML index aims at measuring heterogeneity between statistical units. We believe that SDI is more convenient, because such index maintains the ordinal nature of the data, it is not based on parametric assumptions and its construction is simple to communicate and to interpret.

4 Evidence from Ranking Comparison

Now we show the main results obtained from the application of the two indexes explained in Sect. 3. With regard to ML index, the median has resulted to correspond to evaluation B for each attribute. The normalized version of the Leti index (NLI) has then been used as follows. In the case of maximum homogeneity of the answers ($0 \leq NLI \leq 0.33$) the assigned value is B; in the case of intermediate heterogeneity ($0.33 < NLI \leq 0.66$) the assigned value is BB; in the case of maximum heterogeneity ($0.66 < NLI \leq 1.00$) the assigned value is BBB.

Both indexes have been standardised, the SDI dividing by the number of classes J and the ML by J-1.

Tables 1 and 2 summarise the values obtained by the application of the two methods.

We decided to deepen the analysis on the data by employing the results obtained by means of SDI index, in order to get more information about the customers' satisfaction. Table 2, in fact, shows that higher satisfaction has been expressed for the quality of the products than for website-related attributes, but a closer examination of the data can bring to an understanding of whether there are differences for product category or type of consumer.

In order to understand if the consumers perceive a different quality according to the category of product they are evaluating, Table 3 shows the SDI calculated on the quality differentiated according to the product categories.

It emerges that the perceived quality is high and rather homogeneous across the categories, which means that the Valtellina's products share a homogeneous collective reputation.

Furthermore, Table 4 shows that the SDI for the quality does not differ considerably between the buyers group and the visitors group that have not decided to purchase from the Store, meaning that the perceived quality for the buyers is comparable (precisely slightly higher) to the reputation for the non buyers.

The SDI has been calculated for each combination attribute and product category as well. As Table 5 reports, all categories share a similar average evaluation, although some differences exist for the different attributes and categories. It is interesting to notice, for instance, that visitors interested in purchasing train tickets and ski passes have found higher quality in the information about purchase and

Table 2 SDI for the satisfaction factors of the store

Factors	SDI
Info about products	0.76
Info about purchase mechanism	0.79
Website graphics	0.74
Promotions	0.70
Customer care	0.74
Terms of payement	0.79
Shipping time	0.74
Website user friendliness	0.82
Quality of the products	0.84
Info about producers	0.74
Overall satisfaction	0.80

Table 3 SDI for the quality of the products for product category

Product category	SDI
Handcrafts	0.90
Christmas packages	0.81
Enogastronomy	0.86
Skis	0.90
Ski passes	0.88
Train	0.84
Wine	0.85
Total	0.85

Table 4 SDI for the quality of the products: buyers and non buyers

	Purchase	No purchase
SDI quality of the products	0.85	0.82

payment mechanism with respect to the visitors interested in buying handcrafts, enogastronomic products and skis. On the contrary, visitors interested in buying handcrafts have evaluated the information about producers and shipping time higher than the visitors and buyers of the other product categories.

Finally, the SDI has been calculated according to the amount of money spent for the purchase. It is evident that those who have spent more, are more satisfied both in general and for the single satisfaction factors, with the exception of the information about the producers (Table 6).

Table 5 SDI for product category and satisfaction factors

	Handcrafts	Enogastronomic products	Skis	Train and skipass
Info about products	0.73	0.74	0.80	0.80
Info about purchase	0.80	0.80	0.80	0.87
Website graphics	0.73	0.74	0.80	0.71
Promotion sales	0.60	0.68	0.80	0.78
Customer care	0.73	0.79	0.70	0.76
Terms of payment	0.73	0.81	0.80	0.87
Shipping time	0.87	0.78	0.60	0.76
Website user friendliness	0.87	0.85	0.80	0.82
Quality of the products	0.87	0.86	0.90	0.80
Info about producers	0.87	0.72	0.80	0.67
Overall satisfaction	0.87	0.82	0.80	0.87
Average	0.79	0.78	0.78	0.79

Table 6 SDI for attributes and value of the purchase in euro

	0	<50	50<x<100	>=100
Info about products	0.77	0.74	0.75	0.78
Info about purchase mechanism	0.77	0.80	0.83	0.84
Website graphics	0.77	0.70	0.75	0.82
Promotion sales	0.68	0.65	0.70	0.80
Customer care	0.69	0.76	0.78	0.80
Terms of payement	0.72	0.81	0.83	0.84
Shipping time	0.67	0.77	0.73	0.80
Website user friendliness	0.77	0.83	0.85	0.85
Quality of the products	0.82	0.84	0.83	0.87
Info about producers	0.77	0.73	0.70	0.73
Overall satisfaction	0.73	0.82	0.85	0.85

5 Importance of the Different Satisfaction Factors

A logistic regression has been performed to understand the impact of the different satisfaction factors on the overall satisfaction. The independent variables are the different satisfaction factors (as reported in Table 6), while the "Overall satisfaction" is taken as the dependent variable. All variables have been transformed into binary variables taking the value of 1 for good and high level of satisfaction and 0 otherwise.

The model is the result of variable selection, in particular the comparison of the above with the null model gives a p-value of 3.33e-06.

The only significant variables have resulted to be Website user friendliness and Info about purchase mechanisms. Despite the higher satisfaction for quality of the products, thus, website related attributes have the major impact on overall satisfaction.

Table 7 Results from the logistic analysis on the overall satisfaction using software R

Factors	Estimate	SE	z value	Pr(>|z|)
Constant	−2.601	1.346	−1.932	0.0533
Website user friendliness	4.169	1.177	3.543	0.0004
Info about purchase mechanism	2.497	1.276	1.956	0.0504

These findings confirm the literature (Trepper 2000; Lu 2003) that identified in convenient site design and financial security the critical success factors for an e-commerce application.

6 Conclusions

In this paper we propose a statistical methodology to address the issue of measuring success factors of a website. A research project we are involved in, gave us the chance to monitor a novel website built to sell products of high quality from an Italian region. We could investigate the clients of the website, by proposing them a simple questionnaire made of few questions on the key aspects related to the usability of a website and to the quality of the available products. We proposed to use a descriptive index SDI evaluated comparatively with another descriptive index based on the Leti measure. Those indexes allow us to create rankings on the items of the questionnaire, that is the 11 factors regarding the quality of the products, the website-related attributes and the overall satisfaction. In particular we pay our attention on the SDI index that is a second order stochastic dominance measure, based on the arithmetic mean ideal for comparison purposes. Moreover SDI is more convenient, because the index maintains the ordinal nature of the data, it is not based on parametric assumptions and its construction is simple to communicate and to interpret.

The empirical evidence shows that higher satisfaction has been expressed for the quality of the products than for website-related attributes. Moreover, deepening the analysis by considering categories of available products, we can conclude that perceived quality is high and rather homogeneous across the categories, which means that the Valtellina's products share a homogeneous collective reputation.

Finally a logistic regression has been performed to understand the impact of the different factors on the overall satisfaction. What emerges is that despite the high satisfaction for quality of the products, website-related attributes have the major impact on overall satisfaction. This confirms the literature that identified in convenient site design and financial security the critical success factors for an e-commerce application.

Further investigations will require a more consistent set of data that could be monitored along 1 year of activity of the website, allowing us to track not only new clients but also frequent ones.

Acknowledgments The authors are grateful to prof. Paolo Giudici for scientific supervision and to prof. Silvia Biffignandi for useful comments. Financial support by the Grant *Industria 2015* for the project *@bilita—Nuove Tecnologie per il Made in Italy* is acknowledged.

References

Cerchiello, P., Dequarti, E., Giudici, P., & Magni, C. (2010). Scorecard models to evaluate perceived quality of academic teaching. *In Statistica and Applicazioni, 2*, 145–156.

Cerchiello, P., & Giudici, P. (2012). An integrated statistical model to measure academic teaching quality. *Open Journal of Statistics*, 2(5), 491–497.

Cho, S. (1999). Customer-focused internet commerce at cisco systems. *IEEE Communications Magazine, 37*(9), 61–63.

Galletta, D. F., & Lederer, A. L. (1989). Some cautions on the measurement of user information satisfaction. *Decision Sciences, 20*, 419–438.

Garrity, E. J., & Sanders, G. L. (1998). Introduction to information systems success measurement. In E. J. Garrity & G. L. Sanders (Eds.), *Information systems success measurement*. Series In Information Technology Management: IDEA Group Publishing.

Kurniawan SH. (2000). Modeling online retailer customer preference and stickiness: A mediated structural equation model. In *Fourth Pacific Asia Conference on Information Systems*, Hong Kong, China, June 2000, pp. 238–252.

Leti, G. (1983). *Statistica Descrittiva*. Bologna: Il Mulino.

Liu, C., & Arnett, K. P. (2000). Exploring the factors associated with Website success in the context of electronic commerce. *Information Management, 38*, 23–33.

Lu J. (2001). Assessing web-based electronic commerce applications with customer satisfaction: An exploratory study. In *International Telecommunication Society's Asia-Indian Ocean Regional Conference, Telecommunications and E-Commerce*, Perth, Western Australia, July 2001, pp. 132–144.

Lu, J. (2003). A model for evaluating e-commerce based on cost/benefit and customer satisfaction. *Information Systems Frontiers, 5*(3), 265–277.

Trepper, C. H. (2000). *E-commerce strategies*. Washington: Microsoft.

Component Analysis for Structural Equation Models with Concomitant Indicators

Pietro Giorgio Lovaglio and Giorgio Vittadini

Abstract A new approach to structural equation modelling based on so-called Extended Redundancy Analysis has been recently proposed in literature, enhanced with the added characteristic of generalizing Redundancy Analysis and Reduced-Rank Regression models for more than two blocks. However, in presence of direct effects linking exogenous and endogenous variables, the latent composite scores are estimated by ignoring the presence of the specified direct effects. In this paper, we extend Extended Redundancy Analysis, permitting us to specify and fit a variety of relationships among latent composites and endogenous variables. In particular, covariates are allowed to affect endogenous indicators indirectly through the latent composites and/or directly.

1 Introduction

For fitting Structural Equation Models, despite a number of benefits in Covariance Structure Analysis (CSA) (Jôreskog 1970) and Partial Least Squares (PLS) (Wold 1982), both techniques have some problems. In CSA the occurrence of improper solutions is most likely to interfere with meaningful analysis (Kiers et al. 1996), whereas in PLS the lack of a global optimization criterion (PLS solutions are not optimal in an overall fit) seems to render its use limited (McDonald 1996). Moreover, in applications, often the underlying theory may specify the presence of exogenous covariates that may enter the causal model. Hence, a more comprehensive system would also take into account the possible exogenous factors that may have a causal impact on both observed endogenous variables as well as latent composites. Specifically, exogenous covariates that do not strictly

P.G. Lovaglio (✉) • G. Vittadini
Department of Quantitative Methods, University of Bicocca-Milan, Milan, Italy
e-mail: piergiorgio.lovaglio@unimib.it; giorgio.vittadini@unimib.it

P. Giudici et al. (eds.), *Statistical Models for Data Analysis*, Studies in Classification, Data Analysis, and Knowledge Organization, DOI 10.1007/978-3-319-00032-9_24,

belong to the formative blocks of latent composites (LC), but may have a causal impact on observed endogenous variables and onto latent composites too are called "concomitant indicators". Within the Component Analysis (CA) framework (Millsap and Meredith 1988; Schônemann and Steiger 1976), a few attempts have been made to extend Redundancy Analysis (RA) (van den Wollenberg 1977) and Reduced-Rank Regression model (RRR) (Izenman 1975) to more than two sets of variables. However, they are limited to relationships among three sets of variables (Davies and Tso 1982; Reinsel and Velu 1998) as well as being limited to particular types of models (Bougeard et al. 2008). A new approach to Structural Equation Modeling based on so-called Extended Redundancy Analysis (ERA) (Takane and Hwang 2005), which generalizes RA and RRR for more than two blocks, has been recently proposed in literature. However, in ERA the LC scores are estimated by ignoring the presence of direct effects linking concomitant indicators and the endogenous variables block. In this paper, we propose a new method, called Generalized Redundancy Analysis (GRA) which generalizes ERA and thus, RRR and RA. The proposed method allows fitting diverse complex relationships among variables, including direct effects and concomitant indicators as well.

2 The Extended Redundancy Analysis Model

The ERA model can generally be stated as follows: let $\mathbf{Y}$ denote an n by p matrix consisting of p observed endogenous variables on n subjects. Let $\mathbf{X}$ denote an n by q matrix consisting of q observed exogenous (formative) variables. Assume that all the variables in $\mathbf{X}$ and $\mathbf{Y}$ are standardized. The ERA model can be expressed as $\mathbf{Y} = \mathbf{XWA}' + \mathbf{E} = \mathbf{FA}' + \mathbf{E}$ under the constraint that rank $(\mathbf{WA}') = D \leq \min(q, p)$, where $\mathbf{W}$ denotes a q by D matrix of composite weights, $\mathbf{A}'$ denotes a D by p matrix of composite loadings, $\mathbf{E}$ denotes an n by p matrix of residuals, $\mathbf{F}$ (=$\mathbf{XW}$) denotes an n by D matrix of latent composite scores. For identification, $\mathbf{F}$ is restricted to be $\text{diag}(\mathbf{F}'\mathbf{F}) = \mathbf{I}_\mathbf{D}$. To allow for concomitant indicators, we distinguish two cases: the presence of K concomitant indicators in the model that have a causal impact (by means of the K by p matrix of regression coefficients $\mathbf{A_Y}'$) on observed endogenous variables (Case 1) and onto latent composites, too (Case 2). For Case 2, LC scores are measured by a linear combination of strictly formative ($\mathbf{X}$) and concomitant ($\mathbf{T}$) indicators:

$$\mathbf{F} = \mathbf{XW} + \mathbf{TW_T} \tag{1}$$

where $\mathbf{W_T}$ is a K by D matrix of weights for $\mathbf{T}$. Direct effects could be accommodated ERA (2a) by an equivalent ERA specification (2b) involving three regressors ($\mathbf{X}^\S$) weights ($\mathbf{W}^\S$) and parameters ($\mathbf{A}^\S$) matrices:

$$\mathbf{Y} = \mathbf{TA_Y}' + \mathbf{FA}' + \mathbf{E} \tag{2a}$$

$$\mathbf{Y} = \mathbf{X}^\S\mathbf{W}^\S\mathbf{A}^{\S'} + \mathbf{E} \tag{2b}$$

under the hypothesis that rank($\mathbf{WA}'$) $= D \leq \min(q_1, q_2, p)$ and where $\mathbf{X}^{\$}_{\mathrm{nx(K+q)}} = [\mathbf{T}_{\mathrm{nxK}}|\mathbf{X}_{\mathrm{nxq}}]$, $\mathbf{W}^{\$}_{\mathrm{(K+q)x(K+D)}} = [\mathbf{P}_{\mathrm{KxK}}|\mathbf{W}_{\mathrm{qxD}}]$, $\mathbf{A}^{\$}_{\mathrm{px(K+D)}} = [\mathbf{A}_{\mathbf{Y}pxK}|\mathbf{A}_{pxD}]$, $q = q_1 + q_2$ and $\mathbf{P}$ is a matrix of fixed elements (zeros or unities), where unity in k-th column and a non zero coefficient in the j-th row (j = 1,...,p) of $\mathbf{A_Y}$ selects a causal link between the k-th concomitant indicator and the j-th endogenous indicator. For Case 2, where $\mathbf{F}$ is defined in (1), the associated model is equivalent to model (2b), in which $\mathbf{X}^{\$} = [\mathbf{T}|\mathbf{X}|\mathbf{T}]$, $\mathbf{W}^{\$} = [\mathbf{P}|\mathbf{W}|\mathbf{W_T}]$, $\mathbf{A}^{\$} = [\mathbf{A_Y}|\mathbf{A}|\mathbf{A}]$. Model (2b) allows specifying direct effects and concomitant indicators, but this approach has two noticeable limitations. Firstly, ERA estimates $\mathbf{W}$ in a model between $\mathbf{Y}$ and $\mathbf{X}$, ignoring the presence of concomitant indicators (particularly their direct effects on $\mathbf{Y}$). Secondly, in Case 2 the LC scores are estimated as linear combinations of both formative and concomitant indicators. However, since the indicators embedded in $\mathbf{T}$ (concomitant indicators) and $\mathbf{X}$ (strictly formative indicators) are typically correlated and have different causal effects on endogenous variables ($\mathbf{X}$ has an indirect effect mediated by LC on $\mathbf{Y}$, whereas $\mathbf{T}$ has both a direct effect and an indirect effect mediated by $\mathbf{F}$ on $\mathbf{Y}$) it is impossible to distinguish the separate contribution of two blocks to the determination of the LC scores. A more consistent approach would take into account all specified effects and to examine the separate contribution of strictly formative and concomitant indicators on the LC scores.

3 The GRA Model

For the more general Case 2, we explicitly specify in the model the block of concomitant indicators ($\mathbf{T}$) and we simultaneously estimate all the parameters of model (2a) by an iterative method. In order to separate the contribution of strictly formative and concomitant indicators, instead of measurement model (1) we adopt an equivalent specification, with orthogonal regressors' matrices ($\mathbf{T}^{\circ}$ and $\mathbf{X}$):

$$\mathbf{F} = \mathbf{X}\mathbf{W}^{\circ} + \mathbf{T}^{\circ}\mathbf{W_T} \tag{3}$$

where $\mathbf{T}^{\circ} = \mathbf{T} - \mathbf{X}\ \mathbf{X}^{+}\mathbf{T}$, $\mathbf{W}^{\circ} = \mathbf{W} + \mathbf{X}^{+}\mathbf{T}\ \mathbf{W_T}$ and $\mathbf{X}^{+}$ is the Moore-Penrose generalized inverse of $\mathbf{X}$ obtained by the Singular Value Decomposition (SVD) of $\mathbf{X}$. Substituting (3) into (2a) we obtain:

$$\mathbf{Y} = \left(\mathbf{X}\mathbf{W}^{\circ} + \mathbf{T}^{\circ}\ \mathbf{W_T}\right)\mathbf{A}' + \mathbf{T}\mathbf{A_Y}' + \mathbf{E} = \mathbf{X}\mathbf{W}^{\circ}\mathbf{A}' + \mathbf{T}^{\circ}\ \mathbf{W_T}\mathbf{A}' + \mathbf{T}\mathbf{A_Y}' + \mathbf{E} \tag{4}$$

As has been explained, $\mathbf{W}$, $\mathbf{W_T}$ and/or $\mathbf{A}$ may contain prescribed, fixed (zero) elements, depending on the specified model. Unfortunately, unlike statistical methods based on Singular Value Decomposition (SVD) or Generalized SVD, minimizing the loss function associated to (4) cannot be solved in a closed form, due to the fixed parameters in $\mathbf{W}^{\circ}$, $\mathbf{W_T}$ and $\mathbf{A}'$ containing zeroes. Hence we use an iterative method, employing an Alternating Least Squares (ALS) algorithm (ten Berge 1993). In the algorithm, parameter matrices $\mathbf{A_Y}'$, $\mathbf{W}°$, $\mathbf{W_T}$ and $\mathbf{A}'$ are alternately updated in a four steps algorithm until convergence is reached. In the first step, we update $\mathbf{W}°$, independently from $\mathbf{W_T}$ for fixed $\mathbf{A}$ and $\mathbf{A_Y}'$. In the second step $\mathbf{W_T}$ is updated for fixed, $\mathbf{A}$ and $\mathbf{A_Y}'$ with $\mathbf{W}°$ obtained in the first step. In the third step, $\mathbf{A}$ is updated for fixed $\mathbf{A_Y}'$, $\mathbf{W}°$ and $\mathbf{W_T}$, whereas in the fourth step $\mathbf{A_Y}'$ is updated for fixed $\mathbf{A}$, $\mathbf{W}°$ and $\mathbf{W_T}$. In the first step, $\mathbf{W}°$ is updated independently from $\mathbf{W_T}$, for fixed $\mathbf{A_T}'$ and $\mathbf{A}$. Firstly, with $\mathbf{A_Y}'$ initialized with arbitrary values and defining $\mathbf{Y}^* = \mathbf{Y} - \mathbf{T}\mathbf{A_Y}'$, the loss function associated to (4) is:

$$\mathrm{f} = \mathrm{SS}\left[\mathbf{Y}^* - \mathbf{X}\mathbf{W}°\mathbf{A}' - \mathbf{T}°\mathbf{W_T}\mathbf{A}'\right] \tag{5}$$

where SS[$\mathbf{Z}$] = trace($\mathbf{Z}'\mathbf{Z}$). It is of note that, since (7) involves orthogonal regressors, once weight matrices ($\mathbf{W}°$, $\mathbf{W_T}$) are obtained, regression parameters can be separately resolved. Specifically, let $\mathbf{X} = \mathbf{Q}\mathbf{R}'$ be portion of the QR decomposition of $\mathbf{X}$, pertaining to the $\mathbf{X}$ column space, where $\mathbf{Q}$ is an n by q orthonormal matrix and $\mathbf{R}'$ is a q by n upper-triangular matrix, (5) becomes:

$$\mathrm{SS}\left[\left(\mathbf{Y} * - \mathbf{Q}\mathbf{Q}'\mathbf{Y}^*\right) + \mathbf{Q}\mathbf{Q}'\mathbf{Y}^* - \mathbf{Q}\left(\mathbf{R}'\mathbf{W}°\mathbf{A}' + \mathbf{Q}'\mathbf{T}°\mathbf{W_T}\mathbf{A}'\right)\right] \tag{6}$$

Since $\mathbf{X}$ and $\mathbf{T}°$ are orthogonal, the term $\mathbf{Q}'\mathbf{T}°\mathbf{W_T}\mathbf{A}'$ of (8) becomes null. Further, since the trace of the cross product $[(\mathbf{Y}* - \mathbf{Q}\mathbf{Q}'\mathbf{Y}^*)'\mathbf{Q}(\mathbf{Q}'\mathbf{Y}* - \mathbf{R}'\ \mathbf{W}°\mathbf{A}')]$ is null, (6) can be rewritten as:

$$\mathrm{SS}\left[\mathbf{Y}^* - \mathbf{Q}\mathbf{Q}'\mathbf{Y}^*\right] + \mathrm{SS}\left[\mathbf{Q}\left(\mathbf{Q}'\mathbf{Y}^* - \mathbf{R}'\mathbf{W}°\mathbf{A}'\right)\right] \tag{7}$$

where the first term does not depend on parameters. Hence, minimizing (6) is equivalent to minimizing:

$$\begin{aligned}\mathrm{SS}\left[\mathbf{Q}\left(\mathbf{Q}'\mathbf{Y}^* - \mathbf{R}'\mathbf{W}°\mathbf{A}'\right)\right] &= \mathrm{SS}\left[\mathrm{vec}\left(\mathbf{Q}'\mathbf{Y}^*\right) - (\mathbf{A} \otimes \mathbf{R}')\mathrm{vec}\left(\mathbf{W}°\right)\right] \\ &= \mathrm{SS}\left[\mathrm{vec}\left(\mathbf{Q}'\mathbf{Y}^*\right) - \Omega\mathbf{w}°\right]\end{aligned} \tag{8}$$

where $\boldsymbol{\Omega} = \mathbf{A} \otimes \mathbf{R}'$, $\otimes$ denotes a Kronecker product and $\mathbf{w}° = \mathrm{vec}(\mathbf{W}°)$ denotes a supervector consisting of all columns of $\mathbf{W}°$ one below another. As previously said, $\mathbf{W}°$ has to be estimated without destroying its structure (e.g., zero elements in $\mathbf{w}°$, depending on the existing links between manifest variables and latent composites). Therefore, let $\mathbf{w}°_*$ denote the vector that selects non-zero elements from $\mathbf{w}° = \mathrm{vec}(\mathbf{W}°)$ and $\boldsymbol{\Omega}_*$ denotes the matrix formed by eliminating the columns of $\boldsymbol{\Omega}$ corresponding to the zero elements in $\mathbf{w}°$, we obtain the $\mathbf{w}°_*$ least squares estimate by

$$\mathbf{w}°_* = \left(\boldsymbol{\Omega}_*'\boldsymbol{\Omega}_*\right)^{-1}\boldsymbol{\Omega}_*'\mathrm{vec}\left(\mathbf{Q}'\mathbf{Y}^*\right) \tag{9}$$

Hence, we obtain $\mathbf{w}°$ (reconstructing the zero elements to their original positions), then $\mathbf{W}°$. In the second step $\mathbf{W_T}$ is updated for fixed $\mathbf{W}°$, $\mathbf{A}$ and $\mathbf{A_Y}'$.

This amounts to minimizing (5) which can be expressed as:

$$\mathrm{SS}\left[\left(\mathbf{Y}^{*}-\mathbf{X}\mathbf{W}^{\circ}\mathbf{A}'\right)-\mathbf{T}^{\circ}\mathbf{W}_{\mathbf{T}}\mathbf{A}'\right]=\mathrm{SS}\left[\mathrm{vec}\left(\mathbf{Y}^{**}\right)-\left(\mathbf{A}\otimes\mathbf{T}^{\circ}\right)\mathrm{vec}\left(\mathbf{W}_{\mathbf{T}}\right)\right] \quad (10)$$

where $\mathbf{Y}^{**}=\mathbf{Y}^{*}-\mathbf{X}\mathbf{W}^{\circ}\mathbf{A}'$. Defining $\boldsymbol{\Omega}_{\mathbf{T}}=\mathbf{A}\otimes\mathbf{T}^{\circ}$, $\mathbf{w}_{\mathbf{T}}=\mathrm{vec}(\mathbf{W}_{\mathbf{T}})$ and let $\boldsymbol{\Omega}_{\mathbf{T}*}$, $\mathbf{w}_{\mathbf{T}*}$ denote their non-zero counterparts, we obtain the least squares estimate of $\mathbf{w}_{\mathbf{T}*}$ by

$$\mathbf{w}_{\mathbf{T}*}=\left(\boldsymbol{\Omega}_{\mathbf{T}_*}'\boldsymbol{\Omega}_{\mathbf{T}*}\right)^{-1}\boldsymbol{\Omega}_{\mathbf{T}_*}'\mathrm{vec}\left(\mathbf{Y}^{**}\right) \quad (11)$$

and, reconstructing the zero elements to their original positions, we further obtain $\mathbf{w}_{\mathbf{T}}$, then $\mathbf{W}_{\mathbf{T}}$ and finally, using estimates of $\mathbf{W}_{\mathbf{T}}$ and $\mathbf{W}^{\circ}$, by (3) the $\mathbf{F}$ scores that are normalized it so that $\mathrm{diag}(\mathbf{F}'\mathbf{F})=\mathbf{I}$. Notice that weights to obtain latent composites scores $\mathbf{F}$ are obtained projecting onto the principal directions of exogenous variables, the matrix of endogenous indicators, once the direct effect of $\mathbf{T}$ onto $\mathbf{Y}$ ($\mathbf{W}^{\circ}$) and the effects of $\mathbf{X}$ onto $\mathbf{Y}$ ($\mathbf{W}_{\mathbf{T}}$) are controlled for. In the third step, $\mathbf{A}'$ is updated for fixed $\mathbf{W}$, $\mathbf{W}_{\mathbf{T}}$ and $\mathbf{A}_{\mathbf{Y}}'$. Loss function (5) can be written as

$$\mathrm{SS}\left[\mathbf{Y}^{*}-\mathbf{F}\mathbf{A}'\right]=\mathrm{SS}\left[\mathrm{vec}\left(\mathbf{Y}^{*}\right)-\left(\mathbf{I}\otimes\mathbf{F}\right)\mathrm{vec}\left(\mathbf{A}'\right)\right] \quad (12)$$

Defining $\mathbf{a}=\mathrm{vec}(\mathbf{A}')$, $\boldsymbol{\Gamma}=\mathbf{I}\otimes\mathbf{F}$ and $\mathbf{a}_*$ and $\boldsymbol{\Gamma}_*$ in a way similar to $\mathbf{w}_*$ and $\boldsymbol{\Omega}_*$, the least squares estimate of $\mathbf{a}_*$ is:

$$\mathbf{a}_{*}=\left(\Gamma_{*}'\Gamma_{*}\right)^{-1}\Gamma_{*}'\,\mathrm{vec}\left(\mathbf{Y}^{*}\right) \quad (13)$$

and easily reconstructing the updated $\mathbf{a}$ and $\mathbf{A}'$ from $\mathbf{a}_*$. In the final (fourth) step, $\mathbf{A}_{\mathbf{Y}}'$ is updated with fixed $\mathbf{W}^{\circ}$, $\mathbf{W}_{\mathbf{T}}$ and $\mathbf{A}'$, by Multivariate Regression, regressing $\mathbf{T}$ onto $\mathbf{Y}$–$\mathbf{F}\mathbf{A}'$ scores, where $\mathbf{F}$ (defined in 3) and $\mathbf{A}$ are estimated in the previous three steps. The above four steps are alternated until convergence is reached. For Case 1, since $\mathbf{W}_{\mathbf{T}}$ disappears, in the algorithm we eliminate step 2, whereas the remaining three steps remain the same.

4 Discussion

By (3), we could obtain weights $\mathbf{W}$ in its original specification (1) by the back transformation $\mathbf{W}=\mathbf{W}^{\circ}-\mathbf{X}^{+}\mathbf{T}\ \mathbf{W}_{\mathbf{T}}$. However, the new measurement model (3) offers a benefit in term of meaningful interpretation for latent composites. $\mathbf{F}$ is obtained as a combination of two terms: the combinations of strictly formative indicators $\mathbf{X}\mathbf{W}^{\circ}$, net of contribution of concomitant indicators to the latent scores, and the combinations of concomitant indicators that are orthogonal to exogenous formative indicators $\mathbf{T}^{\circ}\mathbf{W}_{\mathbf{T}}$. In this specification, $\mathbf{X}\mathbf{W}^{\circ}$ are the strictly formative indicators, whereas the sum of two linear combinations $\mathbf{X}\mathbf{W}^{\circ}+\mathbf{T}^{\circ}\mathbf{W}_{\mathbf{T}}$ (=$\mathbf{F}$) are the overall latent scores in the specified model. From (9) to (11), weights defining $\mathbf{F}$

scores have meaningful geometric interpretation. The first set of weights ($\mathbf{W}°$) is the projection of endogenous observed indicators whose scores have been corrected by the effect of $\mathbf{T}$ on $\mathbf{Y}$ ($\mathbf{Q}'\mathbf{Y}^*$) onto the space spanned by the principal directions of the QR decomposition of $\mathbf{X}$ (vec($\boldsymbol{\Omega}_*$)). The second set of weights ($\mathbf{W_T}$) coincides with the projection of $\mathbf{Y}^{**}$ onto the column space spanned by $\mathbf{T}°$, where $\mathbf{Y}^{**}(=\mathbf{Y}^* - \mathbf{X}\mathbf{W}°\mathbf{A}')$ denotes the endogenous observed indicators whose scores have been corrected for the effect of $\mathbf{T}$ on $\mathbf{Y}$ and also for the indirect effect of $\mathbf{X}\mathbf{W}°$ on $\mathbf{Y}$, via $\mathbf{F}$. Local minima, typical of ALS algorithms (ten Berge 1993), may be avoided choosing good initial values for $\mathbf{W}°$ and $\mathbf{W_T}$ since starting values of $\mathbf{A}'$ and $\mathbf{A_Y}'$ are simply obtained by the least squares estimate. Among various alternatives, a recent method, coined Multiblock Redundancy Analysis (MRA) (Bougeard et al. 2008) is used.

Overall, the GRA estimates model parameters by minimizing an overall model fit (the sum of squares of discrepancies between $\mathbf{Y}$ and their predicted values) without any explicit distributional assumptions. The proposed method is simple yet versatile enough to fit various complex relationships among variables, including direct effects of observed variables and/or concomitant variables. Missing values can be inserted in $\mathbf{X}$, $\mathbf{Y}$ and $\mathbf{T}$ as additional parameters, and they are estimated minimizing loss function (5), with estimated parameters (by eliminating non-missing values) and non-missing values fixed. In various applications, the proposed algorithm was found to converge in all cases. It has proved to be efficient and generally results in rapid convergence, also in cases of large sample size, due to the QR-decomposition strategy in the first step ($\mathbf{W^0}$) of the algorithm.

The presented method has two noticeable limitations. Firstly, although GRA may capture reflective relationships among LC and the observed exogenous variables, GRA can accommodate only formative schemes for LC. However, this is the only measurement model consistent with Component Analysis (Millsap and Meredith 1988). A second, more fundamental limitation of the present method is that, although endogenous LCs are allowed in the model, GRA cannot assume any LC for the observed endogenous variables. However, both problems can not be resolved without proposing models outside the area of RA. Potential candidates to resolve this problem may be found in more general methods. Future studies are needed to investigate the feasibility of these additional extensions and to evaluate bias and relative efficiency of GRA estimators, using simulation studies.

References

Bougeard, S., Hanafi, M., Lupo, C., & Qannari, E. M. (2008). From multiblock Partial Least Squares to multiblock redundancy analysis. In *A continuum approach. International Conference on Computational Statistics*. Porto, Portugal.

Davies, P., & Tso, M. (1982). Procedures for reduced-rank regression. *Applied Statistics, 31*, 244–255.

Izenman, A. J. (1975). Reduced-rank regression for the multivariate linear model. *Journal of Multivariate Analysis, 5*, 248–264.

Jôreskog, K. G. (1970). A general method for analysis of covariance structures. *Biometrika, 57*, 409–426.
Kiers, H. A. L., Takane, Y., & ten Berge, J. M. F. (1996). The analysis of multitrait-multimethod matrices via constrained components analysis. *Psychometrika, 61*, 601–628.
McDonald, R. P. (1996). Path analysis with composite variables. *Multivariate Behavioral Research, 31*, 239–270.
Millsap, R. E., & Meredith, W. (1988). Component analysis in cross-sectional and longitudinal data. *Psychometrika, 53*, 123–134.
Reinsel, G. C., & Velu, R. P. (1998). *Multivariate reduced-rank regression*. New York: Springer.
Schônemann, P. H., & Steiger, J. H. (1976). Regression component analysis. *British Journal of Mathematical and Statistical Psychology, 29*, 175–189.
Takane, Y., & Hwang, H. (2005). An extended redundancy analysis and its applications to two practical examples. *CSDA, 49*(3), 785–808.
ten Berge, J. M. F. (1993). *Least squares optimization in multivariate analysis*. Leiden: DSWO.
van den Wollenberg, A. L. (1977). Redundancy analysis: an alternative for canonical analysis. *Psychometrika, 42*, 207–219.
Wold, H. (1982). Soft modeling: The basic design and some extensions. In K. G Jôreskog and H. Wold (Eds.). *Systems under indirect observations: causality, structure, prediction*. Part 2, (pp. 1–54). North-Holland, Amsterdam.

Assessing Stability in NonLinear PCA with Hierarchical Data

Marica Manisera

Abstract Composite indicators of latent variables can be constructed by NonLinear Principal Components Analysis when data are collected by multiple-item scales. The aim of this paper is to establish the stability of the contribution made by each item to the composite indicator, by means of a resampling-based procedure able to take account of the hierarchical structure that often exists in the data, that is when individuals are nested in groups. The procedure modifies the standard nonparametric bootstrap technique and was applied to real data on job satisfaction from the most extensive survey on Italian social cooperatives.

1 Introduction

One of the main goals of social and economic research is to measure individuals' perceptions and attitudes (e.g., customer and job satisfaction), which requires statistical instruments able to manage latent variables, i.e., complex concepts not directly observed. In order to measure latent variables, researchers usually collect data by administrating multiple-item scales, that is, questionnaires with several items referring to the different aspects of the concept being measured. Responses often indicate the degree of agreement with each statement, with higher scores reflecting greater agreement.

When individuals' perceptions and attitudes are measured, data are very often organised in a hierarchical structure. Individuals (i.e., the subjects, or objects, or first-level units) are clustered or nested in groups (second-level units) and these groups can then be grouped in higher-level units. For example, when measuring job satisfaction, workers are nested in organizations (which can be nested in geographical

M. Manisera (✉)
Department of Economics and Management, University of Brescia, C.da S. Chiara, 50 - 25122 Brescia, Italy
e-mail: manisera@eco.unibs.it

P. Giudici et al. (eds.), *Statistical Models for Data Analysis*, Studies in Classification, Data Analysis, and Knowledge Organization, DOI 10.1007/978-3-319-00032-9_25,

areas). The number of individuals within groups is usually not constant, but instead, varies within each cluster (data are said to be unbalanced clustered).

Among the models and techniques conceived to produce quantitative measures of the latent variables underlying a multiple-item scale, we focus on the NonLinear Principal Components Analysis (NL-PCA Gifi 1990; Meulman et al. 2004).[1] NL-PCA is the nonlinear equivalent of classical Principal Components Analysis (PCA; see, among others, Zani and Cerioli 2007) and aims at optimally reducing a large number of categorical variables to a smaller number of composite variables (the principal components), which are useful for representing latent variables. The NL-PCA model is the same linear model as in traditional PCA, but it is applied to nonlinearly transformed data. The variables are transformed by assigning optimal scale values to the categories, resulting in numeric-valued transformed (i.e., quantified) variables. NL-PCA finds category quantifications that are optimal in the sense that the overall variance accounted for in the transformed variables, given the number p of components, is maximized. The variance accounted for is often expressed in percentage (Percentage of Variance Accounted For, PVAF) and is a global measure of the goodness of the NL-PCA solution.

Like PCA, NL-PCA is usually used as a descriptive data analysis technique, because it was developed from an exploratory point of view and does not provide inferential statistics. However, in the literature, there are studies introducing some inferential issues with regard to PCA results (see references in Linting 2007). Regarding NL-PCA, inference has recently been introduced by a nonparametric approach (Linting 2007), which is consistent with the weak distributional assumptions. Following this approach, one may perform either permutation tests or bootstrap studies. Permuting variables or drawing bootstrap samples allows the destruction of the correlational structure in the data to investigate the statistical significance or the stability of the results. In particular, in order to assess the stability of the NL-PCA results (more precisely, the stability to data selection, i.e., the degree of sensitivity of the results to variations in the data), the nonparametric bootstrap procedure (Efron and Tibshirani 1993) has been used (Linting 2007). The bootstrap technique has also been used in the NL-PCA framework to compare groups of subjects with regard to categorical variables by constructing Inferential Confidence Intervals (Goldstein and Healy 1995) around centroids, which represent the group means of quantified variables (Manisera 2011). In addition, in order to establish the statistical significance of the PCA results, permutation tests have been proposed (Linting et al. 2011).

In the construction of composite indicators for latent variables from multiple-item scales, an interesting topic is the contribution made by each item to the definition of the composite indicators. Such a contribution gives, in some sense, the importance of single aspects in composing the measure of the latent variable; analogously, variable importance measures have been proposed in data mining

[1] In the literature (Michailidis and de Leeuw 2000), NL-PCA was also extended to hierarchical data structures to examine how variables are related across groups and how groups vary.

models to select the drivers of latent variables [e.g., in algorithmic models aggregating regression trees, like TreeBost and Random Forests (Carpita and Zuccolotto 2007), or another ensemble learning developed for hierarchical data (Vezzoli and Zuccolotto 2010)].

This paper aims at establishing, by means of a resampling-based procedure, the stability of the contribution of the separate variables to the composite indicator obtained by NL-PCA, in the presence of hierarchical data.

The procedure was applied to real data referring to workers employed in the social cooperatives sampled in the ICSI^{2007} survey on Italian social cooperatives (Carpita 2009). A composite indicator of the workers' job satisfaction was constructed using NL-PCA and the resampling-based strategy was used to assess the stability of the contribution made by the different items (facets of job satisfaction) to the job satisfaction indicator, by taking the clustering of workers within cooperatives into account.

2 Methods

The resampling-based procedure for hierarchical data, based on the nonparametric bootstrap and proposed in Ren et al. (2010), is used in this paper to establish the stability of the contribution made by each item to the composite indicator when data are hierarchical. In setting up the procedure it is worth considering that (*i*) the stability study involves one particular result of NL-PCA, which is the contribution of one variable to the indicator of the latent variable and (*ii*) we intend to take the hierarchical nature of the data into account.

In the NL-PCA framework, the contribution of each separate variable to the composite indicator of the latent variable (given by the nonlinear principal components) is measured by the Variance-Accounted-For (VAF) *per variable*, which is computed as the sum of the squared loadings of that variable across the components. Formally, if the data are in the $(n \times m)$ matrix $\mathbf{X}$, m VAF$_j$, $j = 1, \cdots, m$ can be computed as $\sum_{l=1}^{p} a_{jl}^2$, $l = 1, \cdots, p$, where p is the number of principal components retained in the solution, a_{jl} is the loading of the j-th variable on the l-th component and is given by the correlation coefficient between the j-th quantified variable and the l-th principal component.

In the literature, stability studies on NL-PCA results have been performed using balanced bootstrap procedures involving the entire data set, that is, all the variables together (for example, see Linting 2007). However, when data are hierarchical, simple nonparametric resampling methods are not appropriate because they treat all observations as independent. Nonparametric bootstrapping for hierarchical or clustered data is relatively underdeveloped (Ren et al. 2010), in part because it is not straightforward: in fact, even in the simplest case of hierarchical data with two levels, more than one bootstrap resampling strategy can be defined according to the level chosen to be bootstrapped and the resampling strategy (with or without replacement) on each level.

In order to take the hierarchical structure into account, the resampling-based procedure used in this paper consists in bootstrapping on the highest level, meaning randomly sampling without replacement within the highest level selected by randomly sampling the highest levels with replacement. In other words, if we focus on two-level data (which are very often encountered in practice as well as in the case study reported in Sect. 3), the strategy consists of two steps: (1) to randomly sample second-level units with replacement and (2) to randomly sample, without replacement, first-level units within the second-level units selected at the first step.

Formally, to resample the data matrix $\mathbf{X}$ taking account of the G groups, in which the n observations are clustered (the g-th group counts n_g observations), first a sample with replacement of size equal to G from $\{1, 2, \cdots, g, \cdots, G\}$ is drawn, giving the sequence $\{s_1, s_2, \cdots, s_g, \cdots, s_G\}$ of groups in the resampled data set. The rows of the s_g-th group are then filled in with a random sample (without replacement) drawn from the observed values of $\mathbf{X}$ within that group, with size n_{s_g}. In the presence of unbalanced clustered data, this strategy leads to samples with different sizes: in fact, in the second step, the number n_{s_g} of the first-level units sampled from each selected second-level unit equals the original size of the s_g-th group, and $\sum_{s_g=1}^{s_G} n_{s_g}$ may differ from n.

The resampling-based procedure is repeated B times, and B resampled data sets are obtained. Usually, $B = 1{,}000$ is chosen (Efron and Tibshirani 1993; Linting 2007; Ren et al. 2010). Subsequently, NL-PCA is performed on each of these data sets, which gives B values for each of the VAF_j, forming a distribution from which confidence percentile intervals can be computed. Such intervals can be used to assess the stability of VAF_j, $j = 1, \cdots, m$.

The resampling-based strategy is applied before the optimal quantification of the categorical variables, and no corrections for rotated solutions must be applied, because the VAF *per variable* is not sensitive to rotations of the NL-PCA solution.

Unlike other studies (Linting 2007), in the current work, bootstrap is not balanced. The balanced bootstrap (Efron and Tibshirani 1993) ensures that every subject appears a total of B times in the B bootstrap samples: this is important when the interest is in the stability of the subject scores (on the principal components), while the focus here is on the stability of the VAF *per variable*. In addition, when data are hierarchical, one should decide whether balancing should be performed with reference to first-level or higher-level units, according to the aim of the stability study.

Comparing confidence percentile intervals does not allow graphical study of whether the contributions of two items are statistically different. To achieve this goal, Inferential Confidence Intervals (ICIs; Goldstein and Healy 1995) can be computed. ICIs are a graphic test of statistical mean difference designed to avoid common interpretative problems associated with the null hypothesis statistical testing. Graphed confidence intervals can be used for overlap pairwise comparisons as an inferential graphical tool at the stated significance level only after reducing their widths. Due to the reduction, nonoverlapping ICIs are algebraically equivalent to a null hypothesis statistical test at the stated significance level. In this paper, a modified version of the ICIs, called Bootstrap ICIs (Manisera 2011), is obtained by estimating the standard errors in the ICIs by means of the bootstrap study.

3 Case Study

NL-PCA was applied[2] to construct a job satisfaction indicator that summarizes 11 categorical ordinal variables that measure different aspects of job satisfaction for 2,500 workers employed in 212 social cooperatives.[3] The variables (described in Table 1) refer to the satisfaction of workers with extrinsic aspects (work characteristics, such as variety) as well as intrinsic and relational aspects (such as personal fulfilment and relations with co-workers and the cooperative).

In order to obtain a one-dimensional composite indicator of job satisfaction, NL-PCA was applied with all of the variables scaled ordinally in order to keep in the quantified variables the grouping and the ordering information in the original categorical variables (PVAF $= 45.98$). The 11 VAF_js *per variable* (represented by empty squares in the next Fig. 2) allow to identify which facets mostly contribute to the definition of a global indicator of job satisfaction.[4] The highest contribution is made by the items regarding the relation between the worker and cooperative (`coop.recognition`, `transparency`, `involvement`). Subsequently, the most important items refer to the satisfaction with intrinsic aspects (`growth`, `fulfilment`), followed by a mix of items of satisfaction with individual aspects and relations with superiors, coworkers, and the team. This confirms the results in the literature on job satisfaction in the social service sector (see, for example, Borzaga and Depedri 2005).

According to the resampling-based procedure described in Sect. 2, the entire data set was resampled $B = 2{,}000$ times, obtaining B resampled data sets, and for each of these, NL-PCA was performed. The size of the resampled data sets varied across the replications from 2,170 to 2,806, while the PVAF ranged from 42.63 to 49.09. In both cases (size and PVAF), the resampled distribution closely resembles a Normal distribution, centred on the value of the original sample.

The distributions of the 11 VAF_js *per variable* were obtained (Fig. 1) and the resulting 95% percentile confidence intervals are depicted in Fig. 2.

---

[2]All the computations were performed by R 2.13.1.

[3]Missing values were imputed according to Carpita and Manisera (2011). From the entire sample of the `ICSI`^{2007} survey, the cooperatives with less than 5 workers were removed. The data used in this study result from a preliminary Rasch analysis (Carpita and Golia 2011), which identified the 11 selected job satisfaction items as related to a "global" job satisfaction latent trait, and suggested merging response categories to obtain a 5-point response scale for each item, ranging from 1 = "very dissatisfied" to 5 = "very satisfied", with mid-point 3 = "neither dissatisfied nor satisfied".

[4]Other papers investigated the drivers of job satisfaction by means of regression models with job satisfaction items as independent variables and the overall job satisfaction as a dependent variable (see, for example, Carpita and Zuccolotto 2007, Vezzoli and Zuccolotto 2010). Unlike the current paper, their aim was to identify which facets of job satisfaction drive, from a psychological point of view, the individual perception of the overall job satisfaction, since the latter is measured by a single item in the questionnaire.

Table 1 Job satisfaction items

Item	How satisfied are you with...
`Coop.recognition`	The recognition by the cooperative of your work?
`Transparency`	The transparency in your relation with the cooperative?
`Involvement`	Your involvement in the cooperative decisions?
`Independence`	Your decisional and operative independence?
`Variety`	The variety and creativity of your work?
`Team`	The relations within the team?
`Superiors`	The relations with your superiors?
`Coworkers.recognition`	The recognition by co-workers of your work?
`Growth`	Your vocational training and professional growth?
`Fulfilment`	Your personal fulfilment?
`Career`	Your achieved and prospective career promotions?

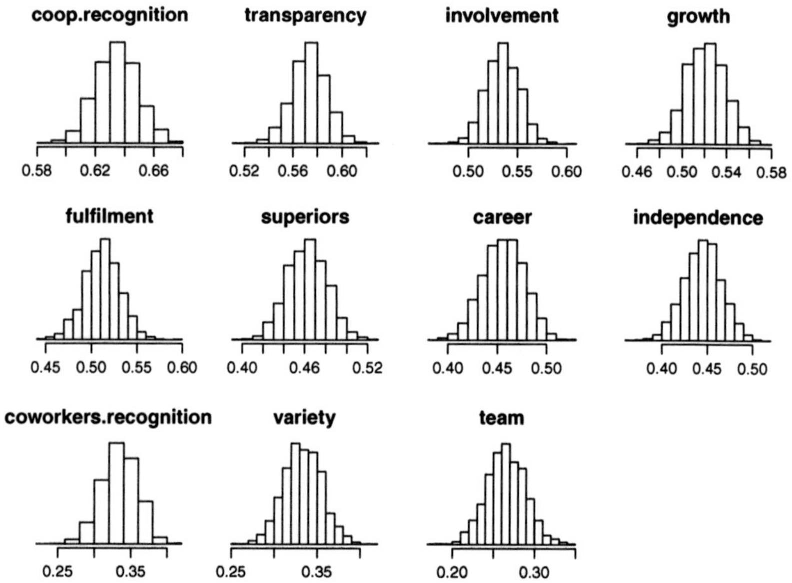

Fig. 1 Distributions of the VAF_js of the 11 job satisfaction items over $B = 2{,}000$ replications of the resampling-based procedure

The bootstrap means, obtained by averaging the VAF_js over the B samples (represented by black circles in Fig. 2), nearly overlap the VAF_js in the original sample (empty squares in Fig. 2), confirming the original ranking of items. With reference to stability, the width of all of the intervals (black lines in Fig. 2) is quite small, suggesting that the contribution of the variables to the job satisfaction measure is quite stable. This is especially true for variables with the highest VAF_js. It is interesting to note that the corresponding loadings on the principal component

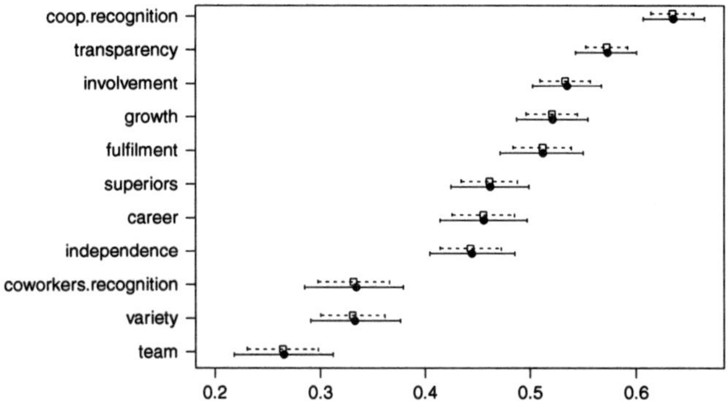

Fig. 2 95% percentile confidence intervals (*black lines*) and 95% Bootstrap Inferential Confidence Intervals (*dotted lines*) for the VAF_js of the 11 job satisfaction items; *black circles* and *empty squares* represent the bootstrap means and the original sample VAF_js, respectively

(VAF_js are squared loadings) were all positive (in each of the B samples); therefore, the corresponding percentile confidence intervals did not contain zero, which indicates that all the considered variables make an important contribution to the job satisfaction indicator.

These results were somewhat expected because in the preliminary Rasch analysis (Carpita and Golia 2011) providing the raw data used in this study, the response scale was reduced by merging some response categories, and this is known in the literature as a procedure to increase stability in the NL-PCA results (Linting 2007).

To statistically compare the contributions of two items, Bootstrap ICIs were computed, after having verified the Normality of the 11 distributions of the VAF_js *per variable* (Fig. 1). Figure 2, representing the Bootstrap ICIs associated with the 11 VAF_js by dotted lines, shows which contributions are statistically different by means of nonoverlapping intervals. For example, the contributions made by the two most important items are statistically different.

The large original sample size made the procedures easier: in the presence of smaller sample sizes, caution should be used when computing Bootstrap ICIs, but also with reference to the varying size of resampled data sets.

4 Conclusions

In this paper, we used a resampling-based procedure for hierarchical data to study the stability of NL-PCA results, in particular the stability of the contribution of items in defining the measure of the latent variable. The application to real data provided interesting insights into the interpretation of job satisfaction in the social cooperatives, by identifying which contribution was made by the different facets to the composite indicator of job satisfaction obtained by NL-PCA. In addition, Bootstrap ICIs allowed the graphical comparison of such contributions.

In this study, the resampling-based procedure allowed the investigation of the absolute stability of the VAF$_j$s *per variable* in the NL-PCA. Future research involves the study of a relative stability, assuming, for example, the stability of linear PCA as a benchmark (as in Linting 2007).

In Ren et al. (2010), it was shown that with the proposed procedure, the original sample information is accurately reflected and the correlational structure within groups is preserved. This issue will be more fully investigated by a next simulation study concerning the stability of by group NL-PCA results.

References

Borzaga, C., & Depedri, S. (2005). Interpersonal relations and job satisfaction: some empirical results in social and community care services. In B. Gui, & R. Sugden (Eds.) *Economics and social interaction: accounting for interpersonal relations* (pp. 132–153). Cambridge: Cambridge University Press.

Carpita, M. (Ed.) (2009). *La qualità del lavoro nelle cooperative sociali, misure e modelli statistici.* Milano: Franco Angeli.

Carpita, M., & Golia, S. (2011). Measuring the quality of work: the case of the Italian social cooperatives. *Quality and Quantity, 46*, 1659–1685. On Line First, DOI 10.1007/s11135-011-9515-0.

Carpita, M., & Manisera, M. (2011). On the imputation of missing data in surveys with Likert-type scales. *The Journal of Classification, 28*, 93–112.

Carpita, M., & Zuccolotto, P. (2007). Mining the drivers of job satisfaction using algorithmic variable importance measures. In L. D'Ambra, P. Rostirolla, & M. Squillante (Eds.) *Metodi, modelli, e tecnologie dell'informazione a supporto delle decisioni* (vol. I, pp. 63–70). Milano: Franco Angeli.

Efron, B., & Tibshirani, R. J. (1993). *An introduction to the bootstrap.* New York: Chapman and Hall.

Gifi, A. (1990). *Nonlinear multivariate analysis.* Chichester: Wiley.

Goldstein, H., & Healy, M. J. R. (1995). The graphical presentation of a collections of means. *Journal of the Royal Statistical Society A, 158*, 175–177.

Linting, M. (2007). *Nonparametric inference in nonlinear principal components analysis: exploration and beyond.* Leiden (NL): Leiden University.

Linting, M., van Os, B. J., & Meulman, J. (2011). Statistical significance of the contribution of variables to the PCA solution: an alternative permutation strategy. *Psychometrika, 76*, 440–460.

Manisera, M. (2011). A graphical tool to compare groups of subjects on categorical variables. *Electronic Journal of Applied Statistical Analysis, 4*, 1–22.

Meulman, J. J., Van der Kooij, A. J., & Heiser, W. J. (2004). Principal components analysis with nonlinear optimal scaling transformations for ordinal and nominal data. In Kaplan, D. (Ed.) *Handbook of quantitative methodology for the social sciences* (pp. 49–70). London: Sage.

Michailidis, G., & de Leeuw, J. (2000). Multilevel homogeneity analysis with differential weighting. *Computational Statistics & Data Analysis, 32*, 411–442.

Ren, S., Lai, H., Tong, W., Aminzadeh, M., Hou, X., & Lai, S. (2010). Nonparametric bootstrapping for hierarchical data. *Journal of Applied Statistics, 37*, 1487–1498.

Vezzoli, M., & Zuccolotto, P. (2010). CRAGGING measures of variable importance for data with hierarchical structure. In S. Ingrassia, R. Rocci, & M. Vichi (Eds.) *New perspectives in statistical modeling and data analysis.* Heidelberg: Springer.

Zani, S., & Cerioli, A. (2007). *Analisi dei dati e data mining per le decisioni aziendali.* Milano: Giuffrè.

Using the Variation Coefficient for Adaptive Discrete Beta Kernel Graduation

Angelo Mazza and Antonio Punzo

Abstract Various approaches have been proposed in literature for the kernel graduation of mortality rates. Among them, this paper considers, as a starting point, the fixed bandwidth discrete beta kernel estimator, a recent proposal conceived to intrinsically reduce boundary bias and in which age is pragmatically considered as a discrete variable. An adaptive variant of this estimator also exists, which allows the bandwidth to vary with age according to the reliability of the data as expressed only by the amount of exposure. This paper presents a further adaptive version, obtained by measuring the reliability via the reciprocal of the variation coefficient, which is function of both the amount of exposure and the observed mortality rates. A simulation study is accomplished to evaluate the gain in performance of the new estimator with respect to its predecessors.

1 Introduction

Mortality rates are age-specific indicators commonly used in demography. Historically, they are also widely adopted by actuaries, in the form of mortality tables, to calculate life insurance premiums, annuities, reserves, and so on. Producing these tables from a suitable set of crude (or raw) mortality rates is called graduation, and this subject has been extensively discussed in the actuarial literature (see, e.g., Copas and Haberman 1983, and Haberman and Renshaw 1996). To be specific, the d_x deaths at age x can be seen as arising from a population, initially exposed to the risk of death, of size e_x. The situation is commonly summarized via the model $d_x \sim \text{Bin}(e_x, q_x)$, where q_x represents the true, but unknown, mortality rate at age x. The crude rate $\mathring{q}_x$ is the observed counterpart of q_x. Graduation is necessary

A. Mazza (✉) · A. Punzo
Dipartimento di Economia e Impresa, Universit di Catania, Italy
e-mail: a.mazza@unict.it; antonio.punzo@unict.it

P. Giudici et al. (eds.), *Statistical Models for Data Analysis*, Studies in Classification, Data Analysis, and Knowledge Organization, DOI 10.1007/978-3-319-00032-9_26,

because crude data usually presents abrupt changes, which do not agree to the dependence structure supposedly characterizing the true rates (London 1985). In fact, a common prior opinion about their form is that each true mortality rate is closely related to its neighbors. This relationship is expressed by the belief that the true rates progress smoothly from one age to the next. So, the next logical step is to graduate the crude rates to produce smooth estimates, $\hat{q}_x$, of the true rates. This is done by systematically revising the crude rates in order to remove any random fluctuations. Nonparametric models are the natural choice if the aim is to reflect this belief. Furthermore, a nonparametric approach can be used: to choose the simplest suitable parametric model, to provide a diagnostic check of a parametric model, or to simply explore the data (see Härdle 1992 for a detailed discussion on the chief motivations that imply their use, and Debn et al. 2006 for an exhaustive comparison of nonparametric methods in the graduation of mortality rates).

Kernel smoothing is one of the most popular statistical methods for nonparametric graduation. Among the various alternatives existing in literature (see Copas and Haberman 1983, Gavin et al. 1993, 1994, 1995 and Peristera and Kostaki 2005), the attention is here focused on the discrete beta kernel estimator proposed by Mazza and Punzo (2011). Roughly speaking, the genesis of this model starts with the consideration that, although age X is in principle a continuous variable, it is typically truncated in some way, such as age at last birthday, so that it takes values on the discrete set $\mathcal{X} = \{0, 1, \ldots, \omega\}$, ω being the highest age of interest. Discretization of age, from a pragmatical and practical point of view, could also come handy to actuaries that have to produce "discrete" graduated mortality tables starting from the observed counterparts. In the fixed bandwidth estimator proposed in Mazza and Punzo (2011), discrete beta distributions (Punzo and Zini 2012, and Punzo 2010) are considered as kernel functions in order to overcome the problem of boundary bias, commonly arising from the use of symmetric kernels. The support $\mathcal{X}$ of the discrete beta, in fact, matches the age range and this, when smoothing is made near the boundaries, allows avoiding allocation of weight outside the support (e.g. negative or unrealistically high ages). Mazza and Punzo (2013) propose an adaptive bandwidth discrete beta kernel estimator, in which the bandwidth is allowed to vary at each age, according to the reliability of the data as expressed by the e_x.

The present paper proposes a new adaptive bandwidth discrete beta kernel estimator, in which the reliability is measured via the reciprocal of the *variation coefficient* (VC). The VC is function of both the amount of exposure and the observed mortality rate. A simulation study will show the performance increase of the new estimator over the two previous approaches.

The paper can be summarized as follows. In Sect. 2 the fixed bandwidth discrete beta kernel estimator of Mazza and Punzo (2011) is briefly illustrated and in Sect. 3 a new adaptive version is provided. In Sect. 4 cross-validation estimation of the adaptive bandwidth is described. In Sect. 5 a simulation study is performed with the aim to ascertain the gain in performance of the proposed estimator with respect to the approaches in Mazza and Punzo (2011) and Mazza and Punzo (2013).

2 Discrete Beta Kernel Graduation

Given the crude rates $\mathring{q}_y$, $y \in \mathcal{X}$, the Nadaraya–Watson kernel estimator of the true but unknown mortality rate q_x at the evaluation age x is

$$\hat{q}_x = \sum_{y \in \mathcal{X}} \frac{k_h\,(y; m = x)}{\sum_{j \in \mathcal{X}} k_h\,(j; m = x)} \mathring{q}_y = \sum_{y \in \mathcal{X}} K_h\,(y; m = x)\,\mathring{q}_y, \quad x \in \mathcal{X}, \tag{1}$$

where $k_h\,(\cdot; m)$ is the *discrete kernel function* (hereafter simply named *kernel*), $m \in \mathcal{X}$ is the single mode of the kernel, $h > 0$ is the (fixed) *bandwidth* governing the bias-variance trade-off, and $K_h\,(\cdot; m)$ is the *normalized kernel*. Since we are treating age as being discrete, with equally spaced values, kernel graduation by means of (1) is equivalent to moving (or local) weighted average graduation (Gavin et al. 1995).

As kernels in (1) we adopt

$$k_h\,(x; m) = \left(x + \frac{1}{2}\right)^{\frac{m+\frac{1}{2}}{h(\omega+1)}} \left(\omega + \frac{1}{2} - x\right)^{\frac{\omega+\frac{1}{2}-m}{h(\omega+1)}}. \tag{2}$$

The normalized version, $K_h\,(x; m)$, corresponds to the discrete beta distribution defined in Punzo and Zini (2012) and parameterized, as in Punzo (2010), according to the mode m and another parameter h that is closely related to the distribution variability. Substituting (2) in (1) we obtain the discrete beta kernel estimator that was introduced in Mazza and Punzo (2011).

Roughly speaking, discrete beta kernels possess two peculiar characteristics. Firstly, their shape, fixed h, automatically changes according to the value of m. Secondly, the support of the kernels matches the age range $\mathcal{X}$ so that no weight is assigned outside the data support; this means that the order of magnitude of the bias does not increase near the boundaries. Further details are reported in Mazza and Punzo (2011).

3 An Adaptive Variant

Rather than restricting h to a fixed value, a more flexible approach is to allow the bandwidth to vary according to the reliability of the data measured in a convenient way. Thus, for ages in which the reliability is relatively larger, a low value for h results in an estimate that more closely reflects the crude rates. For ages in which the reliability is smaller, such as at old ages, a higher value for h allows the estimate of the true rates of mortality to progress more smoothly; this means that at older ages we are calculating local averages over a greater number of observations. This technique is often referred to as a variable or *adaptive kernel estimator* because it is characterized by an adaptive bandwidth $h_x\,(s)$ which depends on the reliability l_x and is function of a further sensitive parameter s.

Although the reliability l_x can be inserted into the basic model (1) in a number of ways (Gavin et al. 1995), here we adopt a natural formulation according to which

$$h_x(s) = hl_x^s, \quad x \in \mathcal{X}, \tag{3}$$

where h is the global bandwidth and $s \in [0,1]$. Reliability decides the shape of the local factors, while s is necessary to dampen the possible extreme variations in reliability that can arise between young and old ages. Naturally, in the case $s = 0$, we are ignoring the variation in reliability, which gives a fixed bandwidth estimator.

Using (3) we are calculating a different bandwidth for each age $x \in \mathcal{X}$ at which the curve is to be estimated, leading model (1) to become

$$\hat{q}_x = \sum_{y \in \mathcal{X}} \frac{k_{h_x}(y; m = x)}{\sum_{j \in \mathcal{X}} k_{h_x}(j; m = x)} \mathring{q}_y = \sum_{y \in \mathcal{X}} K_{h_x}(y; m = x)\, \mathring{q}_y, \quad x \in \mathcal{X}, \tag{4}$$

where the notation h_x is used to abbreviate $h_x(s)$. Thus, for each evaluation age x, the $\omega + 1$ discrete beta distributions $K_{h_x}(\cdot; m = x)$ vary for the placement of the mode as well as for their variability as measured by h_x.

In particular, Mazza and Punzo (2013) consider the reliability a function only of the amount of exposure, according to the formulation

$$l_x = \frac{f_x^{-1}}{\max_{y \in \mathcal{X}} \left\{ f_y^{-1} \right\}}, \quad x \in \mathcal{X}, \tag{5}$$

where

$$f_x = \frac{e_x}{\sum_{y \in \mathcal{X}} e_y}$$

is the empirical frequency of exposed to the risk of death at age x.

According to the model $d_x \sim \text{Bin}\left(e_x, \mathring{q}_x\right)$, where $\mathring{q}_x$ is the maximum likelihood estimate of q_x, a natural index of reliability is represented by the reciprocal of a relative measure of variability. Here, we have chosen to adopt the variation coefficient (VC) which, in this context, can be computed as

$$\text{VC}_x = \frac{\sqrt{e_x \mathring{q}_x \left(1 - \mathring{q}_x\right)}}{e_x \mathring{q}_x}, \quad x \in \mathcal{X}.$$

It is inserted in (3) according to the formulation

$$l_x = \frac{\mathrm{VC}_x}{\sum\limits_{y\in\mathcal{X}} \mathrm{VC}_y}, \quad x \in \mathcal{X}. \tag{6}$$

In (6), VC_x is normalized so that $l_x^s \in (0, 1]$. Note that reliability measured as in (6) takes into account the amount of exposure e_x, but also the crude rate $\mathring{q}_x$.

4 The Choice of *h* and *s*

In (4), two parameters need to be selected: sensitivity, s, and global bandwidth, h. Although s could be selected by cross-validation, we prefer to choose this parameter subjectively, as in Gavin et al. (1995, see also Mazza and Punzo 2013). Once s has been chosen, cross-validation can be still used to select h.

For model (4), the cross-validation statistic or score, $CV(h|s)$, is

$$CV(h|s) = \sum_{x\in\mathcal{X}} \left(\frac{\hat{q}_x^{(-x)}}{\mathring{q}_x} - 1 \right)^2, \tag{7}$$

where

$$\hat{q}_x^{(-x)} = \sum_{\substack{y\in\mathcal{X}\\ y\neq x}} \frac{K_{h_x}(y; m = x)}{\sum\limits_{\substack{j\in\mathcal{X}\\ j\neq x}} K_{h_x}(j; m = x)} \mathring{q}_y$$

is the estimated value at age x computed by removing the crude rate $\mathring{q}_x$ at that age.

Note that, differently from what is done in Mazza and Punzo (2011) and Mazza and Punzo (2013), instead of the standard residual sum of squares, the sum of the squares of the proportional differences is used in (7); this is a commonly used divergence measure in the graduation literature because, since the high differences in mortality rates among ages, we want the mean relative square error to be low (see Heligman and Pollard 1980).

5 Simulation Study

In this section we present the results of a simulation study based on real data, and accomplished with the aim of comparing the performance of three discrete beta kernel estimators. Such estimators differ with respect to the choice made for the bandwidth and specifically:

B0 Corresponds to the fixed bandwidth estimator in (1).
B1 Corresponds to the adaptive bandwidth estimator in (4) with local factor l_x identified in (5).
B2 Corresponds to the adaptive bandwidth estimator in (4) with local factor l_x defined in (6).

Data we consider, composed of the number of exposed to risk e_x and the crude mortality rates $\mathring{q}_x$, with ω set to 85, are referred to the 2008 Italian male population, composed of over 28 million individuals.[1]

5.1 Design of the Experiments

The scheme of the simulations can be summarized as follows:

1. First of all, we have graduated the $\mathring{q}_x$ via the well-known parametric model of Heligman and Pollard (1980). The graduated rates q_x will be hereafter referred to as the "true" mortality rates.
2. For each replication performed and for each age x, the simulated rates are obtained by dividing the d_x generated from a $\text{Bin}\,(ce_x, q_x)$, by ce_x, where $c \in \mathbb{R}_+$ is a multiplying factor added with the aim of reproducing, when $c < 1$, situations of lower smoothness in the generated sequence $d_0, d_1, \ldots, d_{85}$.
3. For each replication and for each age x, once fixed a grid of 11 equally-spaced values for s ranging from 0 to 1, the global bandwidth h in (3), of the estimator $\hat{q}_x$ in (4), is obtained by minimizing the cross-validation statistic $CV\,(h|s)$ in (7). Here it must be noted that *B0* is obtained, as a special case, by taking $s = 0$.
4. For each replication and for each value of s, in line with the divergence defined in (7), the comparison between the smoothed and the "true" mortality rates is dealt via the sum of the squares of the proportional differences

$$S^2 = \sum_{x=0}^{85} \left(\frac{\hat{q}_x}{q_x} - 1 \right)^2.$$

5.2 Results

Table 1 shows simulation results, for the original case $c = 1$, at the varying of s. Here, each subtable corresponds to a (3×3)-matrix having, on the diagonal in bold, the number of times in which each estimator (respectively *B0*, *B1* and *B2*) obtains the minimum S^2 and, outside the diagonal, the number of times in which the

[1] Istat: data available from http://demo.istat.it/.

Table 1 Simulation results for the case $c = 1$

(a) $s = 0.1$

	B0	*B1*	*B2*
B0	**95**	695	106
B1	305	**12**	81
B2	894	919	**893**

(b) $s = 0.2$

	B0	*B1*	*B2*
B0	**116**	725	133
B1	275	**21**	113
B2	867	887	**863**

(c) $s = 0.3$

	B0	*B1*	*B2*
B0	**139**	759	163
B1	241	**24**	139
B2	837	861	**837**

(d) $s = 0.4$

	B0	*B1*	*B2*
B0	**160**	794	184
B1	206	**25**	157
B2	816	843	**815**

(e) $s = 0.5$

	B0	*B1*	*B2*
B0	**195**	827	212
B1	173	**18**	184
B2	788	816	**787**

(f) $s = 0.6$

	B0	*B1*	*B2*
B0	**226**	854	243
B1	146	**17**	201
B2	757	799	**757**

(g) $s = 0.7$

	B0	*B1*	*B2*
B0	**245**	882	261
B1	118	**16**	223
B2	739	777	**739**

(h) $s = 0.8$

	B0	*B1*	*B2*
B0	**278**	910	292
B1	89	**16**	244
B2	707	755	**705**

(i) $s = 0.9$

	B0	*B1*	*B2*
B0	**305**	930	317
B1	69	**13**	267
B2	682	732	**681**

(j) $s = 1.0$

	B0	*B1*	*B2*
B0	**341**	951	352
B1	49	**12**	296
B2	648	704	**647**

Table 2 Simulation results for the case $c = 0.5$

(a) $s = 0.1$

	B0	*B1*	*B2*
B0	**122**	806	132
B1	194	**10**	113
B2	868	887	**868**

(b) $s = 0.2$

	B0	*B1*	*B2*
B0	**138**	821	153
B1	179	**15**	132
B2	847	868	**847**

(c) $s = 0.3$

	B0	*B1*	*B2*
B0	**161**	845	179
B1	155	**19**	151
B2	821	849	**820**

(d) $s = 0.4$

	B0	*B1*	*B2*
B0	**182**	867	197
B1	133	**16**	168
B2	803	832	**802**

(e) $s = 0.5$

	B0	*B1*	*B2*
B0	**199**	883	212
B1	117	**14**	182
B2	788	818	**787**

(f) $s = 0.6$

	B0	*B1*	*B2*
B0	**226**	902	238
B1	98	**12**	203
B2	762	797	**762**

(g) $s = 0.7$

	B0	*B1*	*B2*
B0	**245**	917	260
B1	83	**16**	223
B2	740	777	**739**

(h) $s = 0.8$

	B0	*B1*	*B2*
B0	**275**	934	289
B1	66	**14**	249
B2	711	751	**711**

(i) $s = 0.9$

	B0	*B1*	*B2*
B0	**308**	944	321
B1	56	**15**	269
B2	679	731	**677**

(j) $s = 1.0$

	B0	*B1*	*B2*
B0	**338**	954	346
B1	46	**10**	292
B2	654	708	**652**

estimator on the corresponding row provides a value of S^2 lower than the one of the model on the corresponding column. From the diagonal numbers we can easily see as *B2* is, regardless from s, the best performing estimator, with a gain in performance that slightly increases at the decreasing of s. On the contrary, surprisingly, *B1* is the worst working estimator. Pairwise comparisons among estimators (see the elements outside the diagonals) lead to the same global results. Interestingly, from Table 1e–j, corresponding to the values of s ranging from 0.5 to 1, it can be noted that *B1* beats *B2* more times than *B0*.

Similarly, Table 2 shows simulation results for the case $c = 0.5$. Also in this case, *B2* is clearly the estimator working better (see the values on the diagonals) while *B1* is the one working worse. Global results are preserved with respect to the previous case.

6 Concluding Remarks

In this paper an adaptive version of the discrete beta kernel estimator introduced in Mazza and Punzo (2011) for the graduation of mortality rates has been proposed.

This proposal, differently from the one presented in Mazza and Punzo (2013), allows the bandwidth to vary according to either the amount of exposure and the crude rates themselves, via the well-known variation coefficient. A further sensitivity parameter s has been added, to allow the user to control the degree of emphasis placed on the local factor. The usual bandwidth h is used to control the global level of smoothness. Simulations have confirmed the gain in performance of this new approach with respect to the ones in Mazza and Punzo (2011) and Mazza and Punzo (2013). Finally, it is important to note that the resulting adaptive discrete beta kernel graduation is conceptually simple and so is its implementation.

References

Copas, J. B., & Haberman, S. (1983). Non-parametric Graduation using kernel methods. *Journal of the Institute of Actuaries, 110*(1), 135–156.

Debn, A., Montes, F., & Sala, R. (2006). A comparison of nonparametric methods in the graduation of mortality: application to data from the valencia region (Spain). *International Statistical Review, 74*(2), 215–233.

Gavin, J., Haberman, S., & Verrall, R. (1993). Moving weighted average graduation using kernel estimation. *Insurance: Mathematics and Economics, 12*(2), 113–126.

Gavin, J., Haberman, S., & Verrall, R. (1994). On the choice of bandwidth for kernel graduation. *Journal of the Institute of Actuaries, 121*(1), 119–134.

Gavin, J., Haberman, S., & Verrall, R. (1995). Graduation by kernel and adaptive kernel methods with a boundary correction. *Transactions of Society of Actuaries, 47*, 173–209.

Haberman, S., & Renshaw, A. (1996). Generalized linear models and actuarial science. *The Statistician, 45*(4), 407–436.

Härdle, W. (1992). *Applied nonparametric regression*. Cambridge: Cambridge University Press.

Heligman, L., & Pollard, J. (1980). The age pattern of mortality. *Journal of the Institute of Actuaries, 107*(1), 49–80.

London, D. (1985). *Graduation: the revision of estimates*. Connecticut: Actex publications.

Mazza, A., & Punzo, A. (2011). Discrete beta kernel graduation of age-specific demographic indicators. In S. Ingrassia, R. Rocci, & M. Vichi, (Eds.) *New perspectives in statistical modeling and data analysis, Studies in classification, data analysis and knowledge organization* (pp. 127–134). Berlin-Heidelberg: Springer.

Mazza, A., & Punzo, A. (2013). Graduation by adaptive discrete beta kernels. In A. Giusti, G. Ritter, & M. Vichi (Eds.) *Classification and data mining, Studies in classification, data analysis and knowledge organization* (pp. 243–250). Berlin/Heidelberg: Springer.

Peristera, P., & Kostaki, A. (2005). An evaluation of the performance of kernel estimators for graduating mortality data. *Journal of Population Research, 22*(2), 185–197.

Punzo, A. (2010). Discrete beta-type models. In H. Locarek-Junge, & C. Weihs (Eds.) *Classification as a tool for research, Studies in classification, data analysis and knowledge organization* (pp. 253–261). Berlin/Heidelberg: Springer.

Punzo, A., & Zini, A. (2012). Discrete approximations of continuous and mixed measures on a compact interval. *Statistical Papers, 53*(3), 563–575.

On Clustering and Classification Via Mixtures of Multivariate t-Distributions

Paul D. McNicholas

Abstract The use of mixture models for clustering and classification has received renewed attention within the literature since the mid-1990s. The multivariate Gaussian distribution has been at the heart of this body of work, but approaches that utilize the multivariate t-distribution have burgeoned into viable and effective alternatives. In this paper, recent work on classification and clustering using mixtures of multivariate t-distributions is reviewed and discussed, along with related issues. The paper concludes with a summary and suggestions for future work.

1 Introduction

A finite mixture model is a convex combination of a finite number of probability densities. Formally, a p-dimensional random vector $\mathbf{X}$ arises from a parametric finite mixture distribution if, for all $\mathbf{x} \subset \mathbf{X}$, we can write its density as $f(\mathbf{x} \mid \boldsymbol{\vartheta}) = \sum_{g=1}^{G} \pi_g \gamma_g(\mathbf{x} \mid \boldsymbol{\theta}_g)$, where $\pi_g > 0$, such that $\sum_{g=1}^{G} \pi_g = 1$ are the mixing proportions, $\gamma_1(\mathbf{x} \mid \boldsymbol{\theta}_g), \ldots, \gamma_G(\mathbf{x} \mid \boldsymbol{\theta}_g)$ are the component densities, and $\boldsymbol{\vartheta} = (\boldsymbol{\pi}, \boldsymbol{\theta}_1, \ldots, \boldsymbol{\theta}_G)$ is the vector of parameters with $\boldsymbol{\pi} = (\pi_1, \ldots, \pi_G)$. Note that $f(\mathbf{x} \mid \boldsymbol{\vartheta})$ is called a G-component finite mixture density. Finite mixture models lend themselves naturally to clustering and classification problems, where a class corresponds to one or more mixture components. The component densities are usually taken to be of the same type. Gaussian densities have been predominant since mixture models were first used for clustering (Wolfe 1963); of late, however, there has been a marked increase in the preponderance of non-Gaussian mixture model-based clustering and classification work within the literature. Hereafter, the idiom "model-based clustering" will be used to mean clustering using mixture

P.D. McNicholas (✉)
University of Guelph, Ontario, Canada
e-mail: pmcnicho@uoguelph.ca.

P. Giudici et al. (eds.), *Statistical Models for Data Analysis*, Studies in Classification, Data Analysis, and Knowledge Organization, DOI 10.1007/978-3-319-00032-9_27,

models. The terms "model-based classification" and "model-based discriminant analysis" are used similarly.

Let $\mathbf{z}_i$ denote the component membership of observation i, so that $z_{ig} = 1$ if observation i belongs to component g and $z_{ig} = 0$ otherwise. Suppose n p-dimensional data vectors $\mathbf{x}_1, \ldots, \mathbf{x}_n$ are observed, all of which have unknown group memberships, and we wish to cluster these data into G components. The Gaussian model-based clustering likelihood can be written $\mathscr{L}(\boldsymbol{\vartheta} \mid \mathbf{x}) = \prod_{i=1}^{n} \sum_{g=1}^{G} \pi_g \phi(\mathbf{x}_i \mid \boldsymbol{\mu}_g, \boldsymbol{\Sigma}_g)$, where $\boldsymbol{\mu}_g$ is the mean vector and $\boldsymbol{\Sigma}_g$ is the covariance matrix for component g. In practice, one must often consider the introduction of parsimony. There are $(G-1)+Gp+Gp(p+1)/2$ free parameters in this Gaussian mixture model. Because $Gp(p+1)/2$ of these parameters are contained within the component covariance matrices, the imposition of constraints upon covariance matrices has become a popular solution. The models that result from the imposition of such constraints, together with the unconstrained model, can be collectively referred to as a "family" of mixture models. Families of mixture models will be revisited in Sect. 2 but first we describe model-based classification and discriminant analysis.

In the model-based classification paradigm, we have n observations of which k have known group memberships. Without loss of generality, we can order these n observations so that the first k have known group memberships. Assuming that each of the known groups corresponds to a mixture component, the model-based classification likelihood is

$$\mathscr{L}(\boldsymbol{\vartheta} \mid \mathbf{x}, \mathbf{z}) = \prod_{i=1}^{k} \prod_{g=1}^{G} \left[\pi_g \phi(\mathbf{x}_i \mid \boldsymbol{\mu}_g, \boldsymbol{\Sigma}_g)\right]^{z_{ig}} \prod_{j=k+1}^{n} \sum_{h=1}^{H} \pi_h \phi(\mathbf{x}_j \mid \boldsymbol{\mu}_h, \boldsymbol{\Sigma}_h), \quad (1)$$

for $H \geq G$. The fact that we find a model with a number of groups (H) greater than that already observed (G) gives model-based classification an added flexibility. From the likelihood in Eq. (1), model-based clustering is clearly a special case of model-based classification that arises upon setting $k = 0$.

In the model-based discriminant analysis framework, we again have n observations of which k have known group memberships, ordered so that the first k have known group memberships. Here, rather than using all n observations to estimate the unknown parameters—and so unknown component memberships—we use only the first k. First, we form the likelihood $\mathscr{L}(\boldsymbol{\vartheta} \mid \mathbf{x}, \mathbf{z}) = \prod_{i=1}^{k} \prod_{g=1}^{G} \left[\pi_g \phi(\mathbf{x}_i \mid \boldsymbol{\mu}_g, \boldsymbol{\Sigma}_g)\right]^{z_{ig}}$ based on these k observations. Then, using the maximum likelihood estimates arising from this likelihood, we compute the expected values

$$\hat{z}_{jg} := \frac{\hat{\pi}_g \phi(\mathbf{x}_j \mid \hat{\boldsymbol{\mu}}_g, \hat{\boldsymbol{\Sigma}}_g)}{\sum_{h=1}^{G} \hat{\pi}_h \phi(\mathbf{x}_j \mid \hat{\boldsymbol{\mu}}_h, \hat{\boldsymbol{\Sigma}}_h)},$$

for $j = k+1, \ldots, n$. These expected values play the role of a discriminant rule and the predicted group memberships are given by the maximum *a posteriori* (MAP) classifications MAP$\{\hat{z}_{jg}\}$, where MAP$\{\hat{z}_{jg}\} = 1$ if $\max_g\{\hat{z}_{jg}\}$ occurs at component g

and $\text{MAP}\{\hat{z}_{jg}\} = 0$ otherwise, for $j = k + 1, \ldots, n$. Note that it is also possible to use multiple components for each known group when constructing this discriminant rule (cf. Fraley and Raftery 2002).

The expectation-maximization (EM) algorithm (Dempster et al. 1977) and its variants are usually used for parameter estimation for model-based clustering, classification, and discriminant analysis. The EM algorithm is an iterative procedure used to find maximum likelihood estimates when data are or are taken to be incomplete. The EM algorithm is an iterative procedure based on the "complete-data" likelihood, i.e., the likelihood of the observed plus the missing data. Note that following application of the EM algorithm, it is common to report the MAP classifications. Comprehensive details on EM algorithms, including their application to finite mixture models, are given by McLachlan and Krishnan (2008).

The remainder of this paper is laid out as follows. The notion of a family of mixture models is further developed in Sect. 2, with three Gaussian families used for illustration. Work on mixtures of multivariate t-distributions is then discussed (Sect. 3) before the paper concludes with some discussion (Sect. 4).

2 Three Families of Gaussian Mixture Models

This section expands on the notion of a family of mixture models, using three Gaussian families for illustration. These families were chosen because of the availability of R (R Development Core Team 2012) packages for their implementation and the existence of t-analogues. As suggested in Sect. 1, these families are based on decomposed component covariance matrices. The reader should note that there are very interesting Gaussian families within the literature (e.g., Bouveyron et al. 2007) that are not discussed herein.

The MCLUST family (Banfield and Raftery 1993; Celeux and Govaert 1995; Fraley and Raftery 2002) of Gaussian mixture models is the most well-established family within the literature. The MCLUST family, which is supported by the `mclust` package (Fraley and Raftery 2006) for R, is based on eigen-decomposed component covariance structures so that the gth component covariance structure is $\boldsymbol{\Sigma}_g = \lambda_g \mathbf{D}_g \mathbf{A}_g \mathbf{D}'_g$, where $\mathbf{D}_g$ is the matrix of eigenvectors of $\boldsymbol{\Sigma}_g$, $\mathbf{A}_g$ is the diagonal matrix with entries proportional to the eigenvalues of $\boldsymbol{\Sigma}_g$, and λ_g is the relevant constant of proportionality. Imposing constraints on this covariance structure gives a family of ten multivariate mixture models (cf. Fraley and Raftery 2006, Table 1). The reader should note that the MCLUST family is a subset of the Gaussian parsimonious clustering models of Celeux and Govaert (1995). The MCLUST family is most commonly used for model-based clustering but Fraley and Raftery (2002) also illustrate both discriminant analysis and density estimation applications. Dean et al. (2006) used the MCLUST family for both classification and discriminant analysis.

The second family we will consider is an outgrowth of the mixture of factor analyzers model. The factor analysis model (Spearman 1904; Bartlett 1953) is a well known multivariate statistical technique that models a p-dimensional random vector $\mathbf{X}$ using a q-dimensional vector of latent factors $\mathbf{U}$, where $q \ll p$. The model can be written $\mathbf{X} = \boldsymbol{\mu} + \boldsymbol{\Lambda}\mathbf{U} + \boldsymbol{\epsilon}$, where $\boldsymbol{\Lambda}$ is a $p \times q$ matrix of factor loadings, the latent factors $\mathbf{U} \sim \mathcal{N}(\mathbf{0}, \mathbf{I}_q)$, and $\boldsymbol{\epsilon} \sim \mathcal{N}(\mathbf{0}, \boldsymbol{\Psi})$, where $\boldsymbol{\Psi}$ is a $p \times p$ diagonal matrix with strictly positive entries. From this model, the marginal distribution of $\mathbf{X}$ is $\mathcal{N}(\boldsymbol{\mu}, \boldsymbol{\Lambda}\boldsymbol{\Lambda}' + \boldsymbol{\Psi})$. Ghahramani and Hinton (1997) introduced a mixture of factor analyzers model; the density is that of a finite Gaussian mixture model with $\boldsymbol{\Sigma}_g = \boldsymbol{\Lambda}_g\boldsymbol{\Lambda}_g' + \boldsymbol{\Psi}$. Around the same time, a mixture of probabilistic principal component analyzers model was introduced (Tipping and Bishop 1997, 1999), wherein the $\boldsymbol{\Psi}_g$ matrices are isotropic so that $\boldsymbol{\Sigma}_g = \boldsymbol{\Lambda}_g\boldsymbol{\Lambda}_g' + \psi_g\mathbf{I}_p$ for each component. McLachlan and Peel (2000) then proposed a more general mixture of factor analyzers model with $\boldsymbol{\Sigma}_g = \boldsymbol{\Lambda}_g\boldsymbol{\Lambda}_g' + \boldsymbol{\Psi}_g$. More recently, McNicholas and Murphy (2005, 2008) developed a family of eight Gaussian mixture models by imposing constraints upon the most general covariance structure $\boldsymbol{\Sigma}_g = \boldsymbol{\Lambda}_g\boldsymbol{\Lambda}_g' + \boldsymbol{\Psi}_g$. They used the constraints $\boldsymbol{\Lambda}_g = \boldsymbol{\Lambda}$, $\boldsymbol{\Psi}_g = \boldsymbol{\Psi}$, and $\boldsymbol{\Psi}_g = \psi_g\mathbf{I}_p$, with the resulting family of mixture models called the parsimonious Gaussian mixture model (PGMM) family. McNicholas and Murphy (2010b) modified the factor analysis covariance structure by writing $\boldsymbol{\Psi}_g = \omega_g\boldsymbol{\Delta}_g$, where $\omega_g \in \mathbb{R}^+$ and $\boldsymbol{\Delta}_g$ is a diagonal matrix with $|\boldsymbol{\Delta}_g| = 1$. This modified factor analysis covariance structure, $\boldsymbol{\Sigma}_g = \boldsymbol{\Lambda}_g\boldsymbol{\Lambda}_g' + \omega_g\boldsymbol{\Delta}_g$, along with all legitimate combinations of the constraints $\boldsymbol{\Lambda}_g = \boldsymbol{\Lambda}$, $\omega_g = \omega$, $\boldsymbol{\Delta}_g = \boldsymbol{\Delta}$, and $\boldsymbol{\Delta}_g = \mathbf{I}_p$, give a family of twelve PGMMs (cf. McNicholas and Murphy 2010b, Tables 1 and 2). McNicholas and Murphy (2005, 2008, 2010b) focused on clustering applications, while the PGMM family was used by McNicholas (2010) for model-based classification and by Andrews and McNicholas (2011b) for model-based discriminant analysis. This family is supported by the pgmm package (McNicholas et al. 2011).

The final family we will consider was designed specifically for the model-based clustering of longitudinal data. McNicholas and Murphy (2010a) used the modified Cholesky decomposition (cf. Pourahmadi 1999) to decompose each component precision matrix as $\boldsymbol{\Sigma}_g^{-1} = \mathbf{T}_g'\mathbf{D}_g^{-1}\mathbf{T}_g$, where $\mathbf{T}_g$ is a unique lower unitriangular matrix and $\mathbf{D}_g$ is a unique diagonal matrix with strictly positive diagonal entries. The values of $\mathbf{T}_g$ and $\mathbf{D}_g$ can be interpreted as generalized autoregressive parameters and innovation variances, respectively (cf. Pourahmadi 1999). Constraints are imposed on $\mathbf{T}_g$ and $\mathbf{D}_g$ to give a family of eight mixture models (cf. McNicholas and Murphy 2010a, Table 1). This family is available, for both clustering and classification, within the `longclust` package (McNicholas et al. 2012) for R.

3 Mixtures of Multivariate *t*-Distributions

Model-based clustering using mixtures of multivariate t-distributions has been around for some time (McLachlan and Peel 1998; Peel and McLachlan 2000). The density of a mixture of multivariate t-distributions is

$$f(\mathbf{x} \mid \boldsymbol{\vartheta}) = \sum_{g=1}^{G} \pi_g f_t(\mathbf{x} \mid \boldsymbol{\mu}_g, \boldsymbol{\Sigma}_g, \nu_g),$$

where $f_t(\mathbf{x} \mid \boldsymbol{\mu}_g, \boldsymbol{\Sigma}_g, \nu_g)$ is the density of a p-dimensional multivariate t-distribution with mean $\boldsymbol{\mu}_g$, covariance matrix $\boldsymbol{\Sigma}_g$, and ν_g degrees of freedom. An elegant parameter estimation framework, based on an EM algorithm, is described by McLachlan and Peel (1998). Building on this work, Zhao and Jiang (2006) developed a t-analogue of the probabilistic principal components analysis model and considered mixtures thereof, and McLachlan et al. (2007) introduced a mixture of t-factor analyzers model. Andrews and McNicholas (2011a,b) expanded on this work to introduce a t-analogue of the PGMM family for model-based clustering, classification, and discriminant analysis. They also considered constraining degrees of freedom to be equal across components and observed that this can lead to superior classification performance in some cases. Andrews et al. (2011) illustrated the performance of a four-member family of mixtures of multivariate t-distributions for model-based classification and Steane et al. (2012) studied two specific mixture of t-factors models for model-based classification.

Andrews and McNicholas (2012a) developed a t-analogue of the MCLUST family of models, including two additional covariance structures (i.e., they use 12 of the 14 models of Celeux and Govaert 1995), for model-based clustering, classification, and discriminant analysis. This family of models is supported by the `teigen` package for R (Andrews and McNicholas 2012b). McNicholas and Subedi (2012) developed a t-analogue of the approach of McNicholas and Murphy (2010a) to cluster and classify longitudinal data; they also considered modelling the component means. The `longclust` package for R implements the approach of McNicholas and Subedi (2012) while also allowing the user to select a Gaussian mixture model.

In addition to the aforementioned work on mixtures of multivariate t-distributions, which focuses primarily on the three families of models, there have been several other notable contributions. Space restrictions do not allow for an exhaustive exploration but it seems that some papers deserve special mention. In particular, Shoham (2002) introduce a deterministic agglomeration EM algorithm for mixtures of multivariate t-distributions; and Greselin and Ingrassia (2010a,b) discuss weakly constrained monotone EM algorithms and homoscedastic constraints, respectively, for mixtures of multivariate t-distributions.

4 Discussion

The move from the mixture of multivariate Gaussian distributions to its t-analogue can be considered a more flexible approach. However, there are some downsides as well as some unexpected features. In the latter vein, there is the somewhat surprising effect of constraining the component degrees of freedom to be equal. Although it saves very few parameters, this constraint ($\nu_g = \nu$) can actually lead to improved clustering and classification performance (see Andrews and McNicholas 2011a, for an example). This might be explained, at least in part, by the fact that components might better estimate their degrees of freedom ν_g when borrowing from other components, especially when the sample size is not large. The principal downside of t-mixtures seems to be the extra difficulty involved in parameter estimation; specifically, the greater importance of starting values for the EM algorithm for the t-mixture models over their Gaussian counterparts has been noted, as well as the relatively expensive degrees of freedom update.

Of course, many problems that exist within the Gaussian paradigm carry over into the t-analogues. A glaring example is model selection, where the approach of choice is the Bayesian information criterion (BIC, Schwarz 1978). While it can be effective and there is some theoretical support for its use (cf. Keribin 1998), the BIC leaves room for improvement. There is also the question around whether the best model should be selected at all, as opposed to combining the classification results of the best few models via some model averaging approach. Work is ongoing on both model selection and model averaging for model-based clustering, classification, and discriminant analysis.

The role that work on mixtures of multivariate t-distributions will play, within a historical context, is worthy of consideration. They will almost certainly be regarded as the tip of the non-Gaussian iceberg and, to an extent, the model-based clustering and classification literature has already moved beyond mixtures of multivariate t-distributions. Lin (2010), Lee and McLachlan (2011), and Vrbik and McNicholas (2012) have conducted interesting work on mixtures of skew-t distributions and Karlis and Santourian (2009) discuss non-elliptically contoured distributions, amidst a host of other work on non-Gaussian model-based clustering.

References

Andrews, J. L., & McNicholas, P. D. (2011a). Extending mixtures of multivariate t-factor analyzers. *Statistics and Computing, 21*(3), 361–373.

Andrews, J. L., & McNicholas, P. D. (2011b). Mixtures of modified t-factor analyzers for model-based clustering, classification, and discriminant analysis. *Journal of Statistical Planning and Inference, 141*(4), 1479–1486.

Andrews, J. L., & McNicholas, P. D. (2012a). Model-based clustering, classification, and discriminant analysis via mixtures of multivariate t-distributions. *Statistics and Computing, 22*(5), 1021–1029.

Andrews, J. L., & McNicholas, P. D. (2012b). teigen: Model-based clustering and classification with the multivariate t-distribution. R package version 1.0.

Andrews, J. L., McNicholas, P. D., & Subedi, S. (2011). Model-based classification via mixtures of multivariate t-distributions. *Computational Statistics and Data Analysis, 55*(1), 520–529.

Banfield, J. D., & Raftery, A. E. (1993). Model-based Gaussian and non-Gaussian clustering. *Biometrics, 49*(3), 803–821.

Bartlett, M. S. (1953). Factor analysis in psychology as a statistician sees it. In *Uppsala symposium on psychological factor analysis*, no. 3 in Nordisk Psykologi's Monograph Series (pp. 23–43).

Bouveyron, C., Girard, S., & Schmid, C. (2007). High-dimensional data clustering. *Computational Statistics and Data Analysis, 52*(1), 502–519.

Celeux, G., & Govaert, G. (1995). Gaussian parsimonious clustering models. *Pattern Recognition, 28*(5), 781–793.

Dean, N., Murphy, T. B., & Downey, G. (2006). Using unlabelled data to update classification rules with applications in food authenticity studies. *Journal of the Royal Statistical Society: Series C, 55*(1), 1–14.
Dempster, A. P., Laird, N. M., & Rubin, D. B. (1977). Maximum likelihood from incomplete data via the EM algorithm. *Journal of the Royal Statistical Society: Series B 39*(1), 1–38.
Fraley, C., & Raftery, A. E. (2002). Model-based clustering, discriminant analysis, and density estimation. *Journal of the American Statistical Association 97*(458), 611–631.
Fraley, C., & Raftery, A. E. (2006). MCLUST version 3 for R: Normal mixture modeling and model-based clustering. Technical Report 504, Department of Statistics, University of Washington, minor revisions January 2007 and November 2007.
Ghahramani, Z., & Hinton, G. E. (1997). The EM algorithm for factor analyzers. Tech. Rep. CRG-TR-96-1, University of Toronto, Toronto.
Greselin, F., & Ingrassia, S. (2010a). Constrained monotone EM algorithms for mixtures of multivariate t distributions. *Statistics and Computing, 20*(1), 9–22.
Greselin, F., & Ingrassia, S. (2010b). Weakly homoscedastic constraints for mixtures of t-distributions. In A. Fink, B. Lausen, W. Seidel, & A. Ultsch (Eds.) *Advances in data analysis, data handling and business intelligence, studies in classification, data analysis, and knowledge organization* (pp. 219–228). Berlin/Heidelberg: Springer.
Karlis, D., & Santourian, A. (2009). Model-based clustering with non-elliptically contoured distributions. *Statistics and Computing, 19*(1), 73–83.
Keribin, C. (1998). Estimation consistante de l'ordre de modèles de mélange. *Comptes Rendus de l'Académie des Sciences Série I Mathématique 326*(2), 243–248.
Lee, S., & McLachlan, G. J. (2011). On the fitting of mixtures of multivariate skew t-distributions via the EM algorithm. ArXiv:1109.4706.
Lin, T. I. (2010). Robust mixture modeling using multivariate skew t distributions. *Statistics and Computing, 20*(3), 343–356.
McLachlan, G. J., & Krishnan, T. (2008). *The EM algorithm and extensions* (2nd edn.). New York: Wiley.
McLachlan, G. J., & Peel, D. (1998). Robust cluster analysis via mixtures of multivariate t-distributions. In *Lecture notes in computer science* (vol. 1451, pp. 658–666). Berlin: Springer.
McLachlan, G. J., & Peel, D. (2000). Mixtures of factor analyzers. In *Proceedings of the seventh international conference on machine learning* (pp. 599–606). San Francisco: Morgan Kaufmann.
McLachlan, G. J., Bean, R. W., & Jones, L. B. T. (2007). Extension of the mixture of factor analyzers model to incorporate the multivariate t-distribution. *Computational Statistics and Data Analysis, 51*(11), 5327–5338.
McNicholas, P. D. (2010). Model-based classification using latent Gaussian mixture models. *Journal of Statistical Planning and Inference, 140*(5), 1175–1181.
McNicholas, P. D., & Murphy, T. B. (2005). Parsimonious Gaussian mixture models. Tech. Rep. 05/11, Department of Statistics, Trinity College Dublin, Dublin, Ireland.
McNicholas, P. D., & Murphy, T. B. (2008). Parsimonious Gaussian mixture models. *Statistics and Computing, 18*(3), 285–296.
McNicholas, P. D., & Murphy, T. B. (2010a). Model-based clustering of longitudinal data. *The Canadian Journal of Statistics, 38*(1), 153–168.
McNicholas, P. D., & Murphy, T. B. (2010b). Model-based clustering of microarray expression data via latent Gaussian mixture models. *Bioinformatics, 26*(21), 2705–2712.
McNicholas, P. D., & Subedi, S. (2012). Clustering gene expression time course data using mixtures of multivariate t-distributions. *Journal of Statistical Planning and Inference, 142*(5), 1114–1127.
McNicholas, P. D., Jampani, K. R., McDaid, A. F., Murphy, T. B., & Banks, L. (2011). pgmm: Parsimonious Gaussian Mixture Models. R package version 1.0.
McNicholas, P. D., Jampani, K. R., & Subedi, S. (2012). longclust: Model-Based Clustering and Classification for Longitudinal Data. R package version 1.1.

Peel, D., & McLachlan, G. J. (2000). Robust mixture modelling using the t distribution. *Statistics and Computing, 10*(4), 339–348.

Pourahmadi, M. (1999). Joint mean-covariance models with applications to longitudinal data: unconstrained parameterisation. *Biometrika, 86*(3), 677–690.

R Development Core Team. (2012). R: A Language and Environment for Statistical Computing. R Foundation for Statistical Computing, Vienna, Austria, URL http://www.R-project.org.

Schwarz, G. (1978). Estimating the dimension of a model. *The Annals of Statistics 6*(2), 461–464.

Shoham, S. (2002). Robust clustering by deterministic agglomeration EM of mixtures of multivariate t-distributions. *Pattern Recognition 35*(5), 1127–1142.

Spearman, C. (1904). The proof and measurement of association between two things. *American Journal of Psychology, 15*, 72–101.

Steane, M. A., McNicholas, P. D., & Yada, R. Y. (2012). Model-based classification via mixtures of multivariate t-factor analyzers. *Communications in Statistics – Simulation and Computation, 41*(4), 510–523.

Tipping, T. E., & Bishop, C. M. (1997). Mixtures of probabilistic principal component analysers. Tech. Rep. NCRG/97/003, Aston University (Neural Computing Research Group), Birmingham, UK.

Tipping, T. E., & Bishop, C. M. (1999). Mixtures of probabilistic principal component analysers. *Neural Computation, 11*(2), 443–482.

Vrbik, I., & McNicholas, P. D. (2012). Analytic calculations for the EM algorithm for multivariate skew-mixture models. *Statistics & Probability Letters 82*(6), 1169–1174.

Wolfe, J. H. (1963). *Object cluster analysis of social areas*. Master's thesis, University of California, Berkeley.

Zhao, J., & Jiang, Q. (2006). Probabilistic PCA for t distributions. *Neurocomputing, 69*(16–18), 2217–2226.

Simulation Experiments for Similarity Indexes Between Two Hierarchical Clusterings

Isabella Morlini

Abstract In this paper we report results of a series of simulation experiments aimed at comparing the behavior of different similarity indexes proposed in the literature for comparing two hierarchical clusterings on the basis of the whole dendrograms. Simulations are carried out over different experimental conditions.

1 Introduction

Morlini and Zani (2012) have proposed a new dissimilarity index for comparing two hierarchical clusterings on the basis of the whole dendrograms. They have presented and discussed its basic properties and have shown that the index can be decomposed into contributions pertaining to each stage of the hierarchies. Then, they have obtained a similarity index S as the complement to one of the suggested distance and have shown that its single components S_k obtained at each stage k of the hierarchies can be related to the measure B_k suggested by Fowlkes and Mallows (1983) and to the Rand index R_k. In this paper, we report results of a series of simulation experiments aimed at comparing the behavior of these new indexes with other well-established similarity measures, over different experimental conditions. The first set of simulations is aimed at determining the behavior of the indexes when the clusterings being compared are unrelated. The second set tries to investigate the robustness to different levels of noise. The paper is organized as follows. In Sect. 2 we report the indexes recently proposed in Morlini and Zani (2012) and the similarity indexes used as benchmarks in the simulation studies. We also illustrate some of the properties of these indexes, together with theirs limitations and the

I. Morlini (✉)
Department of Economics, University of Modena and Reggio Emilia, Via Berengario 51, 41100 Modena, Italy
e-mail: isabella.morlini@unimore.it

P. Giudici et al. (eds.), *Statistical Models for Data Analysis*, Studies in Classification, Data Analysis, and Knowledge Organization, DOI 10.1007/978-3-319-00032-9_28,

implied assumptions underlying them. In Sects. 3 and 4 we report results obtained in the simulations. In Sect. 5 we give some concluding remarks.

2 The Indexes

Consider two hierarchical clusterings (or dendrograms) of the same number of objects, n. For measuring the agreement between two non trivial partitions in k clusters ($k = 2, \dots, n-1$) at a certain stage of the procedure, an important class of similarity indexes is based on the quantities T_k, U_k, P_k and Q_k reported in Table 1. This table is a (2×2) contingency table, showing the cluster membership of the $N = n(n-1)/2$ object pairs in each of the two partitions. Among the indexes defined on counting the object pairs on which the two partitions agree or disagree, the most popular ones are perhaps the Rand index:

$$R_k = \frac{N - P_k - Q_k + 2T_k}{N - U_k}, \tag{1}$$

and the criterion B_k suggested by Fowlkes and Mallows (1983):

$$B_k = \frac{T_k}{\sqrt{P_k Q_k}}. \tag{2}$$

The simple matching coefficient, formulated in terms of the quantities in Table 1, is equivalent to the Rand index, while the Jaccard coefficient is $J_k = T_k/(N - U_k$. In Morlini and Zani (2012) we have proposed the following new measure S_k:

$$S_k = \frac{\sum_{j=2}^{n-1} P_j + \sum_{j=2}^{n-1} Q_j - P_k - Q_k + 2T_k}{\sum_{j=2}^{n-1} P_j + \sum_{j=2}^{n-1} Q_j}. \tag{3}$$

The complement to one of S_k, $Z_k = 1 - S_k$, is a metric bounded in [0,1]. This metric takes value 0 if and only if the two clusterings in k groups are identical and value 1 when the two clusterings have the maximum degree of dissimilarity, that is when for each partition in k groups and for each pair i, objects in pair i are in the same group in clustering 1 and in two different groups in clustering 2 (or vice versa). The statistics B_k, J_k and S_k may be thought of as resulting from two different methods of scaling T_k to lie in the unit interval. In these indexes the pairs U_k, which are not joined in either of the two clusterings, are not considered as indicative of similarity. On the contrary, in the Rand index the counts U_k are considered as indicative of similarity. With many clusters U_k must necessarily be large and the inclusion of this count makes R_k tending to 1, for large k. How the treatment of the pairs U_k may influence so much the values of R_k, for different k, is illustrated in Wallace (1983). R_k and S_k may be related to distance measures defined on Table 1, like the

Table 1 Contingency table of the cluster membership of the N object pairs

	Second clustering ($g = 2$)		
First clustering ($g = 1$)	Pairs in the same cluster	Pairs in different clusters	Sum
Pairs in the same cluster	T_k	$P_k - T_k$	P_k
Pairs in different clusters	$Q_k - T_k$	$U_k = N - T_k - P_k - Q_k + 2T_k$	$N - P_k$
Sum	Q_k	$N - Q_k$	$N = n(n-1)/2$

Hamming distance H_k (Mirkin 1996) and the $Z_k = 1 - S_k$ distance (Morlini and Zani 2012). It can be shown that the numerator of Z_k is equal to $N(1-R_k)$ (Morlini and Zani 2012) and $H_k = 2N(1-R_k)$ (Meila 2007). Since the values R_k and S_k are not well spread out over the interval [0,1] for large k, it may be convenient to correct the indexes for association due to chance and to consider the measure (Hubert and Arabie 1985; Albatineh et al. 2006):

$$AS_k = \frac{S_k - E(S_k)}{1 - E(S_k).} \tag{4}$$

It is interesting to note that the adjusted S_k obtained with (4) is equivalent to the Adjusted Rand index (Hubert and Arabie 1985). Indeed, the expectation $E(T_k)$, assuming statistical independence under the binomial distribution for the contingency table showing the cluster membership of the object pairs (Table 1) is (Fowlkes and Mallows 1983; Hubert and Arabie 1985):

$$E(T_k) = P_k Q_k / N \tag{5}$$

Using (5), the expectation $E(S_k)$ is:

$$E(S_k) = \frac{\sum_{j \neq k} P_j + \sum_{j \neq k} Q_j + 2P_k Q_k / N}{\sum_k P_k + \sum_k Q_k} \tag{6}$$

Using (6) in (4), after some algebraic simplification we obtain:

$$AS_k = \frac{2T_k - 2P_k Q_k / N}{P_k + Q_k - 2P_k Q_k / N} \tag{7}$$

which is the same expression of the Adjusted Rand Index.

The most innovative index proposed in Morlini and Zani (2012) is a global measure of similarity which considers simultaneously all the k stages in the dendrograms. In the literature, the only measure that has been presented for measuring the agreement between two whole dendrograms is the γ coefficient of Baker (1974). This criterion is defined as the rank correlation coefficient between stages at which pairs of objects combine in the dendrograms and thus it ranges over the interval $[-1, 1]$ and it is not a similarity index. The global measure of agreement proposed in Morlini and Zani (2012) is:

$$S = \frac{2\sum_k T_k}{\sum_k Q_k + \sum_k P_k}. \tag{8}$$

S does not depend on the number k and thus preserves comparability across clusterings. It has some desirable properties not pertaining to γ. It is a similarity index. Therefore, in a sample of G dendrograms $u_g \in U$, $g = 1, \ldots, G$ it is a function $S(u_g, u_{g'}) = S_{gg'}$ from $U \times U$ into $\mathbf{R}$ with the following characteristics:

- $S_{gg'} \geq 0$ for each $u_g, u_{g'} \in U$ (non negativity).
- $S_{gg} = 1$, for each $u_g \in U$ (normalization).
- $S_{gg'} = S_{g'g}$, for each $u_g, u_{g'} \in U$ (symmetry).

The further additivity property $S_{gg'} = \sum_k V{gg'}_k = \sum_k \frac{2T_k}{\sum_k Q_k + \sum_k P_k}$ permits to decompose the value of the index into contributions pertaining to each stage k of the dendrograms. This makes the values of S more interpretable and comparable.

3 Simulation Experiments: Unrelated Clusterings

For the first study we generate two data sets according to the following steps:

1. For each data set, the sample size is $n = 50$ and the number of variables is $p = 5$.
2. The 50 elements in each set are generated from a multivariate standard normal distribution with a correlation matrix consisting of equal off-diagonal elements ρ_1 (in the first set) and ρ_2 (in the second set). ρ_1 and ρ_2 are chosen randomly in the set $[-0.9, -0.8, \ldots, 0.8, 0.9]$.
3. We repeat steps 1. and 2. 5,000 times. Each time we perform a hierarchical clustering for the two sets with the Euclidean distance and the average linkage and we compute the indexes S_k, R_k, AS_k, B_k, S and the γ coefficient.

The two sets are generated independently and the agreements between clusterings are only due to chance. Since the range of the indices is different, and in these simulations γ takes negative values, we obtain new values of γ, which we call γ^*, lying in the interval $[0, 1]$, with the transformation $\gamma^* = (\gamma + 1)\backslash 2$. Left panel of Fig. 1 shows the boxplots of the values of S (left) and γ^* (right). The median and the mean values of γ^* are approximately 0.5. The boxplots show that S performs better than γ^*, since the median and mean value of S are nearly 0.23 and the index has fewer outliers. In the right panel are reported the mean values of B_k, R_k and AS_K, for $k = 2, \ldots, 49$. With $k = 2$, R_k and B_k have a similar value. Then, the plot shows the tendency of R_k to increase with k and rapidly approaching 1 and the opposite tendency for B_k to decrease with k and assuming values close to 0 for large k. AS_k performs best, showing average values always close to zero, regardless of k.

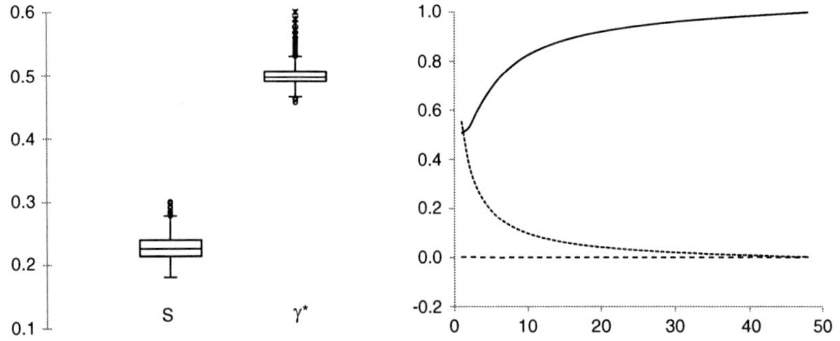

Fig. 1 Results for 5,000 pairs of unrelated samples. *Left panel*: boxplots of S (*left*) and γ^* (*right*). *Right panel*: plots of the mean values of R_k (*solid line*), B_k (*dotted line*) and AS_k (*dashed line*)

Further simulations show that the behavior of all indexes in the case of two unrelated clusterings is robust with respect to the choice of the distance or to the choice of the linkage and also with respect to the size n of the data and to the number of variables p. In several simulations carried out considering the Manhattan distance, different linkages and different values of n (from 50 to 100) and for p (from 2 to 10), boxplots for S and γ^* and plots of B_k, R_k, AS_K are similar to those reported in Fig. 1.

4 Simulation Experiments: Robustness to Noise

In this section simulations are aimed at evaluating the robustness to noise. The first data set is generated as in previous section, setting the sample size $n = 50$, the number of variables $p = 5$ and generating 50 elements from a multivariate standard normal distribution with a correlation matrix consisting of equal elements ρ_1 chosen randomly in the set $[-0.9, -0.8, \ldots, 0.8, 0.9]$. The second data set is obtained by adding to all variables a random normal noise with mean zero and variance σ_e^2. We consider the values $\sigma_e^2 = 0.04, 0.16, 0.36$. Hierarchical clusterings of each data set are carried out using the Euclidean distance and the complete, the single, the average linkages and the Ward method. Since the second data set is just the first one with added noise, indexes should indicate a great similarity between clusterings and the similarity should increase with decrease in σ_e^2. In these simulations γ assumes only positive values, therefore we consider γ instead of the normalized index γ^*. Figures 2 and 3 report the results obtained with $\sigma_e^2 = 0.04$, the single and the complete linkage methods. Results obtained with the average linkage and the Ward methods, not reported for lack of space, are available upon request. For all linkages, the values of S do not exceed 0.9 but are never smaller than 0.4 (for the single linkage, the minimum value obtained in the 5,000 runs is 0.6). On the contrary, γ assumes values greater than 0.9 and close to one but, on the other hand, presents several values smaller than 0.4. If we take the median values for comparing the

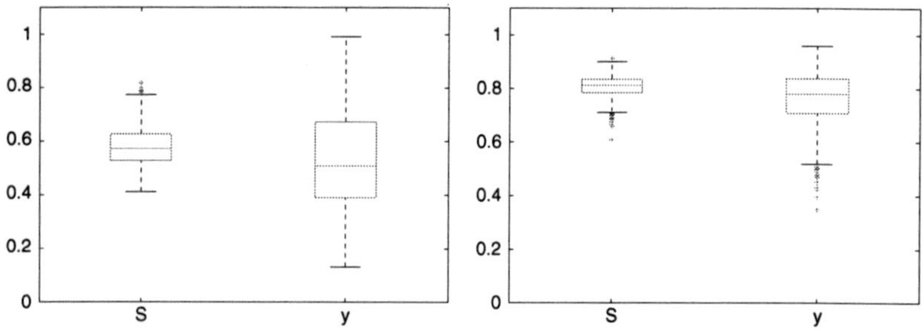

Fig. 2 Boxplots of S and γ using the complete linkage (*left*) and the single linkage (*right*)

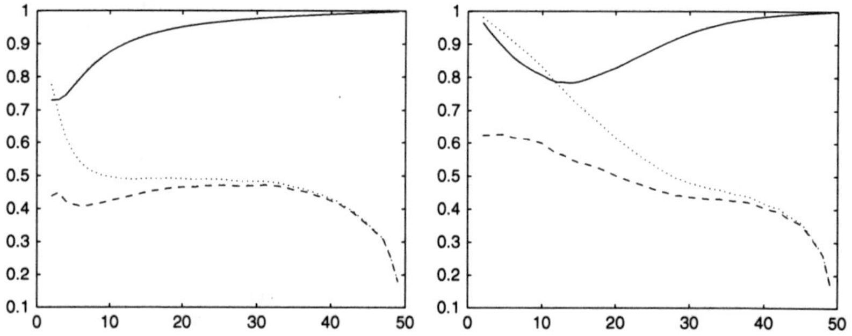

Fig. 3 Plots of the mean values of R_k (*solid line*), B_k (*dotted line*) and AS_k (*dashed line*) using the Euclidean distance and the complete linkage (*left panel*) and the single linkage (*right panel*)

degree of similarity measured by S and γ, we see that S indicates a more marked similarity using the complete and the single linkages. Plots in Fig. 3 show again the opposite tendencies of R_k to approach one and of B_k and AS_k to approach zero as k increases. The plots also show that perturbation affects B_k least for small values of k and greatest for large values of k. This desirable property was just noted in Fowlkes and Mallows (1983). For AS_k this is true for the Ward method, the single and the average linkages, but not for the complete linkage. AS_k shows a relatively more constant pattern with respect to k, without precipitous falloffs. These results show that each index has own desirable properties but also causes for concern and the choice of one index over the others is somehow difficult. That the average values of R_k and B_k are higher in the presence of small perturbation of the sample is reasonable and desirable, but the large values assumed by R_k also in presence of two unrelated clusterings (see Fig. 1) and the greatest variability of B_k across k are causes for concern. For these reasons, a global criterion of similarity like S may be a better choice for measuring the agreement between two hierarchical clusterings.

From Figs. 2 and 3 we may also analyze the stability of the different linkages to small perturbations. Clusterings with the single linkage are less affected by added noise while clusterings recovered by the complete linkage are, in general, less stable.

Figures 4, 5 and 6 show the empirical distribution of S, γ, R_k, B_k, AS_k (with $k = 2, 5, 10, 15, 20, 25, 30, 35, 40, 45$) obtained with $\sigma_e^2 = 0.16$ and $\sigma_e^2 = 0.36$.

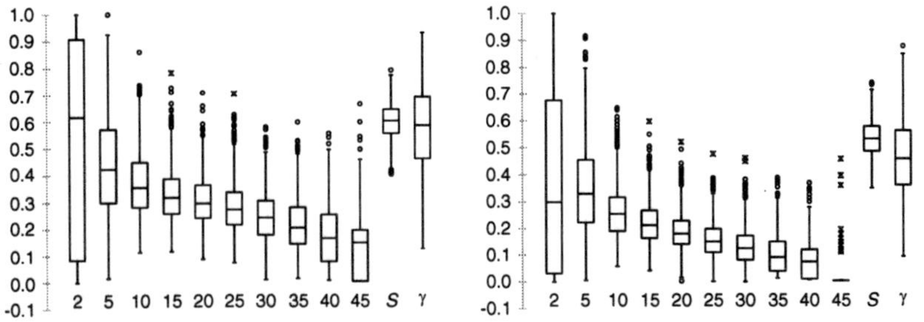

Fig. 4 Boxplots of AS_k ($k = 2, 5, 10, 15, 20, 15, 30, 35, 40, 45$), S and γ. Values are obtained considering pairs of samples where the second one is the first one with added noise with $\sigma_e^2 = 0.16$ (*left panel*), $\sigma_e^2 = 0.36$ (*right panel*)

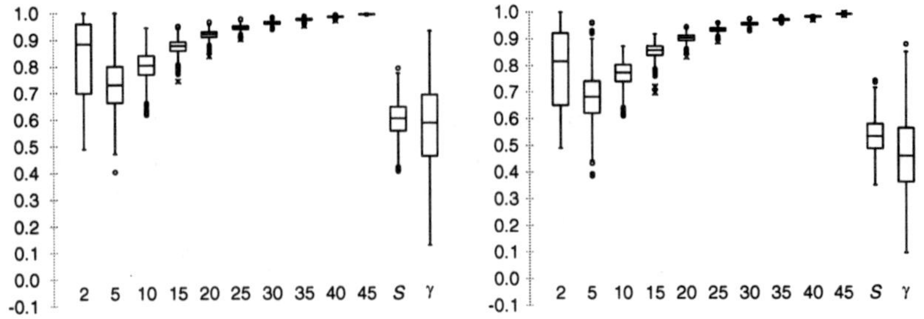

Fig. 5 Boxplots of R_k ($k = 2, 5, 10, 15, 20, 15, 30, 35, 40, 45$), S and γ. Values are obtained considering pairs of samples where the second one is the first one with added noise with $\sigma_e^2 = 0.16$ (*left panel*), $\sigma_e^2 = 0.36$ (*right panel*)

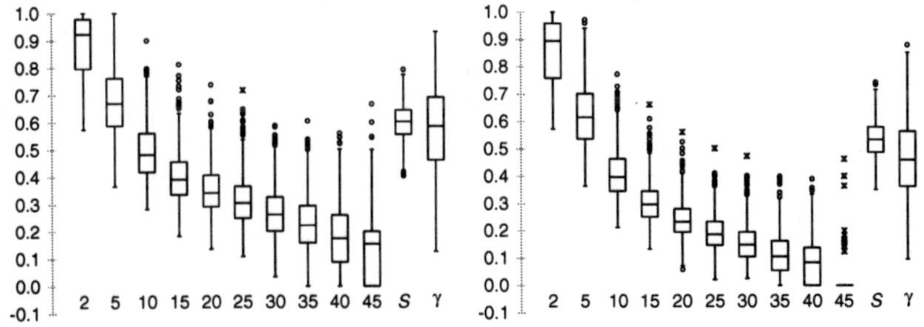

Fig. 6 Boxplots of B_k ($k = 2, 510, 15, 20, 15, 30, 35, 40, 45$), S and γ. Values are obtained considering pairs of samples where the second one is the first one with added noise with $\sigma_e^2 = 0.16$ (*left panel*), $\sigma_e^2 = 0.36$ (*right panel*)

Clusterings are recovered using the Euclidean distance and the average linkage method. The median values of S and γ decrease with increase in σ_e^2. However, this drop is more marked in γ than in S and, for $\sigma_e^2 = 0.36$, the median value of S is substantially higher than the median value of γ. The patterns of the median values of R_k, B_k and AS_k, versus k, do not change across simulations with different σ_e^2. Boxplots show that R_k has a higher variability for small values of k. For $k \geq 30$, R_k is always close to 1 and the values are nearly constant across simulations. The variability of B_k and AS_k, measured by the interquartile range, is more marked for $k = 2$ and $k = 5$.

5 Concluding Remarks

This paper has presented results obtained by simulation studies aimed at comparing the behavior of different similarity indexes used for measuring the agreement between two hierarchical clusterings. In contrast to the well-know criteria like the Rand index and the B_k index of Fowlkes & Mallows, the measure S recently proposed in the literature is not directly concerned with relationship between a single pair of partitions, but depends on the whole set of partitions in the dendrograms. Simulations show that the performances of R_k and B_k strongly depend on the number of groups k. The major drawback of this dependency is that R_k assumes values close to one for large k, even though the two partitions are unrelated. For large k, B_k has improved performances in case of unrelated clusterings but performs worse when the two clusterings are related. There is not a clear best choice between these two competing criteria and thus it is probably meaningless to search for the best criterion. A better goal is to study the behavior of these indexes and their limitations in different experimental conditions. The adjusted version of R_k and S_k, is based on a null model that is reasonable but, nevertheless, artificial. Some authors have expressed concerns at the plausibility of the null model (Meila 2007). However, simulations show that the adjusted version has improved performances and the values of the index are not influenced by k. These results are in agreement with results presented in Albatineh et al. (2006) and Albatineh and Niewiadomska-Bugaj (2011). The new global index S does not depend on k and thus preserves comparability. Simulations show that S has good performances. It takes values close to zero when no clustering structure is present and values close to one when a structure exists.

References

Albatineh, A. N., Niewiadomska-Bugaj, M., & Mihalko, D. (2006). On similarity indices and correction fore chance agreement. *Journal of Classification, 23*, 301–313.

Albatineh, A. N., & Niewiadomska-Bugaj, M. (2011). Correcting Jaccard and other similarity indices for chance agreement in cluster analysis. *Advances in Data Analysis and Classification, 5*, 179–200.

Baker, F. B. (1974). Stability of two hierarchical grouping techniques. Case I: sensitivity to data errors. *JASA, 69*, 440–445.

Fowlkes, E. B. & Mallows, C. L. (1983). A method for comparing two hierarchical clusterings. *JASA, 78*, 553–569.

Hubert, L. J. & Arabie, P. (1985). Comparing partitions. *Journal of Classification, 2*, 193–218.

Meila, M. (2007). Comparing clusterings. An information based distance. *Journal of Multivariate Analysis, 98*(5), 873–895.

Mirkin, B. G. (1996). *Mathematical classificationa and clustering*. Dordrecht: Kluwer Academic.

Morlini, I., & Zani, S. (2012). An overall index for comparing hierarchical clusterings. In W. Gaul, A. Geyer-Schulz, L. Schmidt-Thieme, & J. Kunze (Eds.) *Challenges at the interface of data analysis, computer science and optimization* (pp. 29–36). Berlin: Springer.

Wallace, D. L. (1983). Comment on the paper "A method for comparing two hierarchical clusterings". *JASA, 78*, 569–578.

Performance Measurement of Italian Provinces in the Presence of Environmental Goals

Eugenia Nissi and Agnese Rapposelli

Abstract The widespread of sustainable development concept intimates a vision of an ecologically balanced society, where it is necessary to preserve environmental resources and integrate economics and environment in decision-making. Consequently, there has been increasing recognition in developed nations of the importance of good environmental performance, in terms of reducing environmental disamenities, generated as outputs of the production processes, and increasing environmental benefits. In this context, the aim of the present work is to evaluate the environmental efficiency of Italian provinces by using the non-parametric approach to efficiency measurement, represented by Data Envelopment Analysis (DEA) technique. To this purpose, we propose a two-step methodology allowing for improving the discriminatory power of DEA in the presence of heterogeneity of the sample. In the first phase, provinces are classified into groups of similar characteristics. Then, efficiency measures are computed for each cluster.

1 Introduction

In 1987, a World Commission on Environment and Development report brought the concept of sustainable development into the purview of governments and publics around the world. According to it, sustainable development is development that meets the needs of the present without compromising the ability of future generations to meet their own needs. This concept intimates, therefore, a vision of an ecologically balanced society, where it is necessary to preserve environmental resources whilst integrate economics and environment in decision-making.

E. Nissi • A. Rapposelli (✉)
Dipartimento di Economia, Università "G.d'Annunzio" di Chieti-Pescara,
Viale Pindaro 42, Pescara, Italy
e-mail: a.rapposelli@unich.it

P. Giudici et al. (eds.), *Statistical Models for Data Analysis*, Studies in Classification, Data Analysis, and Knowledge Organization, DOI 10.1007/978-3-319-00032-9_29,

The emergence and the widespread of this idea has intensified the need for indicators which capture the link between the economic, social and environmental dimensions. Hence, there is an increasing need for taking account of the impact of organizations of all kinds on environment, both in terms of reducing environmental disamenities, generated as outputs of the production processes, and increasing environmental benefits.

In this context, the aim of this empirical study is to measure the ecological efficiency of Italian provinces for the year 2008. Our efficiency analysis is undertaken by using the principles of frontier analysis introduced by Farrell (1957), who suggested measuring the efficiency of a productive unit relative to an empirical best practice frontier. The efficiency frontier outlines the technical limits of what an organisation, or a Decision Making Units (DMUs), can achieve: it specifies for a unit the maximum quantities of outputs it can produce given any level of inputs and, for any level of outputs, the minimum quantities of inputs needed for producing the outputs. To this purpose, we use the non-parametric approach to efficiency measurement, represented by Data Envelopment Analysis (DEA) method, which is a very suitable technique for assessing the performance of provinces seen as environmental operating units. DEA, in fact, can easily handle multiple inputs and outputs as opposed to the usual stochastic frontier formulation, represented by Stochastic Frontier (Aigner et al. 1977). Italian provinces, however, are not homogeneous and produce both good outputs and environmental disamenities or "bads" (Scheel 2001). For these reasons, we carry out two main changes to this method.

It is well known, in fact, that DEA method makes a series of homogeneity assumptions about the units under assessment: in general the operating units, or DMUs, have to be similar in a number of ways (Dyson et al. 2001). First of all, the units are assumed to be undertaking similar activities and producing comparable products or services. Second, a similar range of resources is available to all the units. Finally, units have to operate in similar environments, since the external environment generally impacts on the overall performance of units. Hence, we propose a two-step methodology allowing for increasing the discriminatory power of DEA which is limited in the presence of heterogeneity. The idea of this work is firstly to cluster the operating units (the Italian provinces) and then to estimate efficiency for each cluster. Besides, in order to rate the performance of each province, DEA method has to be modified to the field of environmental performance, where there is a joint production of good outputs and bad outputs, such as pollutants emissions (Coli et al. 2011). We try to solve this problem in the traditional measures of efficiency by applying a new model type of DEA, that includes the presence of desired environmental effects and environmental harms (Thore and Freire 2002). The paper is organized as follows. Section 2 reviews the theoretical background, Sect. 3 describes the data used and discusses the results, Sect. 4 contains the main implications for future research and concludes.

2 Methodology: Cluster Analysis and Data Envelopment Analysis

In the first phase of our study we use cluster analysis in order to identify natural groups of provinces based on their structural similarity with regards to the levels of the inputs and outputs they receive and produce (Samoilenko and Osei-Bryson 2008). Provinces of a given cluster may be operating, therefore, in similar environments and share similar input and output mix.

As known, clustering is a statistical multivariate technique that involves the partitioning of a set of objects into a useful set of mutually exclusive clusters such that the similarity between the observations within each cluster (i.e., subset) is high, while the similarity between the observations from the different clusters is low (Johnson and Wichern 2002; Samoilenko and Osei-Bryson 2010). There are numerous algorithms available for doing clustering: they may be categorised in various ways, such as hierarchical or partitional, deterministic or probabilistic, hard or fuzzy. To perform cluster analysis we have employed both hierarchical clustering techniques and partitional clustering techniques. However, hierarchical clustering techniques have not been able to provide satisfactory results. We have obtained, in fact, clusters with a few units, and in this case DEA method cannot be applied with good results, because there would not be a reasonable level of differentiation between DMUs evaluated.[1] Afterwards, we have decided to use a partitional clustering technique based on k-means method,[2] because it represents the best method according to our purpose. The resulting classification, in fact, is more homogeneous and well-differentiated.

Once we have classified the operating units into groups of similar characteristics, we apply the DEA method to assess the different environmental efficiency of provinces for each cluster identified. DEA is a linear-programming based methodology developed by Charnes et al. (1978). It provides a measure of the relative efficiency of a set of homogeneous organisational units in their use of multiple inputs to produce multiple outputs (Cooper et al. 2000). The basic DEA models measure the technical efficiency of a DMU in terms of the maximal radial contraction to its input levels (input orientation) or expansion to its output levels feasible under efficient operation (output orientation). In general, they assume that inputs and outputs are "goods" (Dyckhoff and Allen 2001).

The model used in this paper is the input-oriented one, under the assumption of variable returns to scale (VRS). To present formally this model, known as BCC (Banker et al. 1984), consider a set of n DMUs, indexed by $j = 1, \ldots, n$, each

---

[1] In order to individuate a significant number of efficient organisations, the literature suggests that the number of units has to be greater than 3(m+s), where m+s is the sum of the number of inputs and number of outputs (Dyson et al. 2001).

[2] k-means clustering requires the number of resulting cluster, k, to be specified prior to analysis. Thus, it will produce k different clusters of greatest possible distinction.

producing s different desirable outputs from m different inputs. The technical input efficiency of DMUs under analysis is obtained by implementing the following model:

$$e_0 = \min \theta_0$$

subject to

$$\theta_0 x_{ij0} - \sum_{j=1}^{n} \lambda_j x_{ij} \geq 0, \quad i = 1, \ldots, m \tag{1}$$

$$\sum_{j=1}^{n} \lambda_j y_{rj} \geq y_{rj0}, \quad r = 1, \ldots, s \tag{2}$$

$$\sum_{j=1}^{n} \lambda_j = 1, \tag{3}$$

$$\lambda_j \geq 0, \quad \forall\, j \tag{4}$$

where x_{ij} is the amount of the i-th input to DMU j, y_{rj} is the amount of the r-th output to DMU j, λ_j are the weights of DMU j and θ_0 is the shrinkage factor for DMU j_0. The value of θ_0 obtained is termed the technical output efficiency of DMU j_0 and it is bounded between 0 and 1. DMU j_0 is said to be efficient if it has a score of unity.

However, standard DEA models are not suitable in contexts where at least one of the variables that have to be radially contracted or expanded is not a "good." For example, in the context of environmental performance some production processes may also generate undesirable outputs or "bads" (such as environmental disamenities) which need to be decreased to improve the performance of a unit (Seiford and Zhu 2005). Beside, a symmetric case of inputs which should be maximised may also occur (desired environmental effects). Classical DEA models, therefore, have to be modified in order to extend the analysis by also considering the presence of variables that impact on the environment. To this purpose, we propose a new model type of DEA, that incorporates environmental harms as inputs and environmental benefits as outputs, while also seeking to minimise and maximise them, respectively. Hence, assuming that n DMUs, indexed by $j = 1, \ldots, n$, produce s different desirable outputs and z different undesirable outputs from m different inputs and w environmental benefits, their technical input efficiency is obtained by including the following two more constraint, (5) and (6), into the above linear programming problem:

$$\theta_0 h_{tj0} - \sum_{j=1}^{n} \lambda_j h_{tj} \geq 0, \quad t = 1, \ldots, z \tag{5}$$

$$\sum_{j=1}^{n} \lambda_j e_{vj} \geq e_{vj0}, \quad v = 1, \ldots, w \tag{6}$$

where h_{tj} is the amount of the t-th undesirable output to DMU j and e_{vj} is the amount of the $v-$th environmental benefit to DMU j.

3 Case Study

We apply the proposed methodology to 103 Italian provinces for the year 2008, the most recent for which all required data are available. For evaluating their performance, we focus on economic and environmental aspects of production process, without emphasizing physical inputs and outputs (Kuosmanen and Kortelainen 2005). Hence, we define a model characterized by one input, the number of employees, and a single desirable output, the gross domestic product (GDP) expressed in Euros. Moreover, since the production process of these DMUs may generate desired environmental effects and environmental harmful effects (Dyckhoff and Allen 2001; Scheel 2001), we also include both of these environmental variables. As to data on the undesirable outputs we include two air pollutants—nitrogen dioxide concentration (NO_2)[3] and PM10 concentration[4] (suspended particulate matters)—and nitrates concentration in water. With regard to environmental benefits, we select two variables: percentage of separate refuse collection on total waste produced and public parks and gardens, measured in mq/ha. Finally, we also consider in our analysis the weight of manufacturing sector in the provincial economy. The data required are based on several sources: our principal source of information regarding the three undesirable outputs and the two environmental benefits has been the report "Rapporto 2008 sulla qualità della vita in Italia," while the number of employees and the gross domestic product values have been obtained from the Italian National Institute of Statistics (ISTAT).

As mentioned above, we first apply cluster analysis in order to identify subsets of provinces, heterogenous between the groups, but homogeneous within each group.

To this purpose, we use a partitional clustering technique based on k-means method. In particular, we are able to come up with a solution that partitions the set of 103 DMUs into two clusters. Even if the results from Calinski/Harabasz test, a clustering validity criteria, are very similar for 2 and 3 clusters (Table 1), we have decided to fix k = 2 because the solution with 3 clusters provides a cluster with 13 units, for which it is not possible to individuate a significant number

[3] Current scientific evidence links short-term NO_2 exposures, ranging from 30 min to 24 h, with adverse respiratory effects including airway inflammation in healthy people and increased respiratory symptoms in people with asthma. Nitrogen dioxide also plays a major role in the atmospheric reactions that produce ground-level ozone or smog (Coli et al. 2011).

[4] Suspended particulate matter (SPM) is a mixture of particles of different size and state (solid and liquid) ranging from 0.01 μm to >10 μm in diameter: particles measuring <10 μm (PM10) penetrate into the lower respiratory system and might penetrate into the bloodstream. Particles may contain metals, such as zinc and nickel, organic materials and polycyclic aromatic hydrocarbons, some of which are carcinogenic (Coli et al. 2011).

Table 1 Calinski/Harabasz test

Number of clusters	Calinski/Harabasz (pseudo F)
2	339.93
3	339.61

Table 2 DEA efficiency scores by provinces: Cluster 1

DMU	Score	DMU	Score	DMU	Score	DMU	Score	DMU	Score	DMU	Score
CR	1	AO	1	IM	1	TV	0.94	AP	0.85	NO	0.65
LC	1	GO	1	RE	1	RM	0.92	PR	0.85	AN	0.63
MN	1	UD	1	FE	1	VA	0.92	LO	0.83	GE	0.63
SO	1	BZ	1	IS	1	RO	0.92	PN	0.82	VR	0.63
PV	1	LI	1	TR	0.99	TN	0.91	BG	0.78	AL	0.62
MI	1	LC	1	MO	0.98	PU	0.89	RN	0.78	PG	0.62
AT	1	GR	1	PT	0.97	PI	0.88	BL	0.78	MC	0.60
BI	1	AR	1	TS	0.96	BS	0.88	VE	0.76	PD	0.55
CN	1	SI	1	BO	0.95	RA	0.87	FC	0.75	TO	0.53
VB	1	PO	1	LT	0.95	FI	0.86	CO	0.74		
VC	1	SP	1	SV	0.94	PC	0.86	VI	0.66		

of efficient units.[5] The first one consists of 64 provinces (Northern and Central Italy plus Roma, Latina and Isernia), whilst the second one includes 39 provinces (Southern Italy plus Frosinone, Rieti, Viterbo and Massa Carrara). As mentioned, identification of such clusters can prove useful in understanding the complexion of operating environments and the input/output mix prevailing among the operating units (Thanassoulis 1996). Then, for each cluster, the modified input-oriented BCC model presented in Sect. 2 is carried out. The linear programs associated with the model are solved by using DEAP, a software developed by Tim Coelli (1996). The efficiency ratings obtained are listed in the Tables below.

The results consists of a large number of efficient DMUs in both clusters. In cluster one (Table 2) 26 provinces form the efficiency frontier. Besides, ten provinces have low ratings. In cluster two (Table 3) there are 16 top performers, and two provinces are very close to the best practice frontier. Our findings also show that the efficiency score is higher in cluster one, whereas the scores variability is lower than cluster two (Table 4). In the first cluster Biella, Cuneo, Grosseto and Livorno seem to be the most robustly BCC efficient: they appear very frequently in the peer groups (21, 18, 18 and 16 times, respectively). In the second cluster the most frequent units are Viterbo, Crotone and Massa-Carrara (20, 17 and 10).

[5]We have also tried to fix a larger number of clusters, but we have obtained clusters with a few units, and in this case DEA method cannot be applied with good results.

Table 3 DEA efficiency scores by provinces: Cluster 2

DMU	Score	DMU	Score	DMU	Score	DMU	Score	DMU	Score	DMU	Score
RI	1	BA	1	VV	1	TA	0.90	TP	0.78	PA	0.65
VT	1	LE	1	FR	1	CA	0.85	RC	0.76	FG	0.64
MS	1	OR	1	CE	0.99	CB	0.85	AG	0.75	ME	0.56
CH	1	SA	1	EN	0.98	NU	0.84	BR	0.71	NA	0.56
TE	1	CL	1	SS	0.97	BN	0.84	CS	0.70		
PZ	1	RG	1	CT	0.92	AQ	0.82	AV	0.67		
KR	1	PE	1	MT	0.91	CZ	0.81	SR	0.65		

Table 4 Summary statistics for DEA efficiency scores

	Cluster 1	Cluster 2
Mean	0.885	0.875
Minimum	0.535	0.559
Maximum	1	1
SD	0.140	0.142

4 Conclusions and Future Research

In this paper we have evaluated the environmental performance of Italian provinces for 2008 by means of the non-parametric approach to efficiency measurement, represented by Data Envelopment Analysis. To this purpose, we have adapted this method to the problems at hand, i.e. the heterogeneity of the sample and the presence of positive and negative environmental variables. We have suggested, therefore, to firstly apply a clustering algorithm for obtaining subgroups of homogeneous data, and then to obtain measures of technical efficiency taking into account the presence of both desired environmental effects and environmental harms. The results provided by application of the modified DEA model to Italian provinces show that DMUs are operating at a fairly high level of technical efficiency, although there is room for improvement in several provinces.

However, these results could be improved. First of all, we have applied DEA method because it can easily accomodate multiple inputs and outputs, since a main objective of our study was also to incorporate both undesirable outputs and environmental beneficial effects into the analysis, by augmenting the vectors of inputs and outputs. However, in some cases, it is now possible to include multiple outputs in a parametric analysis. Hence, in our future research we could consider a comparative evaluation of the two alternative techniques to efficiency measurement, DEA and Stochastic Frontier Analysis. Besides, further research in this field could include more variables, such as other undesirable outputs in form of pollutants emissions or other environmentally beneficial variables (recycling of materials, surface for pedestrian walkway, generation of solar energy, energy consumption covered by renewable sources, etc.). Finally, the model employed could be applied in order to compare ecological performance in other territorial systems.

References

Aigner, D. J., Lovell, C. A. K., & Schmidt, P. (1977). Formulation and estimation of stochastic frontier production function models. *Journal of Econometrics, 6*, 21–37.

Banker, R. D., Charnes, A., & Cooper, W. W. (1984). Some models for estimating technical and scale inefficiencies in Data Envelopment Analysis. *Management Science, 30*(10), 78–1092.

Charnes, A., Cooper, W. W., & Rhodes, E. (1978). Measuring the efficiency of decision making units. *European Journal of Operational Research, 2*, 429–444.

Coelli, T. J. (1996). *A guide to DEAP Version 2.1: A Data Envelopment Analysis program*. CEPA Working Paper 96/08, University of New England, Armidale.

Coli, M., Nissi, E., & Rapposelli, A. (2011). Monitoring environmental efficiency: an application to Italian provinces. *Environmental Modelling & Software, 26*, 254–254.

Cooper, W. W., Seiford, L. M., & Tone, K. (2000). *Data envelopment analysis, a comprehensive test with models, applications, references and DEA-solver software*. Boston: Kluwer.

Dyckhoff, H., & Allen, K. (2001). Measuring ecological efficiency with Data Envelopment Analysis (DEA). *European Journal of Operational Research, 132*, 312–325.

Dyson, R. G., Allen, R., Camanho, A. S., Podinovski, V. V., Sarrico, C. S., & Shale, E. A. (2001). Pitfalls and protocols in DEA. *European Journal of Operational Research, 132*, 245–259.

Farrell, M. J. (1957). The measurement of productive efficiency. *Journal of Royal Statistical Society Series A, 120*, 253–281.

Johnson, R. A., & Wichern, D. W. (2002). *Applied multivariate statistical analysis*. Upper Saddle River: Prentice-Hall.

Kuosmanen, T., & Kortelainen, M. (2005). Measuring eco-efficiency of production with data envelopment analysis. *Journal of Industrial Ecology, 9*, 59–72.

Samoilenko, S., & Osei-Bryson, K. (2008). Increasing the discriminatory power of DEA in the presence of the sample heterogeneity with cluster analysis and decision trees. *Expert Systems with Applications, 34*, 1568–1581.

Samoilenko, S., & Osei-Bryson, K. (2010). Determining sources of relative inefficiency in heterogeneous sample: methodology using cluster analysis, DEA and neural networks. *European Journal of Operational Research, 206*, 479–487.

Scheel, H. (2001). Undesirable outputs in efficiency valuations. *European Journal of Operational Research, 132*, 400–410.

Seiford, L. M., & Zhu, J. (2005). A response to comments on modeling undesirable factors in efficiency evaluation. *European Journal of Operational Research, 161*, 579–581.

Thanassoulis, E. (1996). A data envelopment analysis approach to clustering operating units for resource allocation purposes. *Omega-The International Journal of Management Science, 24*, 463–476.

Thore, S. A., & Freire, F. (2002). Ranking the performance of producers in the presence of environmental goals. In S. A. Thore (Ed.), *Technology commercialization: DEA and related analytical methods for evaluating the use and implementation of technical innovation*. Boston: Kluwer.

On the Simultaneous Analysis of Clinical and Omics Data: A Comparison of Globalboosttest and Pre-validation Techniques

Margret-Ruth Oelker and Anne-Laure Boulesteix

Abstract In medical research biostatisticians are often confronted with supervised learning problems involving different kinds of predictors including, e.g., classical clinical predictors and high-dimensional "omics" data. The question of the *added* predictive value of high-dimensional omics data given that classical predictors are already available has long been under-considered in the biostatistics and bioinformatics literature. This issue is characterized by a lack of guidelines and a huge amount of conceivable approaches. Two existing methods addressing this important issue are systematically compared in the present paper. The *globalboosttest* procedure (Boulesteix & Hothorn. (2010). *BMC Bioinformatics, 11*, 78.) examines the additional predictive value of high-dimensional molecular data via boosting regression including a clinical offset, while the *pre-validation* method sums up omics data in form of a new cross-validated predictor that is finally assessed in a standard generalized linear model (Tibshirani & Efron. (2002). *Statistical Applications in Genetics and Molecular Biology, 1*, 1). Globalboosttest and pre-validation are introduced and discussed, then assessed based on a simulation study with survival data and finally applied to breast cancer microarray data for illustration. R codes to reproduce our results and figures are available from http://www.ibe.med.uni-muenchen.de/organisation/mitarbeiter/020_professuren/boulesteix/gbtpv/index.html.

M.-R. Oelker
Department of Statistics, University of Munich, Ludwigstr. 33, 80539 Munich, Germany

Department of Medical Informatics, Biometry and Epidemiology of the Faculty of Medicine, University of Munich, Marchioninistr. 15, 81377 Munich, Germany
e-mail: margret.oelker@stat.uni-muenchen.de

A.-L. Boulesteix (✉)
Department of Medical Informatics, Biometry and Epidemiology of the Faculty of Medicine, University of Munich, Marchioninistr. 15, 81377 Munich, Germany
e-mail: boulesteix@ibe.med.uni-muenchen.de

P. Giudici et al. (eds.), *Statistical Models for Data Analysis*, Studies in Classification, Data Analysis, and Knowledge Organization, DOI 10.1007/978-3-319-00032-9_30,

1 Introduction

While high-dimensional "omics" data such as microarray transcriptomic data have been studied in the context of outcome prediction for more than ten years in biomedical research, the question of the added predictive value of such data given that classical predictors are already available has comparatively focused less attention in the literature (e.g. Boulesteix and Sauerbrei 2011). For a given prediction problem (for example prediction of response to therapy or survival time), we often have two types of predictors. On the one hand, conventional clinical covariates such as, e.g. age, sex, disease duration or tumour stage are available as potential predictors. They have been typically extensively investigated and validated in previous studies. On the other hand, we have a large number of "omics" predictors that are generally more difficult to measure than classical clinical predictors and not yet well-established. In the context of translational biomedical research, researchers are interested in the added predictive value of such omics predictors over classical clinical predictors.

The combined analysis of high-dimensional omics predictors and low-dimensional clinical predictors raises various challenges. How can we build a combined model that is optimal in terms of prediction accuracy? How can we test the added predictive value of high-dimensional omics data over classical clinical predictors and/or assess the respective importance of the two types of predictors? Leaving the first challenge aside, we focus on tests and compare globalboosttest (Boulesteix and Hothorn 2010) and pre-validation (Tibshirani and Efron 2002), two testing approaches. Since omics data are high-dimensional, standard likelihood ratio tests in the framework of Generalized Linear Models (GLM) cannot be performed. The two examined methods tackle this problem based on a two-step procedure, but in different ways: while globalboosttest summarizes clinical predictors as an offset and then fits a regularized regression model to omics data, pre-validation first summarizes omics data as a cross-validated "pseudo predictor" and then tests its significance in a multivariate GLM adjusting for clinical predictors.

2 Globalboosttest

The "globalboosttest" procedure (Boulesteix and Hothorn 2010) aims at testing the additional predictive value of high-dimensional data by combining two well-known statistical tools: GLMs and boosting regression. Suppose we have both high-dimensional omics data as potential predictors on the one hand and a few classical clinical covariates or a well-defined prognostic index on the other hand. The considered null-hypothesis is that "given the clinical covariates the omics data have no added predictive value". To address this testing problem the globalboosttest procedure first builds a clinical model (step 1, also denoted as "internal in this paper") based on clinical covariates only. For example step 1 is based on logistic regression in case of a binary response or on Cox proportional hazard regression in case of a censored survival response. The resulting linear predictor is then considered as an offset in a more complex model involving omics data. As suggested by the procedure's name, the latter model is estimated by boosting regression (step 2, also denoted as external in this paper). Step 2 implies an iterative stepwise selection of the omics predictors while taking the clinical covariates into account in form of the offset. This step 2 is then repeated a large number of times after randomly permuting the omics data (but not the clinical data). A permutation p-value is then derived as the frequency at which the negative binomial log-likelihood of the boosting regression model was smaller in permuted data sets than with the original data set.

Tuning parameters are the number of boosting iterations in the external model and the number of permutations performed in step 2. The number of permutations should be as large as computationally feasible to increase the test's precision. The number of boosting regression steps is a parameter that potentially influences the test result.

Note that, strictly speaking, this permutation procedure tests the joint hypothesis that "omics data have no added predictive value" *and* "omics data and clinical predictors are independent", because by permuting omics data we also destroy the association between omics and clinical predictors. An important feature of the globalboosttest procedure, however, is that the offset is fixed and computed before seeing the omics predictors. Thus, in the case where omics and clinical predictors are strongly correlated, we expect the clinical offset to capture much variability and hence the null-hypothesis to be retained. This issue will be further discussed in Sect. 4.

The fact that the offset is fixed also implies that the coefficients of the clinical predictors fit in step 1 are *not* influenced by the omics predictors added to the model by boosting regression in step 2. On the one hand such an offset can well address the question of the *added* predictive value. The offset can be considered as an artificial but compulsory first predictor that is subsequently completed by the omics predictors selected afterwards. On the other hand the inconvenience is that clinical covariates cannot be tested—either individually or as a whole. The globalboosttest procedure allows to test the omics predictors only.

In principle any type of response variable can be analysed using globalboosttest provided that it can be accommodated into GLMs and boosting regression. This includes normally distributed, binary or censored responses. Furthermore, boosting regression may be essentially replaced by any regularized regression technique allowing an offset, e.g. the Lasso.

3 Pre-validation

The pre-validation method is based on a classical hypothesis testing framework within a GLM including the clinical predictors as well as a "pseudo-predictor" summarizing the omics predictors. This pseudo-predictor can be derived either at the link scale (which is preferred here in the context of survival analysis) or at the predictor scale. In principle all methods that can handle a large number of predictors can be used for this purpose, e.g. boosting regression or Lasso regression. In this study boosting regression is considered for the sake of consistency with the globalboosttest procedure described in Sect. 3.

The obtained pseudo-predictor summarizing the omics data, however, should not be tested in a multivariate regression model based on the data set that were used for its construction. This approach would strongly favor omics data, because the pseudo-predictor constructed from high-dimensional would overfit the data set. To overcome this problem Tibshirani and Efron (2002) suggest "pre-validation". The term pre-validation refers to a cross-validation (CV) performed within the considered data set. At each CV iteration j, a pseudo-predictor is derived from the omics data set $S \setminus S_j$ (where S_j stands for the jth CV fold) and then computed for the observations from S_j. Since the folds S_j form a partition of the data set S, one thus obtains a pseudo-predictor value for each observation. This "pre-validated" pseudo-predictor is not expected to overfit the data set, since at each CV iteration there is no overlap between the "training data" $S \setminus S_j$ and the fold S_j. This pre-validation step is denoted as "internal".

Finally, a multivariate regression model (denoted as "external" model) is fitted using this pre-validated pseudo-predictor and the clinical predictors as predictors. The added predictive value is assessed by testing the significance of the regression coefficient of the pseudo-predictor. However, in a subsequent publication (Höfling and Tibshirani 2008) this test is shown to be biased due to the violation of the i.i.d. assumption in the GLM. Höfling and Tibshirani (2008) address this bias through a permutation procedure which we also use here.

In contrast to globalboosttest, pre-validation considers clinical and omics predictors more symmetrically—in the sense that the coefficients of the clinical predictors are affected by the omics data, which is not the case in globalboosttest. If clinical and omics data are correlated, we thus expect both the clinical predictors and the omics-based pseudo-predictor to capture an important part of the variability. Another difference to globalboosttest is that the pseudo-predictor is computed differently for the K subsets, thus making computation more intensive and interpretation more difficult.

4 Simulation Study

Both methods address the added value of high-dimensional omics data in the same data situation, but essentially ask different questions. While globalboosttest directly focuses on the added predictive value, the rationale behind pre-validation is of more symmetric nature. In this paper, their respective performance is examined in different simulation settings with sample size $n = 200$ and a censored time-to-event as response Y. The partially unobserved survival times T_i ($i = 1, \ldots, 200$) are generated from a Cox–Weibull model (Cox 1972) similarly to Binder and

Schumacher (2008). The cumulative density is given as $F(t) = 1 - \exp(-(\lambda(t) \cdot t/\alpha))$, where $\lambda(t)$ denotes the hazard rate. The survival times T_i ($i = 1, \ldots, 200$) are generated as $T_i = \frac{-\log(U_i)\cdot\alpha}{\lambda(t)}$, whereby U_i is drawn from the uniform distribution $U(0, 1)$ and the shape parameter α is set to 1 in our study. Assuming proportional hazards, $\lambda(t)$ is modeled as $\lambda(t) = \lambda_0 \exp(\eta)$, whereby the baseline hazard rate λ_0 is set to 0.1 and η denotes an additive predictor. The follow-up times F_i are generated independently of the survival times in the same way as T_i but with a constant hazard rate $\lambda(t) = 0.1$ and $\alpha = 1$. Observations are censored if their follow-up time ends before the expected event such that about half of the observations are censored. Finally, the observation times Y_i are obtained as $Y_i = \min(F_i, T_i)$. Further, we generate five standard normal and mutually uncorrelated clinical predictors as well as 1,000 standard normal omics predictors. Ten of these omics predictors are correlated with the first clinical predictor, 10 further omics predictors are correlated with the second clinical predictor, and so on, yielding a total of 50 omics predictors correlated with clinical predictors, whereby the correlation ρ is set either to $\rho = 0$ (no correlation), $\rho = 0.2$ (weak correlation) and $\rho = 0.8$ (strong correlation). The linear predictor η is defined as follows. The regression coefficients of the clinical predictors are chosen to mimick a realistic scenario with predictors of varying strengths: $\beta_{clinic} = (0, 0.5, 2, -1.5, -1)^T$. Out of the 1,000 omics predictors, 20 have non-zero regression coefficients in the linear predictor η. The 20 coefficients are drawn from the uniform distribution $U(0.1, 0.7)$. Importantly, the size of the intersection between the 20 predictive predictors and the 50 correlated predictors is set successively to 0, 5, 10 and 20. Size 0 yields a setting where clinical and informative omics predictors are completely uncorrelated, while size 20 means that all informative omics predictors are correlated with clinical predictors. Table 1 (top) sums up the resulting settings, including an additional "null-scenario" without informative omics predictors (setting I). Moreover, some more extreme situations (bottom of Table 1) are additionally included in the study to complement the considered settings. First, setting VI is enlarged to 5,000 omics covariates (setting XI) and as well reduced to only 20 (setting XII). Settings with more (setting XIII) and fewer (setting XIV) informative omics predictors are also considered. Finally settings with perfect correlation ($\rho = 1$) between the five clinical predictors and the 50 correlated omics predictors are considered, either with completely non-informative omics predictors (setting XV) or with 20 informative predictors included in the 50 correlated omics predictors (setting XVI). For each setting the two methods globalboosttest and pre-validation are evaluated based on 100 randomly generated data sets. For both globalboosttest and pre-validation the number of boosting iteration m_{stop} is set successively to 50, 100, 200, 500, and 1,000. All tests base on 1,000 permutations. As expected, both globalboosttest and pre-validation yield p-values that are approximately uniformly distributed [0, 1] in the absence of informative omics predictors (data not shown). When omics predictors are informative and not perfectly correlated with clinical predictors, globalboosttest tends to yield smaller p-values than pre-validation. This result is illustrated by Fig. 1 that displays the p-value of the two tests for different numbers m_{stop} of boosting

Table 1 Overview on simulation settings with a censored time-to-event as response, 5 clinical covariates and $n = 200$. If so each clinical covariate correlates to 10 omics covariates. For boosting $m_{stop} = (50, 100, 200, 500, 1,000)$ iterations and for testing 1,000 permutations are considered. For each setting globalboosttest/pre-validation are computed on 100 different data sets

Setting	Number of omics covariates	Correlation coefficient	Informative omics covariates	Intersection informative/correlated omics covariates
I	1,000	$\rho = 0$	0	$\{\varnothing\}$
II	1,000	$\rho = 0$	20	$\{\varnothing\}$
III–VI	1,000	$\rho = 0.2$	20	$\{\varnothing, 5, 10, 20\}$
VII–X	1,000	$\rho = 0.8$	20	$\{\varnothing, 5, 10, 20\}$
XI: Many omics predictors	5,000	$\rho = 0.2$	20	$\{10\}$
XII: Few omics predictors	20	$\rho = 0.2$	20	$\{20\}$
XIII: Many informative omics	1,000	$\rho = 0.2$	200	$\{20\}$
XIV: Few informative omics	1,000	$\rho = 0.2$	2	$\{\varnothing\}$
XV: Perfect correlation i	1,000	$\rho = 1$	20	$\{20\}$
XVI: Perfect correlation ii	1,000	$\rho = 1$	0	$\{\varnothing\}$

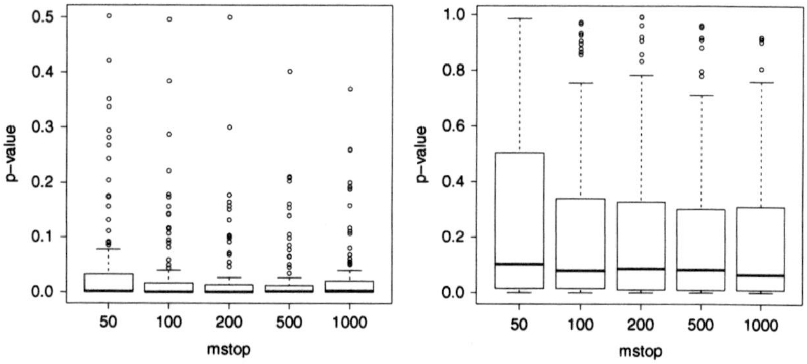

Fig. 1 Results of setting V. *Left:* globalboosttest. *Right:* pre-validation

iterations in Setting V. Moreover, globalboosttest tends to already reach good power for a smaller m_{stop} even in the case of high correlation between omics and clinical predictors. In contrast, pre-validation needs more iterations to capture the added predictive value of omics predictors. That is probably because pre-validation does not take the clinical predictors into account while summarizing the omics predictors and thus first captures information that are already captured by clinical predictors. By considering clinical predictors as an offset, globalboosttest captures the residual variability that is not captured by clinical predictors. Thus, globalboosttest generally needs less boosting iterations to reach good power. An exception is setting X, where *all* informative predictors are strongly correlated with a clinical predictor: globalboosttest then yields uniformly distributed p-values for a small m_{stop}, while a large m_{stop} leads to smaller p-values.This result obtained in setting X is related to the essential goal of globalboosttest. Globalboosttest tests the *added* predictive value of

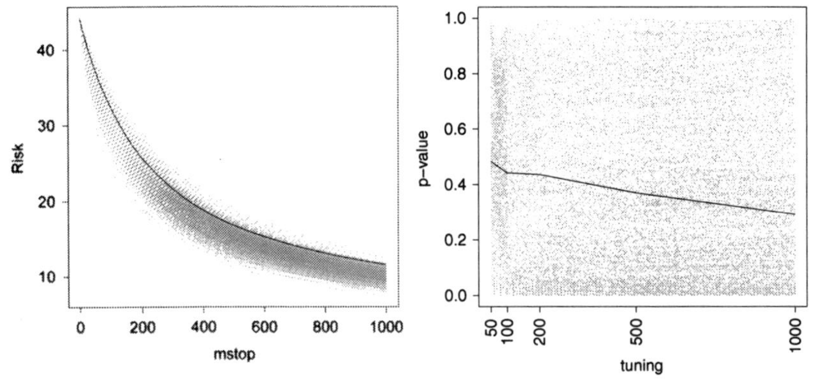

Fig. 2 Permutation procedure on Chin data. *Left:* globalboosttest. *Right:* pre-validation

omics predictors and focuses on the part of the variability that is not captured by the clinical offset. In the unrealistic extreme case where all 50 informative omics predictors are perfectly correlated with a clinical predictor, the p-values are uniformly distributed on [0, 1] with globalboosttest, but not with pre-validation. Note that this result is in contradiction with the theoretical null-hypothesis corresponding to globalboosttest: a strong correlation between omics and clinical predictors does not lead to the rejection of the null-hypothesis. Pre-validation employing the Lasso as "internal model" struggles with some problems related to tuning. As the number of observations is typically small for omics data, there are even less observations in training and test data sets. That makes the choice of λ extremely unstable. Each fold of the pseudo-predictor is based on a different value of λ. In many cases the optimal choice of λ selects no omics predictors at all. The choice of λ is much more crucial as the choice of boosting parameter m_{stop}. Consequently, globalboosttest and pre-validation employing boosting perform substantially better than pre-validation employing the Lasso.

5 Analysis of Breast Cancer Data

For illustration a breast cancer data set (Chin and et. al. 2006) including 77 patients is analyzed using globalboosttest and pre-validation with boosting regression. The response of interest is the censored distal recurrence time in years. The considered data set includes 11 clinical predictors such as age at diagnosis, variables of the TNM staging system or information on estrogen and progesterone receptors, as well as the expression level of 22,215 genes acting as omics predictors.

The permutation-based p-values range from 0.77 to 0.97 for globalboosttest and from 0.29 to 0.48 for pre-validation (depending on m_{stop}). Figure 2 displays the curves representing the negative binomial log-likelihood (for globalboosttest) and the p-value (for pre-validation) obtained with the original data set (black) and the permuted data sets (grey). Both methods suggest that omics predictors do not improve prediction strength.

6 Summary and Outlook

Simulation results suggest that in case of poor to moderate correlation between clinical and omics predictors globalboosttest tends to have a superior power to pre-validation. However, in case of strong correlation globalboosttest becomes more conservative, which reflects its rationale: globalboosttest tests added predictive value, i.e. focuses on the variability that is not already captured by the clinical predictors. Whether it makes more sense to reject or the accept the null-hypothesis in the case of strong correlation depends on the substantive context. Correlation also seems to increase the impact of the number of boosting steps, suggesting that a systematic method for the choice of this parameter should be developed in the future.

In our paper the globalboosttest and pre-validation are assessed with respect to their performance as testing procedures. However, similar approaches may be adopted to derive combined prediction rules based on both clinical and omics predictors, see Boulesteix and Sauerbrei (2011) for an overview of such approaches. Due to its asymmetrical character giving more importance to clinical predictors, we expect the prediction rule derived from globalboosttest to perform poorly when these predictors are weak. A pre-validation approach may be promising, see Boulesteix et al (2008) for an example in the context of binary classification. More research is needed to assess the respective merits of the two methods in terms of predictive accuracy.

Acknowledgements We thank Jutta Engel for helpful advice on the breast cancer data.

References

Binder, H., & Schumacher, M. (2008). Adapting prediction error estimates for biased complexity selection in high-dimensional bootstrap samples. *Statistical Applications in Genetics and Molecular Biology, 7*(1), 12.

Boulesteix, A., Porzelius, C., & Daumer, M. (2008). Microarray-based classification and clinical predictors: on combined classifiers and additional predictive value. *Bioinformatics, 24*(15), 1698–1706.

Boulesteix, A. L., & Hothorn, T. (2010). Testing the additional predictive value of high-dimensional molecular data. *BMC Bioinformatics, 11*, 78.

Boulesteix, A.L., & Sauerbrei, W. (2011). Added predictive value of high-throughput molecular data to clinical data and its validation. *Briefings in Bioinformatics, 12*(3), 215–229.

Chin, K., et al. (2006). Genomic and transcriptional aberrations linked to breast cancer pathophysiologies. *Cancer Cell, 10*(6), 529–541.

Cox, D. R. (1972). Regression models and life-tables. *Journal of the Royal Statistical Society, Series B, 34*(2), 187–220.
Höfling, H., & Tibshirani, R. J. (2008). A study of pre-validation. *The Annals of Applied Statistics, 2*(2), 643–664.
Tibshirani, R. J., & Efron, B. (2002). Pre-validation and inference in microarrays. *Statistical Applications in Genetics and Molecular Biology, 1*, 1.

External Analysis of Asymmetric Multidimensional Scaling Based on Singular Value Decomposition

Akinori Okada and Hiroyuki Tsurumi

Abstract An asymmetric similarity matrix among objects, for example, a brand switching matrix of consumers, can be analyzed by asymmetric multidimensional scaling. Suppose that n brands exist, and that m new brands are introduced. While the brand switching from existing to new brands can be observed, the brand switching from new to existing brands nor that among new brands cannot be observed soon after the introduction of the new brands. The present study analyzed the $n \times n$ similarity matrix by the asymmetric multidimensional scaling based on singular value decomposition. The analysis gives outward and inward tendencies of existing brands. Using the obtained outward tendency of n existing brands, the inward tendency of m new brands is derived. An application to the brand switching data among margarine brands is presented.

1 Introduction

Several procedures of asymmetric multidimensional scaling have been introduced (Borg and Groenen 2005, Chap. 23; Cox and Cox 2001, Sect. 4.8). Most of them deal with one-mode two-way asymmetric similarities among objects. One-mode two-way asymmetric similarities among n objects are represented as an $n \times n$ matrix whose (j, k) element represents the similarity from objects j to k. While one-mode two-way asymmetric similarities among objects are the typical type of data which are dealt with by asymmetric multidimensional scaling, there are other

A. Okada (✉)
Graduate School of Management and Information Sciences, Tama University, Japan
e-mail: okada@rikkyo.ac.jp

H. Tsurumi
College of Business Administration, Yokohama National University, Japan
e-mail: tsurumi@ynu.ac.jp

P. Giudici et al. (eds.), *Statistical Models for Data Analysis*, Studies in Classification, Data Analysis, and Knowledge Organization, DOI 10.1007/978-3-319-00032-9_31,

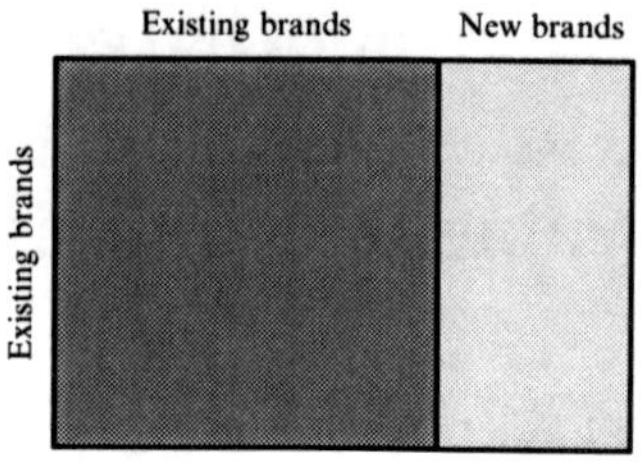

Fig. 1 The brand switching matrix among existing brands (*shaded area*), and the brand switching matrix from the existing brands to new brands (*unshaded area*)

kinds of data which also have been dealt with: two-mode three-way asymmetric similarities (Okada and Imaizumi 1997) or one-mode three-way similarities (de Rooij and Heiser 2000; Nakayama 2005).

In the present study, a procedure of analyzing different sorts of asymmetric similarities is introduced. The similarities are comprised of two parts; one is the $n \times n$ square asymmetric similarity matrix or one-mode two-way asymmetric similarities among n objects, and the other is the $n \times m$ rectangular similarity matrix ($m \geq 1$) whose (j, ℓ) element represents the similarity from objects j to ℓ or two-mode two-way similarities (Okada 2011). The similarities are regarded as an $n \times (n + m)$ similarity matrix where $n \times n$ submatrix (the first part) is asymmetric.

An example of this type of similarities is that suppose n brands already exist in a market, and that m new brands are introduced into the market. We have asymmetric brand switching data of consumers among n existing brands and brand switching data from n existing brands to m new brands. There would not be brand switching among m new brands nor brand switching from new brands to existing brands (Fig. 1) soon after the introduction of the new brands or within the inter-purchase interval, because the brand switching from the new brands to the existing brands nor the brand switching among the new brands do not occur yet.

In the present study, a procedure of analyzing this type of similarities ($n \times (n+m)$ similarity matrix) is presented. And the present procedure is applied to the $n \times (n + m)$ submatrix of similarities of the full $(n + m) \times (n + m)$ similarity matrix. This means that similarities from m objects to n objects as well as those among m objects are available, but they are not analyzed in the present study. The reason for dealing with the present type of data is to validate the effectiveness of the present procedure, because the result derived by the present procedure can be compared with that derived by analyzing the full $(n + m) \times (n + m)$ similarity matrix.

2 The Procedure

In the present study the asymmetric multidimensional scaling based on singular value decomposition (Okada 2012) is used. Let **A** be the $n \times n$ matrix of asymmetric similarities among n objects, and **B** be the $n \times m$ matrix of similarities from n objects to m objects. The (j, k) element of **A** represents the similarity from objects j to k, which is not necessarily equal to the (k, j) element of **A**, and the (j, ℓ) element of **B** represents the similarity from objects j to ℓ. By using singular value decomposition (Eckart and Young 1936; Harville 1997), **A** can be represented as a product of three matrices; the matrix which has left singular vectors as its columns, the diagonal matrix which has singular values as its diagonal elements, and the matrix which has right singular vectors as its columns. By using r $(< n)$ largest singular values and corresponding left and right singular vectors, **A** is approximated by

$$\mathbf{A} \simeq \mathbf{XDY}', \tag{1}$$

where **D** is the $r \times r$ diagonal matrix of r largest singular values in descending order at its diagonal elements, **X** is the $n \times r$ matrix of the corresponding left singular vectors (normalized so that the length is unity), and **Y** is the $n \times r$ matrix of the corresponding right singular vectors (normalized so that the length is unity). The j-th element of the i-th column of **X** (i-th left singular vector), x_{ji}, represents the outward tendency of object j along Dimension i, because rows of **A** represent brands to be switched from. The k-th element of the i-th column of **Y** (i-th right singular vector), y_{ki}, represents the inward tendency of object k along Dimension i, because columns of **A** represent brands to be switched to. We assume that the brand switching from existing brands to new brands can be approximated by

$$\mathbf{B} \simeq \mathbf{XDZ}', \tag{2}$$

where **Z** is the $m \times r$ matrix. The ℓ-th element of the i-th column of **Z**, $z_{\ell i}$, represents the inward tendency of object ℓ along Dimension i. **Z** is derived by

$$\mathbf{Z} = \mathbf{B}'\mathbf{XD}^{-1}. \tag{3}$$

While the inward tendency of the m objects is derived, their outward tendency cannot be derived (Okada 2011).

Equation (1) says that when $r = 2$, a_{jk}, the (j,k) element of **A**, can be approximated by

$$a_{jk} \simeq d_1 x_{j1} y_{k1} + d_2 x_{j2} y_{k2}, \tag{4}$$

where x_{j1} is the outward tendency of brand j and y_{k1} is the inward tendency of brand k along Dimension 1, x_{j2} is the outward tendency of brand j and y_{k2} is the inward tendency of object k along Dimension 2, d_1 is the largest singular value, and d_2 is the second largest singular value. The similarity from objects j to k is approximated by the sum of two terms (a) the similarity from brands j to k along Dimension 1, and (b) the similarity from brands j to k along Dimension 2; (a) the product of the outward tendency of brand j and the inward tendency of brand k multiplied by d_1 along Dimension 1 corresponding to the largest singular value, and (b) the product of the outward tendency of brand j and the inward tendency of brand k multiplied by d_2 along Dimension 2 corresponding to the second largest singular value (Okada 2011). As shown in Fig. 4, term (a) $d_1 x_{j1} y_{k1}$ shows the area along Dimension 1 multiplied by the largest singular value d_1 representing the similarity from brands j to k along Dimension 1. Also as shown in Fig. 5, term (b)

$d_2 x_{j2} y_{k2}$ shows the area with the sign along Dimension 2 multiplied by the second largest singular value d_2 representing the similarity with the sign from brands j to k along Dimension 2. Equation (4) means that a_{jk} is approximated by the sum of two weighted areas (when the second term is negative, the similarity along Dimension 1 is decreased by subtracting the similarity along Dimension 2). Similarly Eq. (2) says that when $r = 2$, $b_{j\ell}$, the (j, ℓ) element of **B**, can be approximated by

$$b_{j\ell} \simeq d_1 x_{j1} z_{\ell 1} + d_2 x_{j2} z_{\ell 2}, \tag{5}$$

where $z_{\ell 1}$ is the derived inward tendency of object ℓ along Dimension 1, and $z_{\ell 2}$ is the derived inward tendency of object ℓ along Dimension 2. The similarity from objects j to ℓ is represented by the sum of two terms; one is the first term of the right hand side of Eq. (5) representing the similarity from objects j to ℓ along Dimension 1 and the other is the second term of the right hand side of Eq. (5) representing the similarity from objects j to ℓ along Dimension 2.

3 The Data

The data analyzed in the present study are brand switching data among margarine brands. The brand switching data among 12 margarine brands (A, B, ..., H, I, J, K, and O) were collected in March through September 2009 at super market stores in the Tokyo metropolitan area. Brand switching data were derived from the purchase records of about 5,500 consumers who purchased margarine twice or more in the period. The details of the data are described in Okada and Tsurumi (2012). In the present study, the brand switching data among nine brands (A, ..., H, and O) and the brand switching data from the nine to three brands (I, J, and K) were dealt with. Brand O consists of all brands other than brands A through K. Brands I, J, and K have the three least market share (Okada and Tsurumi 2012), and are regarded as new brands in the present study. The brand switching matrix among the nine brands and the brand switching matrix from the nine brands to the three brands $(n = 9, m = 3)$ are shown in Table 1.

The submatrix consists of the first nine rows and of the first nine columns is **A** which corresponds to the shaded part of Fig. 1. The submatrix consists of the first nine rows and of the last three columns is **B** which corresponds to the unshaded part of Fig. 1. The full 12×12 brand switching matrix is shown in Okada and Tsurumi (2012).

4 The Analysis

The singular value decomposition of **A** gives nine singular values; 1653.8, 1038.3, 849.4, 792.2, 707.1, 573.3, 479.0, 447.8, and 302.6. The two-dimensional result was chosen as the solution. The reason for adopting the two dimensional result as

Table 1 Brand switching matrix among margarine brands

To brand											
From	A	B	C	D	E	F	G	H	O	I	J	K
Brand A	1,096	458	14	12	12	42	65	16	81	8	27	8
Brand B	312	716	7	5	6	14	52	8	39	3	27	6
Brand C	39	15	837	59	50	13	10	15	104	28	10	40
Brand D	32	18	65	739	15	12	10	17	73	8	9	62
Brand E	25	18	61	15	852	29	6	69	55	4	16	11
Brand F	64	49	18	18	24	505	10	143	64	6	54	18
Brand G	118	63	10	6	3	9	462	3	61	1	7	2
Brand H	31	20	33	17	76	142	4	410	84	3	34	9
Brand O	403	344	134	87	71	84	75	111	1,180	50	96	99

the solution is twofold; the two-dimensional solution was adopted by Okada and Tsurumi (2012), and the proportions of the sum of squared singular value(s) of the largest, the two largest, and the three largest singular values to the sum of squared elements of **A** are 0.42, 0.59 and 0.70 respectively. The increment of the proportion from two- to three-dimensional results is small compared with that from uni- to two-dimensional results. This makes it possible to compare the present result with the result derived by Okada and Tsurumi (2012).

The inward tendency of three brands (I, J and K) along Dimensions 1 and 2 were derived by Eq. (3) using the two-dimensional result. The procedure utilized the outward tendency of nine brands (A, ..., H, and O) which were already derived by the singular value decomposition of the brand switching matrix among the nine brands.

5 Results

The outward and inward tendencies of the nine brands (solid square) and the inward tendencies of the three brands (open diamond) along Dimension 1 are represented in Fig. 2. The horizontal axis or Dimension 1 outward, which corresponds to the first left singular vector corresponding to the largest singular value, represents the outward tendency of nine brands (A, ..., H and O), and the vertical axis or Dimension 1 inward, which corresponds to the first right singular vector corresponding to the largest singular value, represents the inward tendency of the nine brands and three brands (I, J and K). The three brands, which do not have the outward tendency, are represented on the vertical axis.

The outward and inward tendencies of the nine brands (solid square) and the inward tendencies of the three brands (open diamond) along Dimension 2 are represented in Fig. 3. The horizontal axis or Dimension 2 outward, which corresponds to the second left singular vector corresponding to the second largest singular value, represents the outward tendency of nine brands (A, ..., H and O), and the vertical

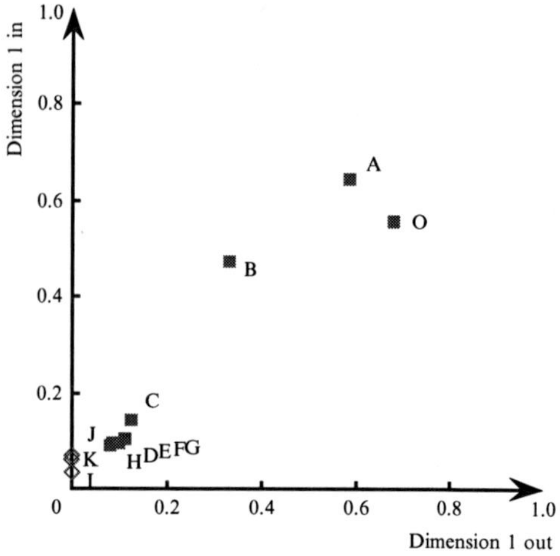

Fig. 2 The outward and inward tendencies along Dimension 1. *Solid squares* represent nine existing brands (A, . . . , H and O). *Open diamonds* represent three new brands (I, J and K)

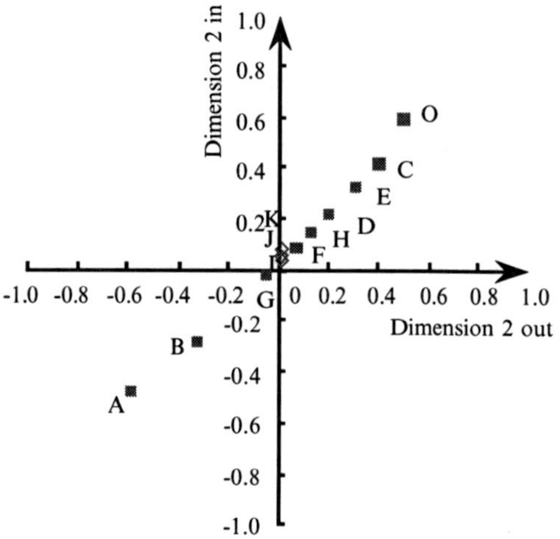

Fig. 3 The outward and inward tendencies along Dimension 2. *Solid squares* represent the nine existing brands. *Open diamonds* represent the three new brands

axis or Dimension 2 inward, which corresponds to the second right singular vector corresponding to the second largest singular value, represents the inward tendency of the nine brands and three brands (I, J and K). Like Fig. 2, the three brands, which do not have the outward tendency, are represented on the vertical axis.

6 Conclusions

Along Dimension 1, brands A, B, and O have large outward and inward tendencies, suggesting they compete strongly each other. From Eq. (4), the similarity from brands A to O along Dimension 1 is $d_1 x_{A1} y_{O1}$; the product of the outward tendency

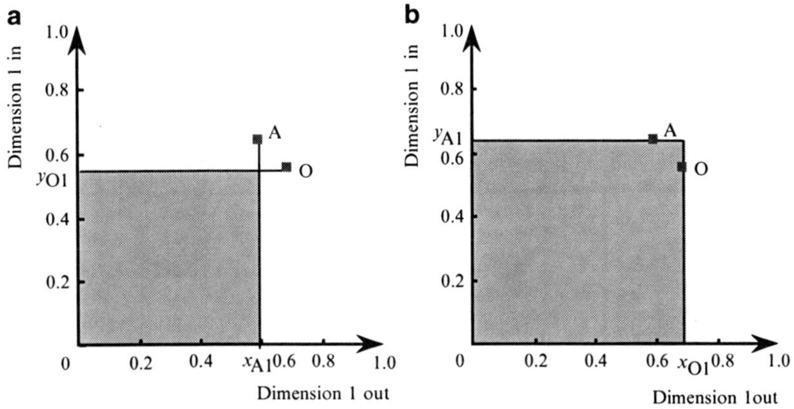

Fig. 4 The similarity along Dimension 1.
(**a**) The area of the *shaded rectangle* $x_{A1}y_{O1}$ corresponds to the similarity from brands A to O along Dimension 1. (**b**) The area of the *shaded rectangle* $x_{O1}y_{A1}$ corresponds to the similarity from brands O to A along Dimension 1

of brand A along Dimension 1, x_{A1}, and the inward tendency of brand O along Dimension 1, y_{O1}, multiplied by the largest singular value d_1. The area of the shaded rectangle of Fig. 4a shows the product $x_{A1}y_{O1}$ which corresponds to the similarity from brands A to O. The area of the shaded rectangle of Fig. 4b shows the product $x_{O1}y_{A1}$ which corresponds to the similarity from brands O to A along Dimension 1. The former is smaller than the latter. Thus the brand switching from brand O to brands A and B is larger than the brand switching from brands A and B to brand O, suggesting brands A and B are dominant over brand O in the interbrand competition among margarine brands along Dimension 1.

Brands C through H have smaller outward and inward tendencies than brands A, B and O have, suggesting the brand switching among them is small. The similarities from brands C through H to brands A and B are similar to those from brands A and B to brands C through H along Dimension 1, and the similarities from brands C through H to brand O are smaller than those from brand O to brands C through H along Dimension 1. This suggests that, while the magnitude of the brand switching is small, brands C through H are dominant over brand O in the interbrand competition among margarine brands along Dimension 1.

Dimension 2 classifies the nine brands into two groups; one group consists of brands in the first quadrant (C, . . . , F, H and O), and the other consists of brands in the third quadrant (A, B and G). The similarity from brands in the third quadrant to those in the first quadrant is negative along Dimension 2. The area of the shaded rectangle in Fig. 5a shows the product $x_{A2}y_{O2}$, which corresponds to the similarity from brands A to O and is negative, because the outward tendency of brand A, x_{A2}, is negative, and the inward tendency of Brand O, y_{O2} is positive. The similarity from brands in the first quadrant to those in the third quadrant is negative along Dimension 2. The area of the shaded rectangle in Fig. 5b shows the product $x_{O2}y_{A2}$,

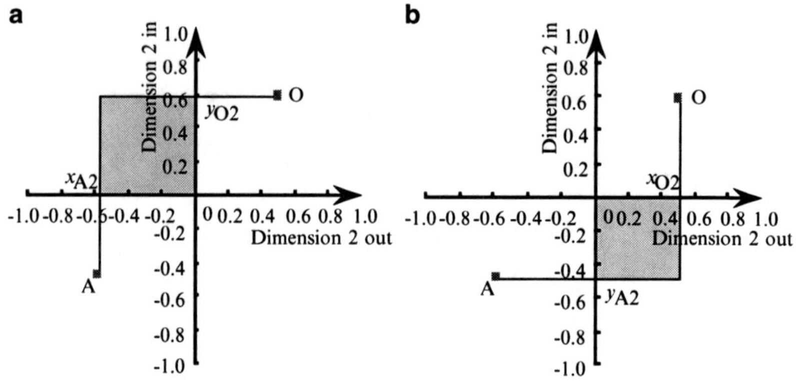

Fig. 5 The similarity along Dimension 2.
(**a**) The area of the *shaded rectangle* $x_{A2}y_{O2}$ corresponds to the similarity from brands A to O along Dimension 2. (**b**) The area of the *shaded rectangle* $x_{O2}y_{A2}$ corresponds to the similarity from brands O to A along Dimension 2

which corresponds to the similarity from brands O to A and is also negative, because x_{O2} is positive and y_{A2} is negative. In Fig. 5, the similarity among brands in the same quadrant (the first or the third quadrants) is positive, because both the inward and the outward tendencies of the brand in the first quadrant are positive, and thus their product is positive, and both the inward and the outward tendencies of the brand in the third quadrant is negative, and thus their product is positive. This suggests that the brand switching between two groups is small, and that within the same group is large along Dimension 2 (Okada 2012). The positive similarity along Dimension 2 increases the similarity along Dimension 1, and the negative similarity along Dimension 2 reduces the similarity along Dimension 1 (Eq. (4)).

All nine brands (A through H) are almost on the 45 degree line in the first and the third quadrants, while brand O is above the 45 degree line. This means that brand O is dominant over brands C through F and H along Dimension 2, because the similarity from brand O to brands C through F and H is smaller than the similarity from the latter to the former along Dimension 2. The dominance of brands A, B and C over brand O along Dimension 1 is reduced by the negative similarity from brand O to brands A and B and by the inferiority of brand C against brand O.

Inward tendencies of three brands (I, J and K) were derived by using Eq. (3). Derived inward tendencies of brands I, J and K along Dimension 1 are 0.028, 0.063 and 0.054, and those along Dimension 2 are 0.033, 0.040 and 0.074. Okada and Tsurumi (2012) derived outward and inward tendencies of all 12 brands by the singular value decomposition of the full 12 × 12 brand switching matrix among brands A, …, H, O, I, J, and K. The inward tendencies of brands I, J and K along Dimension 1 were 0.040, 0.098 and 0.79, and those along Dimension 2 were 0.061, 0.050 and 0.126. The original figures obtained by Okada and Tsurumi (2012) were multiplied by $\sqrt{4/3}$ to derive these figures in order to adjust the length of the right singular vector of **A**, because the singular vector of the present study has nine elements and the singular vector of Okada and Tsurumi (2012) has 12 elements. The magnitude of the inward tendency of the present study is smaller than that of Okada and Tsurumi (2012), but the value varies rather similarly.

Three brands (I, J, and K) are represented on the Dimension 1 inward axis in Fig. 2. In Fig. 2, while the outward tendencies of the three brands are not obtained, when they are above the line connecting a brand and the origin after their outward tendencies are obtained in future, they will be dominant over the brand in the interbrand competition along Dimension 1. On the other hand, when they are below the line in future they will be inferior to the brand in the interbrand competition along Dimension 1. Three brands (I, J, and K) are represented on the positive side of the Dimension 2 inward axis in Fig. 4. This means that they will be in the first or the second quadrants depending on the sign of their outward tendencies obtained in future. If the obtained outward tendency in future is positive, they will be in the first quadrant, and if the obtained outward tendency is negative in future, they will be in the second quadrant. When they are below the line connecting a brand and the origin in the first quadrant, they will be inferior to the brand in the first quadrant in the interbrand competition along Dimension 2. When they are above the line in the first quadrant, they will be dominant over the brand in the first quadrant in the interbrand competition along Dimension 2. As described earlier, when they are in the first quadrant, they have negative similarities with the brands A, B and G in the third quadrant, and vice versa. Thus the three brands and brands A, B and G will not directly compete with each other along Dimension 2. When they are in the second quadrant, they will be dominant over the brands in the first quadrant in the interbrand competition, because the outward tendency of the three brands is negative and the inward tendency of the brand in the first quadrant is positive, suggesting the similarity from the three brands to the brand in the first quadrant is negative, while the inward tendency of the three brands is positive and the outward tendency of the brand in the first quadrant is positive, suggesting the similarity from the brand in the first quadrant to the three brands is positive. The three brands will be inferior to brands in the third quadrant in the interbrand competition for reasons similar to those above.

A procedure for deriving the inward tendency of the newly introduced brands was described, which is based on the outward tendency of the existing brands derived by the asymmetric multidimensional scaling of one-mode two-way asymmetric similarities among existing brands. The present procedure was applied to derive the inward tendency of the three brands where outward and inward tendencies of 12 brands including the three brands had been already been known. This makes it possible to compare the present result with the result which had been obtained by analyzing the asymmetric similarities among 12 brands including the three brands. The purpose of the present application is to show the validity of the present procedure but not to show a practical application. The practical application of the present procedure to the data, where the similarities from the existing brands to the newly introduced brands as well as the similarities among newly introduced brands are unknown, should be done.

The present procedure utilizes the externally given outward tendency of the existing brands derived by analyzing the one-mode two-way asymmetric similarities among existing brands. This means that the procedure is a sort of external analysis of multidimensional scaling (Borg and Groenen 2005, pp. 76–80). The procedure is applied to the brand switching among margarine brands. The inter-purchase interval of two consecutive purchases of the margarine is not long but not so short for most consumers when compared with that of soft drinks. It seems preferable to deal with goods which have a longer inter-purchase interval (e.g., detergent, shampoo, black tea, instant coffee, etc.) so that only a small number of the brand switching from new brands to existing brands and that among new brands occur in the period of collecting brand switching data after the introduction of the new brands.

Acknowledgements The authors would like to express their gratitude to the anonymous referee for her/his constructive review on an earlier version of this paper. They also wish to express their appreciation to Jim Hathaway for his thoughtful help concerning English.

References

Borg, I., & Groenen, P. J. F. (2005). *Modern multidimensional scaling: theory and applications* (2nd edn.). New York: Springer.

Cox, T. F., & Cox, M. A. A. (2001). *Multidimensional scaling* (2nd edn.). Boca Raton, FL: Chapman and Hall/CRC.

de Rooij, M., & Heiser, W. J. (2000). Triadic distance models for the analysis of asymmetric three-way proximity data. *British Journal of Mathematical and Statistical Psychology, 53*, 99–119.

Eckart, C., & Young, G. (1936). The approximation of one matrix by another of lower rank. *Psychometrika, 1*, 211–218.

Harville, D. A. (1997). *Matrix algebra from a statisticians perspective*. New York: Springer.

Nakayama, A. (2005). A multidimensional scaling model for three-way data analysis. *Behaviormetrika, 32*, 95–115.

Okada, A. (2011). Analysis of social affinity for foreigners by spectral decomposition: Asymmetric multidimensional scaling of EASS2008 data. *JGSS Research Series, 8*, 119–128. (in Japanese).

Okada, A. (2012). Analysis of car switching data by asymmetric multidimensional scaling based on singular value decomposition. In W. Gaul, A. Geyer-Schultz, , L. Schmidt-Thieme, J. Kunze (Eds.) *Challenges at the interface of data analysis, computer science, and optimization* (pp. 143–150). Heidelberg: Springer.

Okada, A., & Imaizumi, T. (1997). Asymmetric multidimensional scaling of two-mode three-way proximities. *Journal of Classification, 14*, 195–224.

Okada, A., & Tsurumi, H. (2012). Asymmetric multidimensional scaling of brand switching among margarine brands. *Behaviormetrika, 39*, 111–126.

The Credit Accumulation Process to Assess the Performances of Degree Programs: An Adjusted Indicator Based on the Result of Entrance Tests

Mariano Porcu and Isabella Sulis

Abstract In the frame of the performance indicators this paper aims to consider the bias produced by micro-level Potential Confounding Factors—PCF—by comparing the results observed using adjusted and unadjusted measures of outcome. Results at the university entrance tests together with the previous school experiences have been used as proxies of students' competencies at the beginning of their academic career. The regularity of schooling process has been monitored using as an outcome variable the students' status (drop out, still enrolled) and the number of credits gathered after one academic year. Adjusted indicators of the regularity of the students' career are obtained using the results of *zero-augmented* models to investigate the relationships between the outcome measures and the potential PCF which are not directly associated to the learning process under evaluation.

1 Introduction

Since the second half of the 1980s there has been a growing interest on assessing the effectiveness of educational institutions (Aitkin and Longford 1986; Ball and Wilkinson 1985; Goldstein and Spiegelhalter 1996; Goldstein and Thomas 1996; Bratti et al. 2004; Leckie and Goldstein 2009). Private stakeholders (families and enterprises) and public institutions have strongly demanded to evaluate the performance of formative institutions (such as universities) and to adopt these evaluations in comparative terms in order to support and enhance only those institutions which satisfy specified quality standards in terms of efficiency and efficacy. This way of monitoring the performance, copes on one hand with the need of having measures to make comparisons, while on the other hand, it is

M. Porcu (✉) · I. Sulis
Dip. Scienze Sociali e delle Istituzioni - Università di Cagliari, Cagliari, Italy
e-mail: mrporcu@unica.it; isulis@unica.it

P. Giudici et al. (eds.), *Statistical Models for Data Analysis*, Studies in Classification, Data Analysis, and Knowledge Organization, DOI 10.1007/978-3-319-00032-9_32,

affected by factors which are external to the formative processes: students' socio-cultural characteristics; economic-territorial framework and local job market. From these considerations it arises the demand to assess the effectiveness of formative institutions taking into account, in the process of comparison, not only of the output of the educational process, as for instance, the rate of students who finish on schedule, the regularity of the formative process, or, for secondary schools and universities, the rate of employability, but also of the so called input factors of the process. The latter, may act within a system as Potential Confounding Factors—PCF (Draper and Gittoes 2004) that operate with the same intensity on both macro (institutions) and micro (students) levels. At micro level, it is well known that comparative evaluations are meaningful whenever they are made between students who are homogeneous with respect to their socio-cultural characteristics, whereas, at macro-level, it is important to relate the measures of performance to a territorial framework throughout the evaluation of the influence on indicators of geographical factors (e.g., the local economic system): the presence of enterprises and their vocational sectors; the socio-economic condition of the area; the relevance of the public sector, etc..

This paper makes an attempt to discuss the bias in the indicators of performance generated by micro-level PCF by comparing the results observed using adjusted and unadjusted measures. Results at the university entrance tests together with the previous school experiences have been used as proxies of students' competencies at the beginning of their university career. According to these competencies indicators of the regularity in the teaching process are adjusted in order to assess the *net effect* in terms of efficiency of the institutions (e.g. the degree program).

The adjusted indicator will be obtained using the results of a model-based approach for count data (*zero-augmented* models) to investigate the relationships between the outcome measures and the potential PCF which are not directly attributable to the process under investigation. The modeling approach allows us to assess the role of PCF and to control for them by simulating the composition/structure of a standard population in all the institutions under comparison.

Therefore, this work has two main aims: (i) to assess the informative value of the results in the entrance test to predict students' success during the first year (Häkkinen 2004; Julian 2005; Belfied and Costa 2012); (ii) to consider the effect on the efficiency indicator of the PCF observed at students' level (CNVSU 2010).

2 Efficiency Indicators and Micro-Level PCFs

In Italy, since 2009/2010 academic year, the central government considers for the allocation of the university yearly provision a ranking between academic institutions based on the values of a batch of indicators (CNVSU 2010). The introduction of such allocation system increases, within the universities, the need of employing measuring tools useful in making comparisons, at different levels, across the institutions engaged in the academic educational processes: faculties and "degree committees" for the first and second level DP and departments for the *Ph*D programs. Considering the domain of the teaching, the performances are assessed considering the efficiency of guidance services (monitored using the rate of first year students that get at least 1/2 of the credits), the regularity of formative process (monitored using the credits effectively gained by students on the number of credits that they should have gained), the students' evaluation of university teaching (% of courses surveyed). The first two dimensions are monitored using indicators directly related to the credit accumulation process. It is widely assessed that the regularity of students' careers and their achievement is affected by factors which are external to the process under evaluation (CNVSU 2010). For instance, it is recognized that the relations between students' cultural resources, the choice of educational pathways, and academic performance are the results of previous social and cultural experiences. Thus the cultural resources that students own before to enroll to the university can act as key factor in affecting the transition from school to university and in determining the regularity of academic process and the professional status of individuals (De Graaf 2000; Sullivan 2001). From these considerations, and monitoring the students' achievement in terms of regularity in the credit accumulation process, the purpose of this work is to assess the role played by external factors to the institution under evaluation such as students' competencies at the entrance in influencing students' achievement.

3 Modeling the Credits Accumulation Process

The method here advanced to adjust for PCF the performance indicators related to the credits accumulation process relies on the results a modeling approach for count data and works on the following steps: (1) the observed distribution of credits conditional upon students' covariates is modeled using a count regression model which explicitly considers the excess of zeros in the distribution of credits; (2) the influence of PCF on the expected number of credits is assessed by the estimates of the coefficient parameters; (3) the estimates of the coefficient parameters are used in order to predict the expected number of credits gained by students under the assumption that the "objects" under comparisons (namely, the DP) are homogeneous with respect to students' characteristics.

Zero-augmented models (Zero-inflated and Hurdle) are appropriate in order to model count distributions which show an excessive number of zeros and to assess the role of PCF on the credit accumulation process. They simultaneously model the process of accumulation of credits and the probability to observe a 0 rather than a positive score of credits. This class of models has been already adopted by Boscaino et al. (2007) to study the role played by students' characteristics on the credits distribution. In this context zero-agumented models are adopted to assess the effect of PCF and to use results for predictive purposes: to assess the *net* value of the institution under comparison by comparing adjusted and unadjusted measure of efficiency Zero-inflated models assume the presence of two separate components:

the **zero** one and the **count** one (Cameron and Trivedi 2005; Zeileis et al. 2008). The mixture can be specified as

$$Pr(Y_i = y_i) = \pi_{zero}(0; z, \gamma) I_o(y) + (1 - \pi_{zero}(0; z, \gamma)) f_{count}(y; x, \beta) \quad (1)$$

where, a zero count can contribute to both components (this imply the existence of two kinds of zero observations "***true zeros***" and "***excess zeros***"). π_{zero} (zero *vs* count) can be modeled with a binary model (logit or probit) which considers the probability of zero-inflation (ZI) as a function of the z_i vector of regressors, whereas the f_{count} $(y \geq 0)$ may be modeled as a function of the x_i vector of covariates by specifying a **Poisson** (ZIP) or a **Negative Binomial** (ZINB) distribution. The covariates which influence the two components do not need to be distinct. The expected number of credits is thus estimated as a function of units (individual) covariates $\boldsymbol{x}, \boldsymbol{z}$: e.g. if a Poisson distribution is chosen and the logit link is adopted, the corresponding regression equation for the mean is

$$E(\mu_i) = \pi_i I_0\{y\} + (1 - \pi_i)\exp(\boldsymbol{x}'_i \beta) \quad (2)$$

where, $\pi_i = \exp(z'\gamma)(1 + exp(z'\gamma))^{-1}$. Thus the expected number of credits acquired by students who have the same covariate pattern $\tilde{\boldsymbol{\mu}}_g(\tilde{\boldsymbol{x}}_s, \tilde{z}_s)$ and belong to different DP can be predicted straightforwardly—$\mu_i^e = E(\mu_i | \tilde{\boldsymbol{x}}_s, \tilde{z}_s)$—once that the effect of the covariates is estimated by the parameters γ and β.

3.1 A Proposal to Adjust Efficiency Indicators

We propose to summarize the information on the credits accumulation process by DP (or faculties) $(g = 1, \ldots, G)$ using a re-scaled unadjusted index of efficiency. A gross index of efficiency of the DP could be estimated by comparing the observed distribution of the credits acquired by students with the distribution that we would have observed in the situation of maximum efficiency: if all students would accumulate exactly the number of credits required by the DP (i.e., 60 credits at the end of the first year). The re-scaled unadjusted index to measure the efficiency at DP (or faculties) $(g = 1, \ldots, G)$ level is

$$E_g = [\sum_{i=1}^{n_g} (\mu_i^{max} - \mu_i^o)][\sum_{i=1}^{n_g} \mu_i^{max}]^{-1} \quad (3)$$

where, μ^{max} is the maximum number of credits that could be expected (e.g., 60,120,180); μ^o is the observed number of credits; n_g is the number of students in faculty or DP g. The related adjusted indicator of efficiency is built up by replacing μ_i^o with the "*expected number of credits*" (μ_i^e) that we would observe under the hypothesis that the institutions under comparison match a specific distribution for

PCFs (characteristics of their students), e.g. $\mu_i^e = E(\mu_i|\tilde{\boldsymbol{x}}_s, \tilde{z}_s)$. We propose to use the following bunch of indicators to carry on a comparative analysis of the performances of DP in terms of regularity of the credits accumulation process:

1. The **estimated unadjusted estimated** $\hat{E}_g$ index based on the expected number of credits predicted using ZINB

$$\hat{E}_g = \sum_{i=1}^{n_g} (\mu_i^{max} - \mu_i^e(\boldsymbol{x}, z))(\sum_{i=1}^{n_g} \mu_i^{max})^{-1} \tag{4}$$

2. The **adjusted** $\tilde{E}_g^B$ index based on the results of ZINB model under the assumption that all the DPs were composed by the "best" students

$$\tilde{E}_g = \sum_{i=1}^{n_g} (\mu_i^{max} - \mu_i^e(\tilde{\boldsymbol{x}}_B, \tilde{z}_B))(\sum_{i=1}^{n_g} \mu_i^{max})^{-1} \tag{5}$$

3. The **adjusted** $\tilde{E}_h^g$ index based on the results of ZINB model under the assumption that all the DPs have the "structure" of h.

Specifically, a comparison between indicator (3) and (4) allows us to assess the accuracy of the model-based approach in reproducing the credits accumulation process. The closer the two indexes are the more reliable is the performance indicator.

4 Application

In the application, we use the data on credits accumulation of students enrolled in a faculty of an Italian University in 2008/2009 academic year. We consider the number of credits that students gathered after one academic year (5 proficiency sessions). In Table 1 credits have been classified in 6 classes (from 1–10,. . . , 51–60). The distribution of credits appears to be highly positively skewed and shows a huge number of zeros (147 on 463 observation). The 147 zero counts arise from those students who formally dropout or from those who did not get credits even though they are still enrolled—two kinds of zero observations. An analysis of the distribution of the credits in the three DP reveals that in average students in DP "B" have better performances than students in DP "A": in average they collect a number of credits double with respect to students in DP "A". The credits accumulation process seems to be related to the following students' covariates: sex (mean credits M=18.5, F=23.5), secondary school (mean credits *Liceo*=30, *other*=19—the *Liceo* provides a specific curriculum oriented to university studies), age ($\rho = -0.20$), results in the entrance test ($\rho = 0.27$) and DP (mean credits A=16, B=33, C=22). The composition of the three DP with respect to such

Table 1 Distributions of credits

	Credits							
DP	0	10	20	30	40	50	60	Total
A	55	23	15	22	10	6	4	135
B	7	5	3	9	6	8	8	46
C	85	38	31	48	34	26	20	282
Total	147	66	49	79	50	40	32	463

Table 2 Some descriptive statistics

	School (%)		Sex (%)		Age		Credits	
DP	Liceo	Others	F	M	$\bar{x}$	sd	$\bar{x}$	sd
A	30	70	57	43	26.6	9.9	15.8	17.2
B	37	63	72	28	21.5	6.3	32.6	20.7
C	50	50	56	44	21.3	5.4	22.6	19.8

characteristics is sensibly different (see Table 2). The functional shape of the count part of the model has been selected on the basis of the AIC criterium: the ZINB model, which specifies a negative binomial distribution for the count part and a logit link for the probability to observe a zero count, shows the best goodness of fit (Table 3).

The results of ZINB model are listed in Table 4. Looking at the zero component—which models with a logit function the probability to observe a zero rather than a positive count—arises that an increase of 1 point [min 0; max 40] in the entrance test (`TESTSCORE`) is associated with 7.5% lower odds to fail; further younger students have lower probabilities to fail. Specifically, an increase of 1 year in students' age (`AGE`) is associated with 3% higher odds to fail. The net effect of DP is measured by the coefficient parameters associated to the DP memberships: students enrolled in DP "A" have odds 3 times bigger to fail than students from DP "B". Looking at the effect of gender and secondary school attended, it arises that male students have odds to fail with respect to female equal to 1.76, whereas students from *other* schools have odds to fail 1.5 compared with students from *Liceo*. Regarding to the process which models the expected number of credits, it arises that the score gathered at the entrance test is a relevant predictor for the credit accumulation process. An increase of 1 point in the variable `TESTSCORE` is associated with an increase of 3.39% in the number of expected credits. Looking at the net effect of the DP on the expected number of credits arises that students in DP "B" have an expected number of credits 36.3% higher than students in DP "A". Furthermore, the main "disadvantages" are related to an increase in the variable `AGE`. The adjusted indicators have been build up on the basis of the ZINB results. Table 5 compares the following indexes: the *unadjusted index* (E_g) based on the observed number of credits, the *unadjusted estimated index* ($\hat{E}_g$) based on the expected number of credits predicted using the ZINB model, the *adjusted index*

Table 3 Goodness of fit measures

	ZIP	ZINB
logLik	−2239.30	−1562.00
♯ par.s	14	15
AIC	4506.60	3154.00
♯ zeros	147	147

Table 4 ZINB model

Covariates	Coeff.	SE	*p*-value
Zero component model coefficients (z) (logit link)			
(Intercept)	0.392	0.795	0.621
SEX-M	0.574	0.211	0.006**
TESTSCORE	−0.078	0.024	0.001**
factor(DP-B)	−1.171	0.467	0.012*
factor(DP-C)	−0.128	0.240	0.592
AGE	0.035	0.014	0.012*
SCHOOL-OTHER	0.410	0.225	0.068.
Count model coefficients (x) (NegBin with log link)			
(Intercept)	2.525	0.237	***
SEX-M	−0.019	0.060	0.751
TESTSCORE	0.033	0.006	***
factor(DP-B)	0.312	0.103	**
factor(DP-C)	0.104	0.073	0.153
AGE	−0.008	0.004	0.040
SCHOOL-OTHER	−0.011	0.060	0.844
Log(theta)	1.481	0.090	***

*: p-value < 0.05
**: p-value < 0.01
***: p-value < 0.001

Table 5 Comparison across adjusted and unadjusted indexes

Index %	A	B	C
E_g	26.296	54.348	37.234
$\hat{E}_g$	26.127	53.890	37.337
$\tilde{E}_g^B$	59.317	87.877	66.767
$\tilde{E}_g^A$	26.127	48.292	30.347

$\tilde{E}_g^B$ based on the results of ZINB model under the assumption that all the DPs have just students with the best profile ($age = 19$, *Liceo*, F, total score in the entrance test 40), the *adjusted index* $\tilde{E}_g^A$ based on the results of ZINB model under the assumption that all the DPs have the same composition of DP "A". Results in Table 5 show that if the composition of the three DPs was similar to "A", DP "B" would lose 6.1% points in the assessed efficiency (from 54.34 to 48.29) and the distance between "B" and "C" would change from 11.2% to 4.2%. If all the three DP are composed just by students with the "best" profile it would be a reduction of the differences between "C" and "A" (from 11.2% to 7.4%).

5 Final Remarks

From the comparison between the unadjusted indicator 3 and the expected unadjusted indicator 4 it arises a good accuracy of the modeling approach based on zero-inflated models in reproducing the credits distribution. The method allows us to assess the role played by some key factors that we consider in order to define adjusted indicators of efficiency. Such indicators take into account not just the final output of the process (e.g. total number of credits and dropout rates) but also of the differences in the input factors (secondary school attended and students' competencies at the enrollment). Further research are still in progress in order to perform the analysis in a multilevel framework. The use of Multilevel-Zero-Augmented Models will enable us to apply the method on all DP of a university and to control in the analysis also for the second-level covariates such as number of students enrolled or the field of studies to which the degree program belongs to.

References

Aitkin, M., & Longford, N. (1986). Statistical modeling issues in school effectiveness studies. *Journal of the Royal Statistical Society A, 149*(1), 1–443.

Ball, R., & Wilkinson, R. (1985). The use and abuse of performance indicators in UK higher education. *Higher Education, 27*(4), 417–427.

Belfied, C. L., & Costa, P. M. (2012). Predicting success in college: The importance of placement tests and high school transcripts. *CCRC Working Paper, 42.*

Boscaino, G., Capursi, V., & Giambona, F. (2007). La "sofferenza formativa": unánalisi per coorte di immatricolati. *DSSM Working Paper, 1.*

Bratti, M., McKnight, A., Naylor, R., & Smith, J. (2004). Higher education outcomes, graduate employment and university performance indicators. *Journal of the Royal Statistical Society A, 167*(3), 475–496.

Cameron, A. C., & Trivedi, P. K. (2005). *Microeconometrics: methods and applications.* Cambridge: Cambridge University Press.

CNVSU. (2010). Sistema di indicatori per la misura dell'efficienza della formazione universitaria. Technical Report RdR 3/10, MIUR.

De Graaf, N. D., De Graaf, P. M., & Kraaykamp, G. (2000). Parental cultural capital and educational attainment in the netherlands: A refinement of the cultural capital perspective. *Sociology of Education, 73*(2), 92–111.

Draper, D., & Gittoes, M. (2004). Statistical analysis of performance indicators in UK higher education. *Journal of the Royal Statistical Society A, 167*(3), 449–474.

Goldstein, H., & Spiegelhalter, D. J. (1996). League tables and their limitations: statistical issues in comparisons of istitutional performance. *Journal of the Royal Statistical Society A, 159*, 385–443.

Goldstein, H., & Thomas, S. (1996). Using examination results as indicators of school and college performance. *Journal of the Royal Statistical Society A, 159*, 149–163.

Häkkinen, I. (2004). Do university entrance exams predict academic achievement? *Working paper - Department of Economics - University of Uppsala*, (16).

Julian, E. R. (2005). Validity of the medical college admission test for predicting medical school performance. *Academic Medicine Research Report, 80*(10), 910–917.

Leckie, G., & Goldstein, H. (2009). The limitation of using school league tables to inform school choice. *Journal of the Royal Statistical Society A, 174*, 835–851.
Sullivan, A. (2001). Cultural capital and educational attainment. *Sociology, 35*(4), 893–912.
Zeileis, A., Kleiber, C., & Jackman, S. (2008). Regression models for count data in R. *Journal of Statistical Software, 27*(8), 1–25.

The Combined Median Rank-Based Gini Index for Customer Satisfaction Analysis

Emanuela Raffinetti

Abstract The quality assessment represents a relevant topic especially with regard to several real contexts. Currently, firms and services suppliers pay particular attention to customer satisfaction surveys in order to investigate about the "perceived quality" feature. Typically, a useful tool to obtain information about the customer satisfaction degree is represented by the quality questionnaires. The use of quality questionnaires implies that the collected data mostly assume ordinal nature.

A contribution in dealing with ordinal data is provided by this paper. Here, we propose a novel Gini measure built on ranks. By combining it with the median index, one can depict the customer satisfaction degree by exploiting information coming from the responses given to the quality questionnaires items.

1 Background on Quality Assessment and Current Proposal

Ordinal data are assuming a relevant role in many application areas since they provide information about phenomena which are not directly observable. In fact, in real contexts many interesting aspects seem not to be evaluable through the employment of quantitative variables and this condition implies many difficulties when the main purpose regards the specification of dependence relations among the involved variables, as discussed in Ferrari and Raffinetti (2012), Raffinetti and Giudici (2012, 2011) and Giudici and Raffinetti (2011). Such critical issue typically appears when measuring the "perceived quality", since data are only available at ordinal level. This occurs in particular when considering the customer satisfaction problem which can affect many fields such as economics, health and education.

E. Raffinetti (✉)
Department of Economics, Management and Quantitative Methods, Università degli Studi di Milano, Via Conservatorio 7, Milano, Italy
e-mail: emanuela.raffinetti@unimi.it

P. Giudici et al. (eds.), *Statistical Models for Data Analysis*, Studies in Classification, Data Analysis, and Knowledge Organization, DOI 10.1007/978-3-319-00032-9_33,

In literature, several approaches providing measures able to assess, through based-quality questionnaires, the perceived quality have been proposed (see e.g. Allison and Foster 2004). Such proposals employ partial orderings based on a median-preserving spread of distributions, analogous to partial orderings based on a mean-preserving spread provided for instance by the Lorenz curves comparison, as described in Muliere and Petrone (1992). However, these measures only define a partial ordering and there may be instances when the underlying conditions do not hold leading to the inability of ranking different ordinal data distributions.

An alternative approach to compute quality indices, when dealing with ordinal data, is to transform ordinal variables into cardinal ones and then calculating standard (mean-based) quality indices. Van Doorslaer and Jones (2003) present a review and an assessment of all the possible methods for such transformation. Further contributions in this area can be found also in Abul Naga and Yalcin (2008) and in Madden (2010).

Here a novel statistical procedure in treating ordinal variables will be discussed following a different *modus operandi* with respect to suggestions existing in literature. Through the Lorenz curve (for more details, see e.g. Gastwirth 1972) extension to ordinal data, we define a new quality index, called "*rank-based Gini measure*", whose main properties will be described in Sect. 2. By the rank-based Gini measure, one can obtain information about the agreement or disagreement condition related to customers' opinion with regard to a service or a product perceived quality. Furthermore, the proposed measure combined with the median index is suitable in detecting the opinions dissimilarities concerning the evaluated topics. All the details related to this new measure, hereafter called *combined median rank-based Gini index*, will be illustrated in the following sections.

2 Our Proposed Approach

The focus of this contribution is formalizing a novel quality measure able to specify the different customers' opinions concerning the evaluated items of a service or a product.

The idea is based on Lorenz curves and Gini measure employment, according to which analyzing the disagreement or agreement status among the interviewed subjects, whose opinions are summarized by the responses given to the questionnaire items. Since ordinal variables representing the customer satisfaction degree measured in terms of "perceived quality" are involved, an appropriate approach for the Lorenz curve construction is needed.

Many definitions of quality indices can be found in economical literature. Examples are presented in Allison and Foster (2004), where the problem concerning the scales arbitrary choice for ordered categories has been deeply discussed. The use of different scales can lead to subjectivity and this implies troubles in results interpretation. Our proposal overcomes the restrictions arising with an arbitrary chosen scale, by resorting to ranks tool. Through ranks employment, one specifies

Table 1 Results

"Perceived Quality"	Absolute Frequency	Rank ($r_j, j = 1, \dots, 3$)	$F(r)$	$Q(r)$
Poor	$n_1 = 5$	$1\ (r_1 = 1)$	5/11	5/49
Average	$n_2 = 4$	$6\ (r_2 = r_1 + n_1)$	9/11	29/49
Good	$n_3 = 2$	$10\ (r_3 = r_2 + n_2)$	1	1

a more homogeneous interpretation of ordered categories. More precisely, if Y is an ordinal variable, the y_i's do not correspond to numerical values since they do not represent the assumed ordered categories. For this reason, our purpose is in substituting the assumed ordered categories with values able to represent their real impact. As well known, in a quantitative context, the Lorenz curve construction is obtained by ordering all the variable values in increasing sense, as described in Muliere and Petrone (1992). In an ordinal framework, supposing that k ordered categories are considered, we propose to assign rank 1 to the lowest assumed ordered category and value $(r_{k-1}+n_{k-1})$ to the highest one, where r_{k-1} and n_{k-1} correspond to the rank and absolute frequency associated to the $(k-1)-th$ ordered category. Thus, given a variable Y characterized by k ordered categories, the set of points defining the corresponding Lorenz curve is obtained by the following pairs of values

$$\left(\frac{\sum_{j=1}^{i} n_j}{\sum_{j=1}^{k} n_j}, \frac{\sum_{j=1}^{i} r_j n_j}{\sum_{j=1}^{k} r_j n_j}\right), \text{ with } i = 1, \dots, k, \tag{1}$$

where r_j specifies the rank assigned to the $j-th$ category. More in detail, $r_1 = 1$ for the first ordered category, $r_2 = (r_1 + n_1)$ for the second ordered category and $r_k = (r_{k-1}+n_{k-1})$ for the last ordered category. Let us now denote the x-axis values with $F(r)$ and the y-axis values with $Q(r)$, where $F(r)$ represents the *cumulative frequency percentage* and $Q(r)$ represents the *cumulative rank percentage*.

To illustrate our proposal we introduce an example. Let us suppose that 11 individuals have been asked to express their personal opinion with regard to the "perceived quality" towards a service. Let Y be a variable describing each interviewed individual's perceived quality. Furthermore, let the Y variable assumed ordered categories be specified in terms of poor, average and good. Table 1 reports the results and Fig. 1 provides the ordinal Lorenz curve graphical representation.

As well known, according to the classical income distribution hypothesis, a useful measure able to summarize information about homogeneity or heterogeneity is the Gini measure. If each individual owns the same percentage of income, the Gini measure is null and the set of points characterizing the Lorenz curve lies on the egalitarian curve. On the other hand, if only an individual owns the total percentage of income, the corresponding Gini measure assumes its maximum value. When focusing on the ordinal context, the maximum homogeneity, corresponding to the minimum dispersion degree, is achieved when all the statistical units are located in an unique ordered category. In this case, the concentration area is null.

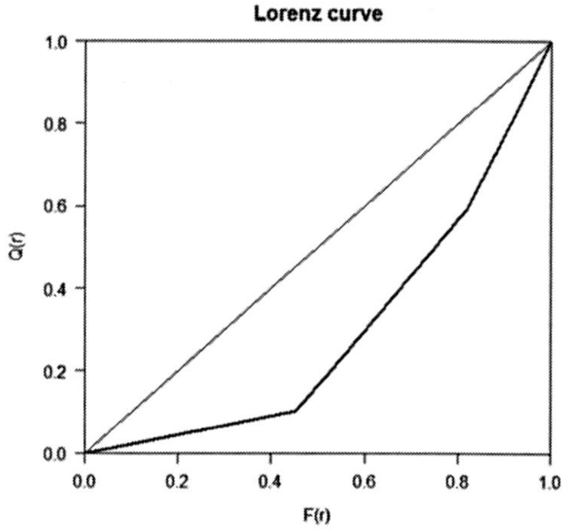

Fig. 1 The ordinal Lorenz curve

A novel Gini measure, related to the ordinal Lorenz curve and named "*rank-based Gini measure*", can be defined as follows:

$$G = 1 - \sum_{r=1}^{k} (Q(r-1) + Q(r))(F(r) - F(r-1)). \tag{2}$$

Through the rank-based Gini measure application, the agreement or disagreement level of individuals about each evaluated topic is specified. In particular, if the rank-based Gini measure $G(r)$ increases up to assume values closer to one, then there is a great disagreement among the interviewed subjects. On the other hand if $G(r)$ assumes values very close to 0, then this corresponds to a great consensus. However, in order to establish the consensus degree among individuals, one has to take into account the median index. As already discussed, Allison and Foster (2004) showed how standard measures of the spread of a distribution, which use the mean as a benchmark, are inappropriate when dealing with ordinal data. This because one needs a measure independent on the arbitrarily chosen scale applied to different categories (see e.g. Madden 2010). For this reason, a more appropriate benchmark, is the median index. Through the median index one describes the evaluation position of the 50% of individuals with regard of an evaluated item, providing a central measure able to detect the interviewed subjects' assessment distribution. In general, if the median index has the same position for all the evaluated items, then one has to consider the corresponding $G(r)$ value to measure the individuals' agreement or disagreement level about each specific item. For instance, if $G(r)$ takes values close to 0, then a relevant consensus among the customers' opinions is achieved.

The procedure based on the joint use of the median index and the rank-based Gini measure allows to define a novel customer satisfaction index which will be called "*combined median rank-based Gini index*". The rank-based Gini index role is stressed by the following property.

Property 1. In case of a median index located at the same ordered category for all the evaluated items, the rank-based Gini index allows to catch the agreement or disagreement degree with regard to each provided customers' assessment.

Through Property 1, an appropriate *ranking* among the evaluated items can be obtained in terms of individuals' agreement or disagreement.

3 Evaluation of a Chain of Restaurants in the UK

In this section we discuss about the quality assessment issue, expressed in terms of customers satisfaction, with regard to a restaurants chain in the United Kingdom (UK). The chain is composed by four restaurants. In particular, in order to validate our proposed methodological approach we focus on the analysis of information collected through a quality-based questionnaire submitted to people visiting each restaurant.

To answer to an ordered scale level questionnaire, the involved subjects specify their personal assessment by responding to a question of this kind: "*What is your personal opinion about . . .*

- *The menu content?"* (Question 1).
- *The proposed food quality?"* (Question 2).
- *The staff behavior?"* (Question 3).
- *The restaurant rooms?"* (Question 4).

The variables of interest, representing the perceived customers satisfaction degree related to the evaluated items, assume the following five ordered categories:

- Poor.
- Fair.
- Average.
- Good.
- Excellent.

Due to the five considered ordered categories, our proposed methodological procedure assigns rank 1 to the lowest ordered category ("poor") and rank $(r_4 + n_4)$ to the highest one ("excellent").

As already anticipated in Sect. 2, if $G(r)$ is close to 0 then a relevant consensus among the interviewed subjects is achieved. On the other hand, if $G(r)$ moves aways from 0 by increasing its value then a disagreement among the interviewed subjects occurs. Moreover, if the customers responses provide a positive median opinion to the specific quality-based questions and $G(r)$ has a low value, this implies a good evaluation. On the contrary, a negative median position with regard to the specific quality-based questions associated to a $G(r)$ low value implies a bad evaluation.

Table 2 Restaurant 1

	$G(r)$	Median Index
Question 1	0.1680189	Excellent
Question 2	0.1788070	Excellent
Question 3	0.0648627	Excellent
Question 4	0.0674505	Excellent

Table 3 Restaurant 2

	$G(r)$	Median Index
Question 1	0.2149883	Excellent
Question 2	0.2225268	Excellent
Question 3	0.1012204	Excellent
Question 4	0.0670991	Excellent

Table 4 Restaurant 3

	$G(r)$	Median Index
Question 1	0.1861510	Excellent
Question 2	0.2385738	Excellent
Question 3	0.0812033	Excellent
Question 4	0.0781560	Excellent

Table 5 Restaurant 4

	$G(r)$	Median Index
Question 1	0.1691841	Excellent
Question 2	0.1714296	Excellent
Question 3	0.0725393	Excellent
Question 4	0.0541574	Excellent

All the combined median-ranks based Gini index values, corresponding to each restaurant evaluated feature, are represented in Tables 2, 3, 4 and 5.

Let us now point out some of the obtained results, recalling that the related interpretation is strictly linked to median index values. Since in all restaurants the evaluated topics achieve the same median index value ("Excellent"), the following conclusions can be provided. Restaurant 1 and Restaurant 4 present similar $G(r)$ values about Question 1 and Question 2 meaning that these restaurants achieve the same customer satisfaction degree with regard to menu content and food quality. Restaurant 2 shows the highest $G(r)$ values respectively in Question 1 and Question 3: this result means that Restaurant 2 has to improve the provided services in terms of menu content and staff behavior. Furthermore, the two restaurants that achieve the best results in terms of customer satisfaction degree with regard to almost all the evaluated items are Restaurant 1 and Restaurant 4 since they are characterized by the smallest $G(r)$ values and the maximum median values.

All these aspects stress how our proposed index gives detailed information about products or services critical features which have to be monitored through appropriate marketing or re-organization policies aimed at improving the quality aspect.

4 Final Remarks and Conclusions

This paper deals with the description of a novel statistical approach for quality assessment. In particular, our presented research contribution proposes to investigate about the "perceived quality" in order to catch information about the achieved customer satisfaction degree.

Through the median rank-based Gini index one can specify the existing dissimilarities among the single evaluated items of a service or a product, allowing to assign a rating to services or products supplier.

The combined median rank-based Gini index can be compared with other agreement measures, such as for instance the so-called Kappa-type indices (see e.g. De Mast 2007 for more details). Typically, Kappa-type indices use the concept of agreement to express the reproducibility of nominal measurements, however De Mast (2007) provides a novel definition by considering them as measures of predictive association, rather than absolute measures of reproducibility. Even if our proposal and Kappa-indices are both based on the agreement concept, some relevant dissimilarities have to be highlighted. In fact our contributed index is built on an ordinal measurement and it is dealt here only as a descriptive approach without any predictive meaning, as instead it happens for Kappa-indices. Furthermore, Kappa-indices are computed in terms of two different probabilities: the one expressing the probability of agreement for the measurement system under study and the other one representing the probability of agreement for a "chance" measurement system (a completely uninformative measurement system that assigns measurement values to objects randomly). Despite that, some specific similarities can be detected. In fact, both measures assume values in range $[0, 1]$. In our case, the $G(r)$ assumed values provide a descriptive information about the agreement or disagreement condition among the interviewed individuals, and Kappa-indices are measures of predictive association based on the Gini measure of dispersion.

In conclusion, our proposal is particularly suitable when considering customer satisfaction surveys since it gives more detailed information about the positive or negative features towards a particular service or product when all the evaluated items are characterized by the same median ordered category.

References

Abul Naga, R., & Yalcin, T. (2008). Inequality measurement for ordered response health data. *Journal of Health Economics, 27*, 1614–1625.

Allison, R. A., & Foster, J. E. (2004). Measuring health inequality using qualitative data. *Journal of Health Economics, 23*, 505–524.

De Mast, J. (2007). Agreement and kappa-type indices. *The American Statistician, 61*(2), 148–153.

Ferrari, P. A., & Raffinetti, E. (2012). An extension and a new interpretation of the rank-based concordance index. In: Analysis and modeling of complex data in behavioural and social sciences. Cleup, Padova.

Gastwirth, J. L. (1972). The estimation of the Lorenz curve and Gini index. *The Review of Economics and Statistics, 54*(3), 306–316.

Giudici, P., & Raffinetti, E. (2011). On the Gini measure decomposition. *Statistics and Probability Letters, 81*(1), 133–139.

Madden, D. (2010). Ordinal and cardinal measures of health inequality: an empirical comparison. *Health Economics Letters, 19*, 243–250.

Muliere, P., & Petrone, S. (1992). Generalized Lorenz curve and monotone dependence orderings. *Metron, L*(3–4), 19–38.

Raffinetti, E., & Giudici, P. (2011). Model selection based on Lorenz zonoids. *Book of papers ASMDA 2011 (applied stochastic models and data analysis) international conference* (pp. 1145–1152). Rome, 7–10 June 2011. CD-Rom, Ed. ETS ISBN 97888467-3045-9.

Raffinetti, E., & Giudici, P. (2012). Multivariate ranks-based concordance indexes. In A. Di Ciaccio, M. Coli, I. Angulo, M. Jose (Eds.) *Advanced statistical methods for the analysis of large data-sets, Series, "Studies in Theoretical and Applied Statistics"*, (pp. 465–473), Berlin/Heidelberg: Springer.

Van Doorslaer, E., & Jones, A. M. (2003). Inequalities in self-reported health: validation of a new approach to measurement. *Journal of Health Economics, 22*, 61–87.

A Two-Phase Clustering Based Strategy for Outliers Detection in Georeferenced Curves

Elvira Romano and Antonio Balzanella

Abstract A two-phase clustering method for the detection of geostatistical functional outliers is proposed. It first, clusters data by a modified version of a Dynamic Clustering algorithm for geostatical functional data and then detects groups of outliers according to a cut-off value defined by a measure of spatial deviation in a minimum spanning tree. The performance of the proposed procedure is analyzed by several simulation studies.

1 Introduction

In an increasing number of applied sciences, like agronomy, meteorology, ecology, spatially located sensors collect curves. Traditionally, the analysis of curves has been performed through Functional Data Analysis (Ramsay and Silverman 2005). However, in the last years in order to consider the further information provided by the spatial locations, recent contributions have introduced the new research field of Spatial Functional Data Analysis (SFDA) (Delicado et al. 2010). Methods in this new field are based on the assumption that the variability among the curves depends on their spatial distance so that at near locations correspond similar curves while, at opposite, far locations are characterized by very different curve behaviors.

In this context, we focus on the problem of anomaly detection. In particular we aim at discovering groups of curves, from the georeferenced functional dataset, whose the variability depends in an anomalous way on the spatial location. A basic example of this kind of outliers is a fire in a forest. A set of sensors monitors the temperature over a wide geographic area however, due to the fire, a group of such

E. Romano (✉) · A. Balzanella
Second University of Naples, Caserta, Italy
e-mail: elvira.romano@unina2.it; antonio.balzanella@unina2.it

P. Giudici et al. (eds.), *Statistical Models for Data Analysis*, Studies in Classification, Data Analysis, and Knowledge Organization, DOI 10.1007/978-3-319-00032-9_34,

sensors records curves which are anomalous in terms of spatial dependence and magnitude of the temperature.

The problem we analyze is different from what is addressed in Sun and Genton (2012) and Romano and Mateu (2012). The former, founds the detection of outliers on the functional boxplot tool: as in classical boxplot, outliers are observations which deviates from the central region identified by the IQR. The latter, is based on extending the ordering criterion defined in modal depth functions to account for the spatial-functional variability. In this framework, the outliers will be the curves having a very low modal depth.

In both cases, outliers are isolated curves which deviate from a single typical behavior; in our case, outliers are groups of similar curves which deviate, strongly, from a set of typical behaviors.

Our proposal is a clustering-based outliers detection strategy. It is made by two steps: the first one, extends the method proposed in Romano et al. (2010) for clustering geostatistical functional data, to support the detection of clusters of potential outliers; the second step identifies groups of outliers by selecting the small clusters characterized by anomalous spatial variability behaviors according to the schema in Jiang et al. (2001).

The clustering algorithm we refer to, reveals a partition of curves into a chosen apriori number of clusters optimizing a criterion of spatial variability and discovers a set of variogram prototypes which describe the spatial variability structure of each cluster.

We modify this strategy introducing an heuristic which allows to optimize the number of cluster according to the following rule "if one new input curve is far, in terms of spatial variability, from all existing clusters prototypes, then it is generated a new cluster and this curve is allocated to it".

The second step is run on the results of the first one. The detection of the clusters of outliers is performed building a minimum spanning tree (MST) on the clusters and on cutting the longest edges.

The paper is organized as follows: Sect. 2 provides a formal description of the analyzed data and of the tool used as reference for modeling spatial variability in the functional data setting. Section 3 introduces the details of the proposed method. Section 4 presents an application on simulated data. Finally, Sect. 5 closes the paper with future perspectives.

2 Clustering Geostatistical Functional Data

The method we introduce aims at analyzing a set $\chi = (\chi_{s_1}(t), \ldots, \chi_{s_i}(t), \ldots, \chi_{s_n}(t))$ of curves observed at the n spatial sites $(s_1, \ldots, s_i, \ldots, s_n)$ in $D \subseteq R^d$ (with positive volume). The measurements on each curve are part of a single underlying continuous spatial functional process defined as

$$\{\chi_s : s \in D \subseteq \mathbb{R}^d\} \quad (1)$$

In particular each function $\chi_{s_i}(t)$ is defined on $T = [a,b] \subseteq \mathbb{R}$ and assumed to belong to an Hilbert space

$$L_2(T) = \{f : T \to \mathbb{R}, \text{ such that } \int_T f(t)^2 dt < \infty\}.$$

with the inner product $\langle f, g\rangle = \int_T f(t)g(t)dt$.

For each t, the random process is assumed to be second order stationary and isotropic: that is, the mean and variance functions are constant and the covariance depends only on the distance between sampling sites.

Formally we have:
$\mathbb{E}(\chi_s(t)) = m(t)\ \forall\ t \in T,\ s \in D$,
$\mathbb{V}(\chi_s(t)) = \sigma^2(t),\ \forall\ t \in T,\ s \in D$, and
$\mathbb{C}ov(\chi_{s_i}(t), \chi_{s_j}(t)) = \mathbb{C}(h,t)$ where $h = \|s_i - s_j\|\ \forall\ s_i, s_j \in D$.

Moreover, since we are assuming that the mean function is constant over D, the semivariogram function $\gamma(h,t) = \gamma_{s_i s_j}(t) = \frac{1}{2}\mathbb{V}(\chi_{s_i}(t) - \chi_{s_j}(t))$ where $h = \|s_i - s_j\|\ \forall\ s_i, s_j \in D$ can be expressed by

$$\gamma(h,t) = \gamma_{s_i s_j}(t) = \frac{1}{2}\mathbb{V}(\chi_{s_i}(t) - \chi_{s_j}(t)) = \frac{1}{2}\mathbb{E}\left[\chi_{s_i}(t) - \chi_{s_j}(t)\right]^2 \tag{2}$$

By considering the integral on T of this expression, using Fubini's theorem and following Delicado et al. (2010), a measure of spatial variability can be considered

$$\gamma(h) = \frac{1}{2}\mathbb{E}\left[\int_T (\chi_{s_i}(t) - \chi_{s_j}(t))^2 dt\right], \text{ for } s_i, s_j \in D \text{ with } h = \|s_i - s_j\|$$

which is the so called trace-variogram. This can be estimated as

$$\hat{\gamma}(h) = \frac{1}{2|N(h)|} \sum_{i,j \in N(h)} \int_T (\chi_{s_i}(t) - \chi_{s_j}(t))^2 dt, \tag{3}$$

where $N(h) = \{(s_i, s_j) : \|s_i - s_j\| = h\}$, and $|N(h)|$ is the number of distinct elements in $N(h)$. For irregularly spaced data there are generally not enough observations separated by exactly h. Then $N(h)$ is modified to $\{(s_i, s_j) : \|s_i - s_j\| \in (h - \varepsilon, h + \varepsilon)\}$, with $\varepsilon > 0$ being a small value.

In this framework, we define as outlier, a subset of χ having a spatial-functional variability structure, described by its trace-variogram function, which deviates so much from the other variability structures of the data to be considered as part of a different spatial functional process.

We propose to detect this subset by running, at first, an appropriately extended clustering algorithm which partitions the set of curves χ into clusters which are homogeneous in terms of spatial-functional variability and then, by pruning the clusters which highlight the anomalous trace-variogram functions.

The clustering strategy we use as reference is based on a Dynamic Clustering Algorithm (Diday 1971). It simultaneously searches for a partition P of the curves into K clusters and a set of prototypes $L = \left(\gamma_1^*(h), \ldots, \gamma_k^*(h), \ldots, \gamma_K^*(h)\right)$ which describe the spatial variability behavior of each cluster C_k (with $k = 1, \ldots, K$). To reach this aim, the following criterion is optimized:

$$\Delta(P, L) = \sum_{k=1}^{K} \sum_{\chi_{s_i}(t) \in C_k} \sum_{h} (\upsilon_k^{s_i}(h) - \gamma_k^*(h))^2 \tag{4}$$

The criterion $\Delta(P, L)$ measures the fitting between the partition of curves and the prototypes. In particular, the prototype function $\gamma_k^*(h)$ is the trace-variogram for the cluster C_k which can be estimated by the method of moments as follows:

$$\gamma_k^*(h) = \frac{1}{|2N_k(h)|} \sum_{i,j \in N_k(h)} \int_T \left(\chi_{s_i}(t) - \chi_{s_j}(t)\right)^2 dt \tag{5}$$

where $N_k(h) = \{(s_i; s_j) \in C_k : \|s_i - s_j\| = h\}$.

Then, $\upsilon_k^{s_i}(h)$ is the centered variogram function which can be estimated by:

$$\upsilon_k^{s_i}(h) = \frac{1}{2\left|N_k^i(h)\right|} \sum_{j \in N_k^i(h)} \int_T \left(\chi_{s_i}(t) - \chi_{s_j}(t)\right)^2 dt * \left|N_k^i(h)\right| \tag{6}$$

with $N_k^i(h) \subset N_k(h), \ |N_k(h)| = \sum_i \left|N_k^i(h)\right|$.

Unlike to the trace-variogram function, $\upsilon_k^{s_i}(h)$ considers the curve χ_{s_i} as a pivot so that rather than measuring the whole spatial variability of the curves in a cluster C_k, it measures the variability of a single curve with regard to the other curves in C_k.

In order to optimize the criterion $\Delta(P, L)$ we follow a k-means like schema where starting from a random partition, a step of representation and a step of allocation are executed until the convergence to a local minimum. In the *representation* step, the theoretical variogram $\gamma_k^*(h)$ of the set of curves $\chi_{s_i}(t) \in C_k$, for each cluster C_k is estimated (4). This involves the computation of the empirical variogram and its model fitting by the Ordinary Least Square method.

In the *allocation* step, the function $\upsilon_k^{s_i}$ is computed for each curve $\chi_{s_i}(t)$. Then a curve $\chi_{s_i}(t)$ is allocated to a cluster C_k by evaluating its matching with the spatial variability structure of the clusters according to the following rule:

$$d(\upsilon_k^{s_i}(h); \gamma_k^*(h)) < d(\upsilon_l^{s_i}(h); \gamma_l^*(h)) \ \ \forall k \neq l \tag{7}$$

3 Outliers Detection from Geostatistical Functional Data

The method we propose is based on extending the clustering algorithm introduced above, to support a number of clusters K', not defined apriori, which is larger that the initial value K. This will allow to carry out clusters of potential outliers from which the second phase of the procedure will detect the final outliers.

With this aim, we modify the optimized criterion $\Delta(P, L)$ introducing a heuristic which generates a new cluster if a curve is far away from all the existing clusters. This is obtained adding a further step between the allocation and the representation step and modifying the allocation criterion.

The new step consists in computing the minimum min_d of the Euclidean distances $d(.)$ among the variograms prototype of the clusters as follows:

$$min_d = min\; d(\gamma_k^*(h); \gamma_l^*(h)) \;\; for \;\; k, l = 1, \ldots, K',\; k \neq l \tag{8}$$

As consequence, the allocation of the curves to the clusters is modified so that a curve $\chi_{s_i}(t)$ is allocated to a cluster C_k only if the following rule is true:

$$d(\upsilon_k^{s_i}(h); \gamma_k^*(h)) < d(\upsilon_l^{s_i}(h); \gamma_l^*(h)) \;\; \forall k \neq l \cap d(\upsilon_k^{s_i}(h); \gamma_k^*(h)) < min_d. \tag{9}$$

otherwise, a new cluster is generated and $\chi_{s_i}(t)$ is allocated to it.

This procedure could, in some extreme case, generate $K' >> K$ clusters, so that we constraint the maximum value of K' to K_{max}. Especially, if the generation of a new cluster involves that $K' > K_{max}$ then two clusters C_k, C_l such that $d(\gamma_k^*(h); \gamma_l^*(h)) < d(\gamma_{k'}^*(h); \gamma_{l'}^*(h)) \forall \{k, l\} \neq \{k', l'\}$ (with $k \neq l$) are merged into a single cluster.

In this way, the total criteria to be optimized can be expressed as:

$$\delta(w, \gamma_k^*(h)) = \sum_{s_i=1}^{n} \sum_{l=1}^{K'} w_{s_i,l} (\upsilon_k^{s_i}(h) - \gamma_k^*(h))^2 \tag{10}$$

subject to $\sum_{l=1}^{K'} w_{s_i,l} = 1,\;\; s_i = 1 \ldots n$ where $w_{s_i,l} = 1$ if the curve s_i is allocated to the cluster l, $\forall s_i = 1 \ldots n,\; l = 1, \ldots, K'$.

Starting from the obtained clustering results, the second phase, selects the clusters which are anomalous in terms of spatial-functional variability.

It is based on constructing a complete, undirected graph $G = (V, E)$ where the nodes set V is made by the K' variogram functions which describe the spatial functional variability of the clusters while the edges E record the Euclidean distance between each couple of variograms.

We propose an iterative procedure which processes the graph G in order to select the nodes representing the outliers clusters:

- Repeat until the number of nodes in G reaches the desired number of clusters of outliers

  - Split G into two sub-graphs by choosing as cutting edge the one corresponding to the maximum value of the distance between the nodes.
  - Select the sub-graph having the lowest number of nodes and regard it as the new G

- End repeat

As result, this procedure returns a set of nodes representing the clusters characterized by the most anomalous behavior in terms of spatial variability structure.

4 Simulation Study

The developed two phase clustering method is evaluated through a simulation study on four different spatio-functional data sets. The datasets have been generated according to a separable and a non separable spatio-functional covariance function contaminated with outlier models. Original data are drawn from a zero-mean, stationary spatio-functional Gaussian random field, $\chi_s(t)$ (with $s \in D$ and $t \in T$) whose covariance function $C(h,u) = cov\{\chi_{s_i}(t_1), \chi_{s_j}(t_2)\}$ depends (for any couple of s_i,s_j and t_1, t_2) on the spatial distance $h = s_i - s_j$ and on the functional distance $u = t_1 - t_2$.

We consider the following separable and non separable covariance functions:

- A separable covariance function:

$$C_{SEP}(h,u) = cov\{\chi_{s_i}(t_1), \chi_{s_j}(t_2)\} = C_s(h)\,C_T(u) \tag{11}$$

  with $C_s(h) = (1-\nu)\exp(-c\,|h|) + \nu I\{h=0\}$ a spatial covariance function and $C_T(u) = \left(u + a\,|u|^{2\alpha}\right)^{-1}$ a purely stationary functional covariance functions of the Cauchy type having with a time span $u = |t_1 - t_2|$. Here $a > 0$ is the scale parameter in time, that is fixed to $a = 1$ in all the tests, and α, is the parameter that controls the strength of the functional variability.
- A Symmetric but generally non-separable correlation function:

$$C_{Sim}(h,u) = \frac{1-\nu}{1+au^{2\alpha}}\left[\exp\left\{-\frac{c\,\|h\|}{(1+au^{2\alpha})^{\frac{\beta}{2}}}\right\} + \frac{\nu}{1-\nu} I\{h=0\}\right] \tag{12}$$

  where the parameter $0 \le \beta \le 1$, we set to 0.9 controls the degree of non-separability.

The four generated datasets are composed respectively by two and three different clusters, each of them is characterized by different values of the parameters defining the Spatial Functional covariance function as in Table 1.

In all the cases one of them is contaminated by a group of $n_o = 66$ outliers following a model contaminated by peaks

Table 1 The 4 datasets and their parameters

		α-parameters			c-parameters		
Dataset Id	Cov. Model	α_1	α_2	α_3	c_1	c_2	c_3
D_1	Separable	0.2	0.2		0.1	0.5	
D_2	Separable	0.2	0.2		0.3	0.5	
D_3	Non Separable	0.2	0.2	0.2	0.1	0.5	0.7
D_4	Non Separable	0.2	0.2	0.2	0.3	0.5	0.7

Table 2 Rand Index for the Two phase (TP) and the Dynamic(DC) clustering methods

Dataset Id	RI_{TP}	RI_{DC}
D_1	0.88	0.60
D_2	0.75	0.59
D_3	0.80	0.48
D_4	0.79	0.49

Table 3 Percentage of the detected outliers A_{TP} for the Two phase clustering methods

Dataset Id	A_{TP}
D_1	0.88
D_2	0.85
D_3	0.77
D_4	0.82

$$\phi_{s_i}(t) = \chi_{s_i}(t) + c_{s_i}\sigma_{s_i}K, \;\; i = 1, \ldots, n_o \tag{13}$$

where $c_{s_i}(t)$ is 1 with probability 0.1 and 0 with probability 0.9, $K = 6$ is a contamination size constant, σ_{s_i} is a sequence of random variables independent of $c_{s_i}(t)$ taking values 1 and -1 with probability $1/2$.

Each cluster has been generated with locations $s_1, \ldots, s_{196}$ on a grid of size 14×14 in the unit square with the grid spatial spacing $1/13$, and 50 equally spaced time points in $[0, 1]$. Thus two datasets of 392 curves with two clusters C_1, C_2 and two datasets of 588 curves with three clusters of data C_1, C_2, C_3 were generated.

Our experimentations consist in comparing the results of the two phase clustering strategy with the DCA for geostatistical functional data on the generated datasets. At first, we evaluate the clustering results computing the well known Rand Index *RI* (Rand 1971) between the real partition of data and the one obtained by the two strategies. Then, we monitor the performance of the strategy by measuring the percentage of detected outliers. Table 2 contains the RI for both the strategies. It shows that the real clustering structures are discovered by the two phase procedure at an higher rate for all the datasets.

This is a consequence of the detection of the outlier cluster. At the opposite the lower rates of the RI_{DCA} are due to the misclassification of outliers. By analyzing the cluster of outliers we have measured the proportion of curves inside the cluster that follows the model (13). Table 3 shows the results.

5 Conclusion

In this paper, we introduced a cluster-based outlier detection approach based on Dynamic Clustering algorithm. In our method, we give attention to the group outliers rather than a single outlier. The implementation of our algorithm on various datasets shows successful results. For future work, we need to improve our approach to make it more time efficient. A further modification will consist in extending the validity of the algorithm to make it applicable for streaming time series also.

References

Delicado, P., Giraldo, R., Comas, C., & Mateu, J. (2010). Statistics for spatial functional data: some recent contributions. *Environmetric, 21*, 224–239.

Diday, E. (1971). La methode des Nuees dynamiques. *Revue de Statistique Appliquee, 19*(2), 19–34.

Jiang, M. F., Tseng, S. S., & Su, C. M. (2001). Two-phase clustering process for outliers detection. *Pattern Recognition Letters, 22*(6), 691–700.

Ramsay, J. E., & Silverman, B. W. (2005). *Functional data analysis* (2nd edn.). New York: Springer.

Rand, W. M. (1971). Objective criteria for the evaluation of clustering methods. *Journal of the American Statistical Association, 66*(336), 846–850.

Romano, E., Balzanella, A., & Verde, R. (2010). A new regionalization method for spatially dependent functional data based on local variogram models: an application on environmental data. In *Atti delle XLV Riunione Scientifica della Societá Italiana di Statistica Universitá degli Studi di Padova*. ISBN/ISSN:978 88 6129 566 7. Padova: CLEUP.

Romano, E., & Mateu, J. (2012). Outlier detection for geostatistical functional data: an application to sensor data. In A. Giusti, G. Ritter, M. Vichi (Eds.) *Classification and data mining*, (pp. 131–138). Springer. ISBN: 978-3-642-28893-7.

Sun, Y., & Genton, M. G. (2012). Adjusted functional boxplots for spatio-temporal data visualization and outlier detection. *Environmetrics, 23*, 54–64.

High-Dimensional Bayesian Classifiers Using Non-Local Priors

David Rossell, Donatello Telesca, and Valen E. Johnson

Abstract Common goals in classification problems are (i) obtaining predictions and (ii) identifying subsets of highly predictive variables. Bayesian classifiers quantify the uncertainty in all steps of the prediction. However, common Bayesian procedures can be slow in excluding features with no predictive power (Johnson & Rossell. (2010). In certain high-dimensional setups the posterior probability assigned to the correct set of predictors converges to 0 (Johnson and Rossell 2012). We study the use of non-local priors (NLP), which overcome the above mentioned limitations. We introduce a new family of NLP and derive efficient MCMC schemes.

1 Introduction

Two common goals in classification problems are (i) predicting the class that an individual belongs to and (ii) identifying variables with high predictive power. Oftentimes, it is also important to measure the confidence both in the obtained predictions and variable subsets. Bayesian classifiers are appealing in that they quantify the uncertainty associated to all steps in the prediction process in a natural manner.

D. Rossell (✉)
Institute for Research in Biomedicine of Barcelona, Barcelona, Spain
e-mail: david.rossell@irbbarcelona.org

D. Telesca
University of California, Los Angeles, USA
e-mail: dtelesca@ucla.edu

V.E. Johnson
University of Texas MD Anderson Cancer Center, Houston, USA
e-mail: vejohnson@mdanderson.org

P. Giudici et al. (eds.), *Statistical Models for Data Analysis*, Studies in Classification, Data Analysis, and Knowledge Organization, DOI 10.1007/978-3-319-00032-9_35,

Let $y_i \in \{0, 1\}$ be the class of individual i for $i = 1, \ldots, n$, $\mathbf{y}' = (y_1, \ldots, y_n)$, $\mathbf{x}_i' = (x_{i1}, \ldots, x_{ip})$ be a vector with p predictors and $\mathbf{X}$ the $n \times p$ matrix with i^{th} row equal to $\mathbf{x}_i$. We partition $\mathbf{X} = (\mathbf{X}_1, \mathbf{X}_2)$, where $\mathbf{X}_i$ is $n \times p_i$ for $i = 1, 2$ and $p_1 + p_2 = p$. The idea is to use all variables in $\mathbf{X}_2$ but only the most relevant ones in $\mathbf{X}_1$ for prediction purposes. We denote an arbitrary subset of predictors by $\mathbf{k} = \{k_1, \ldots, k_{|\mathbf{k}|}\}$, where $k_j \in \{1, \ldots, p_1\}$ and $|\mathbf{k}| = \dim(\mathbf{k})$. Let $\mathcal{M}$ be the collection of all 2^{p_1} subsets, denote by $\mathbf{t}$ the predictors truly related to $\mathbf{y}$, and let $P(\mathbf{k}|\mathbf{y})$ be the posterior probability assigned to model $\mathbf{k}$. The goals are to predict the class y_{n+1} for a new individual $n + 1$ given the observed predictors $\mathbf{x}_{n+1}$ and to obtain large $P(\mathbf{t}|\mathbf{y})$.

In modern applications with large p, one expects that a relatively small proportion of them hold predictive power. Johnson and Rossell (2010) showed that variable selection based on non-local prior (NLP) distributions (Sect. 2) discards spurious covariates at a faster rate than local priors (LP). Further, Johnson and Rossell (2012) showed that under linear model setups, when p grows at rate faster than $\sqrt{n}$, $P(\mathbf{t}|\mathbf{y})$ converges in probability to 0 when LP are used. In contrast, under the same assumptions, for NLP $P(\mathbf{t}|\mathbf{y})$ converges to 1 as long as $p < n$. Despite these undesirable properties, most current Bayesian procedures are either based on or asymptotically equivalent to LP. Here we explore the application of NLP to classification problems, with emphasis in large p. We adopt a probit regression model with latent variables (Albert and Chib 1993), i.e. we let $y_i = \mathrm{I}(z_i > 0)$, where $\mathbf{z} \sim N(\mathbf{X}_1\boldsymbol{\theta}_1 + \mathbf{X}_2\boldsymbol{\theta}_2, 1)$. Our proposal remains valid for other binary regression models, e.g. $\mathbf{z} \sim \text{Logistic}(\mathbf{X}_1\boldsymbol{\theta}_1 + \mathbf{X}_2\boldsymbol{\theta}_2, 1)$ gives a logistic regression. Prior elicitation and implementation details in Sects. 3, 4 focus on probit models and would need to be appropriately adjusted for other models. We compute the predictive probability for y_{n+1} via Bayesian model averaging as

$$\hat{p}(y_{n+1} = 1) = P(y_{n+1} = 1|\mathbf{x}_{n+1}, \mathbf{y}, \mathbf{X}) = \sum_{\mathbf{k} \in \mathcal{M}} P(y_{n+1} = 1|\mathbf{k}, \mathbf{x}_{n+1}, \mathbf{y}, \mathbf{X}) P(\mathbf{k}|\mathbf{y}). \tag{1}$$

2 Prior Formulation

Let $\mathbf{k}$ denote the model with $\theta_{1j} \neq 0$ for $j \in \mathbf{k}$ (and $\theta_{1j} = 0$ for $j \notin \mathbf{k}$). A LP under model $\mathbf{k}$ is any distribution π_L such that $\pi_L(\boldsymbol{\theta}_1|\mathbf{k}) \to c > 0$ as $\theta_{1j} \to 0$ for some $j \in \mathbf{k}$. That is, LP assign non-vanishing prior density to a neighbourhood of $\theta_{1j} = 0$, even though model $\mathbf{k}$ assumes that $\theta_{1j} \neq 0$. In contrast, a NLP distribution under model $\mathbf{k}$ satisfies $\pi_L(\boldsymbol{\theta}_1|\mathbf{k}) \to 0$ as $\theta_{1j} \to 0$ for any $j \in \mathbf{k}$. Johnson and Rossell (2011) defined two families of NLP: product moment (pMOM) and product inverse moment (piMOM) densities. Let $\boldsymbol{\theta}_1^{(k)}$ be the non-zero coefficients under model $\mathbf{k}$ and $|\mathbf{k}| = \dim(\boldsymbol{\theta}_1^{(k)})$. The normal pMOM prior density of order r and prior dispersion τ_1 is defined as

$$\pi_M(\boldsymbol{\theta}_1^{(k)}|\tau_1) = \left(\frac{1}{(\tau_1\phi)^r(2r-1)!!}\right)^{|\mathbf{k}|} \prod_{j\in\mathbf{k}} (\theta_{1j}^{(k)})^{2r} N(\theta_{1j}^{(k)}; 0, \tau_1\phi), \tag{2}$$

where !! denotes the double factorial and ϕ is a dispersion parameter (for our probit model $\phi = 1$). Following Johnson and Rossell (2011), we set $r = 1$ as a default choice. Regarding τ_1, default and decision theoretical choices are discussed in Sect. 3.

A heavy-tailed pMOM prior can be obtained by placing a hyper-prior $\tau_1 \sim \mathrm{IG}(\frac{a_\tau}{2}, \frac{b_\tau}{2})$. The corresponding marginal prior for $\boldsymbol{\theta}_1^{(k)}$ is

$$\pi_T(\boldsymbol{\theta}_1^{(k)}) = \left(\prod_{j\in\mathbf{k}} \frac{\theta_{1j}^{2r}}{(\frac{b_\tau}{2}\phi)^r(2r-1)!!}\right) \frac{\Gamma\left(r|\mathbf{k}| + \frac{|\mathbf{k}|+a_\tau}{2}\right)}{(\pi\phi b_\tau)^{\frac{|\mathbf{k}|}{2}}\Gamma(\frac{a_\tau}{2})} \left(1 + \frac{\boldsymbol{\theta}_1'\boldsymbol{\theta}_1}{\phi b_\tau}\right)^{-(r|\mathbf{k}| + \frac{|\mathbf{k}|+a_\tau}{2})}. \tag{3}$$

Notice that (3) is essentially a polynomial penalty term times a multivariate T density with $r|\mathbf{k}| + \frac{a_\tau}{2}$ degrees of freedom, location $\mathbf{0}$ and scale matrix $\phi b_\tau I$.

An attractive feature of pMOM priors is that computing univariate marginal densities is immediate, which makes them amenable for use in MCMC schemes (Sect. 4). Unfortunately, for piMOM priors no closed form expressions are available. Although it is possible to obtain approximations (Johnson and Rossell 2011), this poses a challenge when exact calculations are desired and p is large. To address this issue we introduce the product exponential moment prior densities (peMOM)

$$\pi_E(\boldsymbol{\theta}_1^{(k)}) = c\exp\left\{-\sum_{j\in\mathbf{k}} \frac{\tau_1\phi}{\theta_{1j}^{2r}}\right\} \prod_{j\in\mathbf{k}} N(\theta_{1j}|0, \tau_1\phi), \tag{4}$$

where r is the prior order, τ_1 a dispersion parameter and c the normalization constant with the simple form $c = e^{|\mathbf{k}|\sqrt{2}}$ for $r = 1$. The ratio of peMOM to piMOM prior densities near the origin is bounded by a finite constant, which guarantees equal learning rates when $\boldsymbol{\theta}_1 = \mathbf{0}$ under a wide class of models. The key advantage of peMOM over piMOM priors is that under certain MCMC setups they provide closed-form expressions, i.e. they are computationally appealing. Both pMOM and peMOM fall within the generalized moment prior family (Consonni and La Rocca 2010).

For $\boldsymbol{\theta}_2$ we use a conjugate $N(\mathbf{0}, \phi\tau_2 A)$ prior, where A is an arbitrary matrix and τ_2 a dispersion parameter. By default we set $\tau_2 = 10^6$, $A = \mathrm{I}$ and $\pi(\mathbf{k}) = \text{Beta-Binomial}(|\mathbf{k}|; p_1, 1, 1)$ (Scott and Berger 2010).

3 Prior Parameter Setting

Different prior settings may influence the finite sample performance of the proposed Bayesian classifier. τ_1 can be used to determine the magnitude of $|\theta_{1j}|$ that has practical relevance for the problem at hand.

To propose default τ_1 we recall that θ_{1j} is the z-score associated to a + 1SD increase in covariate j (assuming that covariates are standardized), and consider that changes in $P(y_i = 1)$ less than 0.05 lack practical relevance. The largest possible change in $P(y_i = 1)$ produced by covariate j is $\Phi(\theta_{1j}) - 0.5$. Since $\Phi(0.126) - 0.5 = 0.05$, we regard $|\theta_{1j}| < 0.126$ as practically irrelevant. π_M in (2) with $\tau_1 = 0.139$ and π_E in (4) with $\tau_1 = 0.048$ assign 0.01 prior probability to $|\theta_{1j}| < 0.126$. For π_T in (3) we set $0.5a_\tau = 1$ and $0.5b_\tau = .278$, i.e. a fairly non-informative prior with prior mode at $\tau_1 = 0.139$.

The proposed default τ_1 seem reasonable for classification problems, and results are usually insensitive to moderate changes in τ_1 (Sect. 5). For practitioners seeking to set τ_1 in a data-adaptive manner, we describe a decision-theoretic procedure that aims to provide optimal predictions. Let $\mathbf{y}^p$ be a draw from the posterior predictive distribution. Following Gelfand and Ghosh (1998), the deviance posterior predictive loss (PPL) equally penalizing deviations from $\mathbf{y}$ and $\mathbf{y}^p$ for a fixed Bernoulli sampling model $\mathbf{k}$ is

$$D_{\mathbf{k}}(\tau_1) = \sum_i \{h_i - h(\mu_i^p)\} + 2\sum_i \left\{ \frac{h(\mu_i^p) + h(y_i)}{2} - h\left(\frac{\mu_i^p + y_i}{2}\right) \right\}$$

where $h(x) = (x + 1/2)\log(x + 1/2) + (1.5 - x)\log(1.5 - x)$, $h_i = E\{h(y_i^p \mid \mathbf{y}; \mathbf{k}, \tau_1)\}$ and $\mu_i^p = E(y_i^p \mid \mathbf{y}; \mathbf{k}, \tau_1)$.

A decision about the optimal value of τ_1 is made integrating over the model space $\mathscr{M}$ and, provided a Monte Carlo sample $(\mathbf{k}_1, \ldots, \mathbf{k}_M)$ is available from $p(\mathbf{k} \mid \mathbf{y}; \tau_1)$, the deviance PPL associated with τ_1 may be approximated by $\bar{D}(\tau_1) = \frac{1}{M}\sum_{j=1}^{M} D_{\mathbf{k}_j}(\tau_1)$. A simple grid search can be used to find τ_1 minimizing $D(\tau_1)$.

We assessed the procedure via 100 simulations with $n = 500$, $p = 50$ in a sparse setup with $\theta_{11} = 0.2$, $\theta_{12} = 0.4$, $\theta_{13} = 0.6$, $\theta_{1j} = 0$ for $j > 3$ and $\mathbf{X}_1$ with $\rho = 0.25$ as in Sect. 5. We also simulated from a non-sparse setup with $\theta_{1j} = 0.1$ for $1 \leq j \leq 10$, $\theta_{1j} = 0.2$ for $11 \leq j \leq 20$ and $\theta_{1j} = 0$ for $j > 20$. The average selected τ_1 for π_M was 1.19 and 0.05 in the sparse and non-sparse setup, respectively, and the average posterior model sizes were 15.9 and 2.5. That is, the chosen τ_1 captured the degree of sparsity in the data.

4 Model Fitting

We use a Metropolis-Hastings (MH) within Gibbs scheme to sample from the joint posterior of $(\boldsymbol{\theta}_1, \boldsymbol{\theta}_2, \mathbf{k}, \mathbf{z})$ given $\mathbf{y}$. We re-parameterize $\mathbf{k}$ with $\boldsymbol{\delta}' = (\delta_1, \ldots, \delta_{p_1})$, where $\delta_j = 1$ if $\theta_{1j} \neq 0$ and $\delta_j = 0$ otherwise. We propose updates $(\delta_j^*, \theta_{1j}^*)$

sequentially for $j = 1, \ldots, p_1$ and evaluate each proposal via MH. The distribution of θ_{1j} is dominated by the σ-finite measure $\delta_0(\cdot) + \mathcal{L}(\cdot)$, where δ_0 is a point mass at 0 and $\mathcal{L}$ the Lebesgue measure, and hence the MH technical conditions are met (Gottardo and Raftery 2009). The resulting chain allows for computationally efficient updates and typically shows a reasonable mixing. This agrees with Gottardo and Raftery (2009), who found that a similar Gibbs sampling scheme outperformed more sophisticated reversible jump strategies.

Let $\boldsymbol{\theta}_{1(-j)}$ and $\boldsymbol{\delta}_{(-j)}$ denote $\boldsymbol{\theta}_1$ and $\boldsymbol{\delta}$ after removing the jth element, respectively. Further denote by $\mathbf{x}_1^{(j)}$ the jth column in X_1, and let $X_1^{(-j)}$ be X_1 after removing $\mathbf{x}_1^{(j)}$. We provide a generic scheme for (2)–(4). Step 2 is skipped for π_M and π_E, as τ_1 is fixed. To initialize $(\boldsymbol{\theta}_1, \boldsymbol{\theta}_2, \boldsymbol{\delta})$ we start with the null model and sequentially consider $\mathbf{x}_1^{(1)}, \ldots, \mathbf{x}_1^{(p_1)}$. We deterministically accept moves increasing $\hat{p}(\mathbf{k}) = L(\mathbf{y}|\hat{\boldsymbol{\theta}}_1^{(k)}, \hat{\boldsymbol{\theta}}_2)\pi(\hat{\boldsymbol{\theta}}_1^{(k)})\pi(\mathbf{k})$, where $(\hat{\boldsymbol{\theta}}_1^{(k)}, \hat{\boldsymbol{\theta}}_2)$ is the MLE under model $\mathbf{k}$, until no moves are made after a full covariate pass. The MCMC scheme is outlined below. Steps 1–4 are repeated until a sufficient number of updates are obtained (10,000 by default).

1. Sample $z_i \sim N(\mathbf{x}_{1i}'\boldsymbol{\theta}_1 + \mathbf{x}_{2i}'\boldsymbol{\theta}_2, 1)$ for $i = 1, \ldots, n$, truncating at $z_i > 0$ when $y_i = 1$ and $z_i < 0$ when $y_i = 0$.
2. Sample $\tau_1 \sim \text{IG}\left(\frac{1}{2}\left(a_\tau + (2r+1)\sum_{j=1}^{p_1}\delta_j\right), \frac{1}{2}\left(b_\tau + \boldsymbol{\theta}_1'\boldsymbol{\theta}_1\right)\right)$.
3. Sample (θ_{1j}, δ_j). Let $\mathbf{e} = \mathbf{z} - X_1^{(-j)}\boldsymbol{\theta}_{1(-j)} - X_2\boldsymbol{\theta}_2$ be the partial residuals for $\mathbf{x}_1^{(j)}$, $s = (\mathbf{x}_1^{(j)})'\mathbf{x}_1^{(j)} + \tau_1^{-1}$ and $m = \mathbf{e}'\mathbf{x}_1^{(j)}/s$. Let $m_0(\mathbf{e}) = N(\mathbf{e}; 0, I)$ and $m_1(\mathbf{e}|\tau_1) = \frac{\exp\{\frac{1}{2}(m^2 s - \mathbf{y}'\mathbf{y})\}}{(2\pi)^{n/2}(s\tau_1)^{1/2}} A(m,s)$. For the pMOM priors (2)–(3) $A(m,s) = \frac{\text{E}(\psi^{2r})}{\tau_1^r(2r-1)!!}$, where $\psi \sim N(m, s^{-1})$, For the eMOM prior (4)

$$A(m,s) = \exp\{\sqrt{2} - \frac{1}{2}m^2 s\}2^{\frac{3}{4}} \sum_{\nu=0}^{\infty} \frac{(\tau_1 s)^{\frac{1+2\nu}{4}}(m^2 s)^\nu}{2^{\frac{3\nu}{2}}\Gamma(\nu+\frac{1}{2})\nu!} \text{K}_{\nu+\frac{1}{2}}\left(\sqrt{2s\tau_1}\right), \quad (5)$$

where $\text{K}(\cdot)$ is the modified Bessel function of the second kind. We define

$$q = \left(1 + \frac{m_0(\mathbf{e})P(\delta_j = 0)}{m_1(\mathbf{e}|\tau_1)P(\delta_j = 1)}\right)^{-1}. \quad (6)$$

Propose a new value $\delta_j^* \sim \text{Bern}(q)$. If $\delta_j^* = 0$ set $\theta_{1j}^* = 0$, otherwise propose $\theta_{1j}^* \sim T_\nu(m, \phi/s)$ with degrees of freedom $\nu = \sqrt{n}$. That is, δ_j^* is proposed from the exact $\pi(\delta_j|\mathbf{z}, \boldsymbol{\theta}_{1(-j)}, \boldsymbol{\theta}_2, \boldsymbol{\delta}_{(-j)}, \tau_1)$ and θ_{1j}^* from an asymptotic approximation to $\pi(\theta_{1j}|\delta_j^*, \mathbf{z}, \boldsymbol{\theta}_{1(-j)}, \boldsymbol{\theta}_2, \boldsymbol{\delta}_{(-j)}, \tau_1)$. The acceptance probability is equal to $\min\{1, \lambda\}$, where λ is equal to 1 if $\delta_j^* = \delta_j = 0$ and

$$\frac{N(\mathbf{e};\theta_{1j}^{*}\mathbf{x}_1^{(j)},I)\pi(\theta_{1j}^{*}|\tau_1)}{N(\mathbf{e};\theta_{1j}\mathbf{x}_1^{(j)},I)\pi(\theta_{1j}|\tau_1)}\frac{T_\nu(\theta_{1j};m,s^{-1})}{T_\nu(\theta_{1j}^{*};m,s^{-1})}, \text{ if } \delta_j^{*}=1,\delta_j=1$$

$$\frac{N(\mathbf{e};\theta_{1j}^{*}\mathbf{x}_1^{(j)},I)\pi(\theta_{1j}^{*}|\tau_1)}{T_\nu(\theta_{1j}^{*};m,s^{-1})m_1(\mathbf{e}|\tau_1)}, \text{ if } \delta_j^{*}=1,\delta_j=0$$

$$\frac{T_\nu(\theta_{1j};m,s^{-1})m_1(\mathbf{e}|\tau_1)}{N(\mathbf{e};\theta_{1j}\mathbf{x}_1^{(j)},I)\pi(\theta_{1j}|\tau_1)}, \text{ if } \delta_j^{*}=0,\delta_j=1. \quad (7)$$

4. Sample $\boldsymbol{\theta}_2 \sim N(\mathbf{m}, \phi S^{-1})$, where $S = X_2'X_2 + \frac{1}{\tau_2}I$, $\mathbf{m} = S^{-1}X_2'\mathbf{e}$ and $\mathbf{e} = \mathbf{z} - X_1\boldsymbol{\theta}_1$.

5 Results

We assess our proposed classifiers on simulated and experimental data, and compare them with Zellner's prior π_z (normal kernel in (2); $\tau_1 = 0.139$ as for π_M), maximum likelihood estimation (MLE), BIC-based probit regression, SCAD (Fan and Li 2004) and diagonal linear discriminant analysis (DLDA). For NLP, $\hat{p}(y_{n+1} = 1)$ in (1) is estimated by averaging $\Phi(\mathbf{x}_{1,n+1}\boldsymbol{\theta}_1 + \mathbf{x}_{2,n+1}\boldsymbol{\theta}_2)$ over the MCMC output, where $\Phi(\cdot)$ is the standard normal cdf. Similarly, for BIC we run a Gibbs scheme using the approximate $P(\mathbf{k}|\mathbf{y}) \propto e^{-0.5BIC(\mathbf{k})}P(\mathbf{k})$, and average $\Phi(\mathbf{x}_{1,n+1}\hat{\boldsymbol{\theta}}_1^{(k)} + \hat{\mathbf{x}}_{2,n+1}\hat{\boldsymbol{\theta}}_2)$, where $(\hat{\boldsymbol{\theta}}_1(\mathbf{k}), \hat{\boldsymbol{\theta}}_2)$ is the MLE under model $\mathbf{k}$. For MLE we obtain $(\hat{\boldsymbol{\theta}}_1, \hat{\boldsymbol{\theta}}_2)$ under the full model and set $\hat{p}(y_{n+1} = 1) = \Phi(\mathbf{x}_{1,n+1}\hat{\boldsymbol{\theta}}_1 + \mathbf{x}_{2,n+1}\hat{\boldsymbol{\theta}}_2)$. For SCAD we use the logit model implementation in R package `ncvreg` to set the penalty parameter via ten fold cross-validation, and for DLDA we use the `supclust` package. Following a 0-1 loss rule, in all cases we predict $y_{n+1} = 1$ when $\hat{p}(y_{n+1} = 1) > 0.5$.

In the simulations we considered a sparse scenario with 5 true predictors and 90 spurious covariates, and a non-sparse scenario with 5 and 5 (respectively). In both cases we set $n = 100$, $\boldsymbol{\theta}_1 = (0.4, 0.8, 1.2, 1.6, 2.0)$ and generated $\mathbf{X}_1$ from a multivariate normal with mean $\mathbf{0}$, unit variance and all pairwise correlations $\rho = 0.25$. $\mathbf{X}_2$ is the intercept with $\theta_2 = 0$. For each simulated training sample, we evaluated the correct classification rate in 1,000 independent test samples. We simulated 250 training datasets, obtaining a standard error for all reported rates below 0.003. For comparison we include the oracle classifier obtained by plugging in the true $(\boldsymbol{\theta}_1, \boldsymbol{\theta}_2)$. SCAD and DLDA are not included in this comparison, as they are not based on a probit model and hence are at a disadvantage.

Table 1 shows the out-of-sample correct classification rates. For $p = 10$ all methods achieve rates comparable to the oracle. For $p = 100$ the MLE performance decreases sharply, whereas the rates for pMOM (π_M, π_T) remain high. Relative to $p = 10$ the default π_M rate decreases 0.026 (0.037 for its LP counterpart π_z).

Table 1 Out-of-sample correct classifications in 250 simulations with 1,000 independent test samples each (π_M: normal pMOM, π_Z: Zellner; π_T: T pMOM; π_E: eMOM)

		Default τ_1				PPL τ_1			
p	Oracle	π_M	π_Z	π_T	π_E	π_M	π_E	BIC	MLE
10	0.921	0.898	0.899	0.900	0.898	0.900	0.898	0.899	0.891
95	0.921	0.872	0.862	0.876	0.843	0.858	0.864	0.826	0.577

Table 2 Correct classification rates (ten fold cross-validated) in colon cancer microarray data

Default τ_1				PPL τ_1					
π_M	π_Z	π_T	π_E	π_M	π_E	BIC	SCAD	MLE	DLDA
0.754	0.746	0.754	0.743	0.761	0.761	0.754	0.754	0.557	0.657

The BIC drop is more pronounced at 0.073. We set $\tau_1 = 0.05, 0.1, \ldots, 0.4$ in π_M to assess sensitivity, obtaining rates between 0.834 and 0.878. For π_E we set $\tau_1 = 0.025, \ldots, 0.2$ and obtain rates between 0.810 and 0.866. Results are fairly insensitive to moderate τ_1 changes.

We now consider the 135 intestinal stem cell gene markers which Merlos-Suárez et al. (2011) showed to be related to colon cancer recurrence after surgery. The authors combined the GEO (Edgar et al. 2002) microarray data GSE17537 (Smith et al. 2010) and GSE14333 (Jorissen et al. 2009), obtaining 280 patients with recorded recurrence status. Our goal is to predict recurrence based on gene expression and to unravel potential mechanisms relating gene expression to recurrence.

Table 2 reports the out-of-sample correct classification rates, as estimated by tenfold cross-validation. NLPs show good predictive ability, similar to π_Z, BIC and SCAD and substantially better than DLDA and MLE.

To assess the explanatory potential of each method, i.e. its ability to identify promising models, we compute the proportion of MCMC visits to the 10 most visited models. For the default NLPs π_M, π_T, π_E it was 0.796, 0.780 and 0.362, for π_Z 0.199 and for BIC 0.734. In contrast, the top 10 models selected by the PPL π_M and π_E were visited less than 0.01 of the iterations. The top 10 default π_M, π_T and BIC models contain combinations of the genes ARL4C, PXDN, PRICKLE1, ASRGL1, RASSF5 and NAV1. These genes are known to be associated with leukemia and multiple kinds on cancer, including colorectal, liver, kidney, lung and brain cancer (Rhodes et al. 2007). The top 10 π_E models select the same genes, except for RASSF5 and PRICKLE1. Additionally, SCAD selected ST3GAL3 and TACC1.

The results suggest that pMOM priors concentrate the posterior probability on a reduced set of models. We assessed this option with a simulation study trained on the experimental data. We set $\boldsymbol{\theta}_{1i} = 0$ for all genes except ARL4C and PXDN (the most often selected genes by π_M, π_T, π_E and BIC). We focused on 2 genes as all top 10 models contain ≤ 2 covariates. We set their coefficients to the MLE from a probit model only containing ARL4C, PXDN and the intercept, obtaining

0.278, 0.272 and −0.781 (respectively). We simulated 100 datasets with $n = 1,000$ observations each, with $\mathbf{X}_2$ being the intercept, $\mathbf{X}_1 \sim N(\mathbf{0}, \hat{\Sigma})$ the 135 genes and $\hat{\Sigma}$ the empirical covariance matrix. The average posterior probability assigned to the data-generating model by the NLPs was 0.339 (π_M) and 0.269 (π_T). π_Z and BIC assigned lower probabilities at 0.129 and 0.230.

Our results illustrate the advantage of using feature selection methods to penalize unnecessary model complexity. DLDA and MLE do not perform feature selection and showed the lowest out-of-sample classification rates. NLPs provided good predictions in sparse setups, pMOM being particularly robust to spurious covariates. In non-sparse setups, all methods provided good predictions.

Regarding explanatory potential, default NLPs favor focusing the posterior probability on a small subset of models and a reduced number of variables. pMOM showed an improved ability in assigning a larger posterior probability to the data-generating model, which agrees with the findings in Johnson and Rossell (2010, 2011). In contrast, PPL trained NLPs provide good out-of-sample predictions but they may fail to focus the posterior probability on a small set of models. This is not surprising, as the loss function targets good predictive, rather than explanatory, performance. In summary, NLPs provide important advantages for explanatory purposes while maintaining a good predictive ability, especially in sparse setups.

Acknowledgments This research was partially funded by the NIH grant R01 CA158113-01.

References

Albert, J., & Chib, S. (1993). Bayesian analysis of binary and polychotomous response data. *Journal of the American Statistical Association, 88*, 669–679.

Consonni, G., & La Rocca, L. (2010). On moment priors for Bayesian model choice with applications to directed acyclic graphs. In J. M. Bernardo, M. J. Bayarri, J. O. Berger, A. P. Dawid, D. Heckerman, A. F. M Smith, & M. West (Eds.) *Bayesian statistics 9 - proceedings of the ninth Valencia international meeting* (pp. 63–78). Oxford: Oxford University Press.

Edgar, R., Domrachev, V., & Lash, A. E. (2002). Gene expression omnibus: NCBI gene expression and hybridization array data repository. *Nucleic Acids Research, 30*, 207–210.

Fan, J., & Li, R. (2004). New estimation and model selection procedures for semiparametric modeling in longitudinal data analysis. *Journal of the American Statistical Association, 99*, 710–723.

Gelfand, A. E., & Ghosh, S. K. (1998). Model choice: A minimum posterior predictive loss approach. *Biometrika, 85*, 1–11.

Gottardo, R., & Raftery, A. E. (2009). Markov chain Monte Carlo with mixtures of singular distributions. *Journal of Computational and Graphical Statistics, 17*, 949–975.

Johnson, V. E., & Rossell, D. (2010). Prior densities for default Bayesian hypothesis tests. *Journal of the Royal Statistical Society B, 72*, 143–170.

Johnson, V. E., Rossell, D. (2012) Bayesian model selection in High-dimensional settings. *Journal of the American Statistical Association, 107*(498), 649–660; doi: 10.1080/01621459.2012.682536.

Jorissen, R. N., Gibbs, P., Christie, M., Prakash, S., Lipton, L., Desai, J., Kerr, D., Aaltonen, L. A., Arango, D., & Kruhoffer, M., et al. (2009). Metastasis- associated gene expression changes predict poor outcomes in patients with dukes stage b and c colorectal cancer. *Clinical Cancer Research, 15*, 7642–7651.

Merlos-Suárez, A., Barriga, F. M., Jung, P., Iglesias, M., Céspedes, M. V., Rossell, D., Sevillano, M., Hernando-Momblona, X., da Silva-Diz, V., Muñoz, P., Clevers, H., Sancho, E., Mangues, R., & Batlle, E. (2011). The intestinal stem cell signature identifies colorectal cancer stem cells and predicts disease relapse. *Cell Stem Cell, 8*, 511–524.

Rhodes, D. R., Kalyana-Sundaram, S., Mahavisno, V., Varambally, R., Yu, J., Briggs, B. B., Barrette, T. R., Anstet, M. J., Kincead-Beal, C., Kulkarni, P., Varambally, S., Ghosh, D., & Chinnaiyan, A. M. (2007). Oncomine 3.0: genes, pathways, and networks in a collection of 18,000 cancer gene expression profiles. *Neoplasia (New York), 9*, 166–180.

Scott, J. G., & Berger, J. O. (2010). Bayes and empirical Bayes multiplicity adjustment in the variable selection problem. *The Annals of Statistics, 38*(5), 2587–2619.

Smith, J. J., Deane, N. G., Wu, F., Merchant, N. B., Zhang, B., Jiang, A., Lu, P., Johnson, J. C., Schmidt, C., & Edwards, C. M., et al. (2010). Experimentally derived metastasis gene expression profiles predicts recurrence and death in patients with colon cancer. *Gastroenterology, 3*, 958–968.

A Two Layers Incremental Discretization Based on Order Statistics

Christophe Salperwyck and Vincent Lemaire

Abstract Large amounts of data are produced today: network logs, web data, social network data... The data amount and their arrival speed make them impossible to be stored. Such data are called streaming data. The stream specificities are: (i) data are just visible once and (ii) are ordered by arrival time. As these data can not be kept in memory and read afterwards, usual data mining techniques can not apply. Therefore to build a classifier in that context requires to do it incrementally and/or to keep a subset of the information seen and then build the classifier. This paper focuses on the second option and proposed a two layers approach based on order statistics. The first layer uses the Greenwald and Khanna quantiles summary and the second layer a supervised method such as MODL.

1 Introduction

Many companies produce today large amounts of data. Sometimes data can be kept into a database, sometimes their arrival speed makes them impossible to be stored. In that specific case mining data is called stream mining. The stream specificities are: (i) data are just visible once and (ii) are ordered by arrival time. Such an amount of data leads to the impossibility to keep them in memory and to read them afterwards. Therefore to build a classifier in that context requires doing it incrementally and/or to keep a subset of the information seen and then build the classifier. In this paper the focus is on the second option. This can be achieved by: keeping a subset of the stream examples, calculating a density estimation or having order statistics. The work presented in this paper focuses on numeric attributes discretization based on order statistics. The discretization is then used as a pretreatment step for a supervised classifier.

C. Salperwyck (✉) · V. Lemaire
Orange Labs, Lannion, France – LIFL, Université de Lille 3, Villeneuve d'Ascq, France
e-mail: christophe.salperwyck@orange.com; vincent.lemaire@orange.com

P. Giudici et al. (eds.), *Statistical Models for Data Analysis*, Studies in Classification, Data Analysis, and Knowledge Organization, DOI 10.1007/978-3-319-00032-9_36,

2 Related Works

Incremental discretization is mainly used in two fields: (i) data mining field to be able to discretize large data set or to discretize data on the fly; (ii) Data Base Management Systems (DBMS) to have order statistics (quantiles estimates) on tables for building efficient query plans. This section gives a brief state of art of the main incremental discretization methods used in these two fields to have order statistics.

2.1 Data Mining Field

Gaussian Density Approximation: The main idea of this method relies on the hypothesis that the observed data distribution follows a Gaussian law. Only two parameters are needed to store a Gaussian law: the mean and the standard deviation. The incremental version required one more parameter: the number of elements. An improved version for supervised classification on stream can be found in Pfahringer et al. (2008) but it needs a parameter to set up the number of bins derived from the Gaussian. This method has one of the lowest memory footprints.

PiD: Gama and Pinto in 2006, proposed a two layers incremental discretization method. The first layer is a mix of a discretization based on the methods named "Equal Width" and "Equal Frequency" (algorithm details: Gama and Pinto 2006, p. 663). This first layer is updated incrementally and needs to have much more bins than the second one. The second layer uses information of the first one to build a second discretization. Many methods can be used on the second layer such as: Equal Width, Equal Frequency, Entropy, Kmeans... The advantage of this method is to have a fast first layer which can be used to build different discretizations on it (second layer).

Online Histogram: Ben-Haim and Tom-Tov (2010), presented an incremental and online discretization for decision trees. Their algorithm is based on three methods: (i) UPDATE—add a new example. It can be done by inserting the new example directly in an existing histogram or create a new bin with it and then do a merge, (ii) MERGE—merge two bins in one, (iii) UNIFORM: use a trapezoid method to build the final Equal Frequency bins. This method has a low computational requirement and is incremental but it introduces some errors. In case of skewed distributions the authors recommend to use bound error algorithms.

2.2 DBMS Field

MLR: Manku et al. (1998) developed an algorithm to approximate quantiles based on a pool of buffers. Their approach has three operation: (i) NEW—takes an empty buffer and fill it with new values from the stream, (ii) COLLAPSE—when all buffers are full, some need to be merged to get new empty buffers—this operation takes at least two buffers and merges them to have just one full at the end, (iii) OUTPUT—this operation collapses all the buffers into one and returns the quantile value for the given parameter. This method has a theoretical bound on the error ϵ and on the required space: $\frac{1}{\epsilon}log^2(\epsilon N)$, where N is the size of stream.

GK: Quantiles provide order statistics on the data. The ϕ-quantile, with $\phi \in [0, 1]$ is defined as the element in the position $\lceil \phi N \rceil$ on a sorted list of N values. ϵ is the maximum error on the position of the element: an element is an ϵ approximation of a $\phi - quantile$ if its rank is between $\lceil (\phi - \epsilon)\, N \rceil$ and $\lceil (\phi + \epsilon)\, N \rceil$. It corresponds to an "Equal Frequency" discretization; the number of quantiles being in that case the number of intervals.

The GK quantiles summary, proposed by Greenwald and Khanna (2001) is an algorithm to compute quantiles using a memory of $O(\frac{1}{\epsilon}log(\epsilon N))$ in the worst case. This method does not need to know the size of the data in advance and is insensitive to the arrival order of the examples. The algorithm can be configured either with the number of quantiles or with a bound on the error. Its internal structure is based on a list of tuples $< v_i, g_i, \Delta_i >$ where:

- v_i is a value of an explanatory feature of the data stream
- g_i corresponds to the number of values between v_{i-1} and v_i
- Δ_i is the maximal error on g_i

2.3 Summary

This subsection aims to present a synthetic overview of the methods described above. This overview uses two criteria taken from Dougherty et al. (1995). The first criterion: *global/local* corresponds to the way methods use data to build intervals. A method using all data for building all bins is considered as *global*. A method splitting data into subset and doing local decision is considered as *local*. The second criterion: *supervised* corresponds to methods using class labels to build the discretization. We added two other criteria: *parametric*—a non-parametric method finds the number of intervals automatically, *online/stream*—evaluate the ability to work online and to deal with data streams.

Table 1 presents the comparison of all methods seen above versus these four criteria. The second part of the table reports the widely used offline methods *Equal Width* and *Equal Frequency*, and also two competitive supervised methods used in the next section: MDLP and MODL.

Table 1 Discretization methods comparison

Method	Global/Local	Parametric	Supervised	Online/Stream
Gaussian	Global	Yes	No	Yes
PID (Layer 1)	Global	Yes	No	Yes
Online histogram	Global	Yes	No	Yes
MLR	Global	Yes	No	Yes
GK	Global	Yes	No	Yes
Equal Width/Freq	Global	Yes	No	No
MDLP	Local	No	Yes	No
MODL	Global	No	Yes	No

3 Our Proposal

3.1 Objective

Our proposal aims to used at best the data mining field and the DBMS field to propose an incremental discretization method which intrinsically realizes a compromise between the error ϵ and the memory used. This method will also have to be robust and accurate for classification problems.

3.2 Proposal

The idea is to use a two layer incremental discretization method as PiD (Gama and Pinto 2006) but in our case bounds in memory. The first layer summarizes (using counts per class) the input data, using much more intervals than required, in a single scan over the data stream. The second layer processes the first layer summary and produces the final discretization. The memory is used at best to have the lowest error.

For the first layer, the Greenwald and Khanna quantiles summary (GK) suits this requirement the best and provides order statistics. We adapted the GK summary to store directly the class counts in tuples. For the second layer, among methods using order statistics two are particularly interesting considering their performances: Recursive Entropy Discretization (MDLP) (Fayyad and Irani 1993) and Minimum Optimized Description Length (MODL) (Boullé 2006). They both use an entropy based criterion to build the discretization and the MDL (Minimum Description Length) criterion to stop finding intervals. They are supervised and known to be robust. The choice to use GK for the first layer and either MDLP or MODL in the second layer is coherent since the complete structure is based on order statistics. The errors on cut points depends mainly on the number of bins used the first layer. Because the second layer used the MODL approach and since MODL is a discretization method based on counts, the error on a split position of our two level

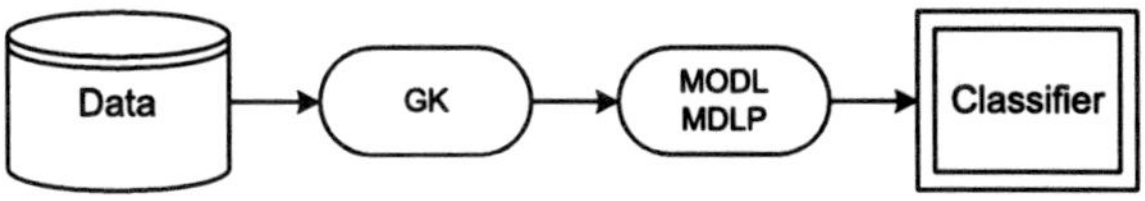

Fig. 1 Two layers discretization

method is related at worst to: $\arg\max_i (g_i + \Delta_i)/2$, where i is the index of the interval int the first level. This indicates that when the number of intervals of the first level increases, the error decreases.

Figure 1 shows how our method proceeds. A GK summary is created for each feature and updated after the arrival of a new example. When the model is needed, GK summaries provide univariate contingency tables to the discretization method (MODL or MDLP). A new contingency table (expected smaller) is built and returned by these methods. Using GK quantile values and the contingency table after discretization give at the same time cut points, density estimations per interval and conditional density estimation per interval for this numeric feature. Finally a classifier based on order statistics is built, as for example a naive Bayes.

4 Experiments

4.1 Large Scale Learning Challenge

The Delta training dataset from the large scale learning challenge[1] is used for a first experiment. This dataset contains 500,000 examples; each example consists of 500 numerical features and a boolean label. 100,000 examples were kept for a test set and train examples were taken from the 400,000 remaining instances. We adapted the MODL discretization so that it uses GK quantiles summary as an input and built a naive Bayes classifier on this discretization. The GK quantiles summary is set up with 10 and 100 quantiles: GK10 (200K bytes) and GK100 (2M bytes). The reservoir sampling approach (Vitter 1985) is also used as a baseline method to compare performances with a bounded memory technique: GK10, GK100 corresponds respectively to a reservoir of 50 and 500 examples.

Figure 2 shows the comparison of theses approaches using the AUC (Area Under learning Curve) performance indicator. The two reservoir sizes used are equivalent to the memory consumed by GK10 and GK100. With limited memory GK methods are performing much better than reservoir sampling. Compared to the naive Bayes classifier using all data into memory, the GK performances with 100 quantiles (GK100) are almost the same.

This first experiment shows that with a small given amount of memory our method performances are similar to the one loading all the data into memory.

[1] http://largescale.ml.tu-berlin.de.

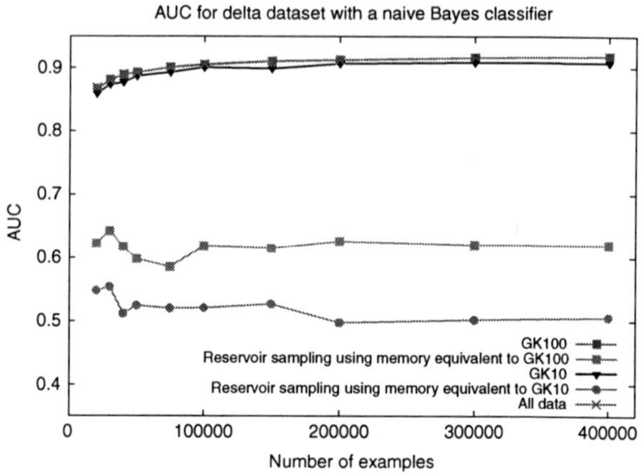

Fig. 2 Naive Bayes AUC performances with all data, GK and reservoir sampling

4.2 *ICML Exploration and Exploitation Challenge*

The ICML 2011 challenge[2] aims to show the online ability and robustness of our two levels discretization. This challenge data are very unbalanced and contain a high level of noise. The dataset contains 3 millions examples; each example consists of 100 numerical and 19 nominal features labels by a boolean (click/no-click). The purpose of the challenge is to evaluate online content selection algorithms. Each algorithm has to perform a sequence of iterations. For each iteration, a batch of six visitor-item pairs is given to the algorithm. The goal is to select the instance which is most likely to provoke a click. This challenge has strong technical constraints: (i) a time limit (100ms per round), (ii) a limited space (1.7GB of memory). These challenge data contain nominal features; we dealt with them using a hash based solution: nominal values are hashed and put into a fixed number of buckets. The click/no-click counts are stored for each bucket. These buckets are used as numerical bins so that we can deal with nominal features as numerical ones.

Due to the challenge specificities our two layers approach was adapted. This dataset is very unbalanced: clicks are very scarce—0.24%. Methods as MODL are known to be robust but this robustness with noisy data leads to a late discovery of cut points as shown by the MODL curve on Fig. 3. As the challenge score was cumulative rewards, the model has to make decision even with just few clicks. Waiting to make decision could provide a better classifier at the end but a lower score on this challenge evaluation. To be more reactive probability estimation tree (PETs: Provost and Domingos 2003) were built on our first level summaries. A tree can be seen as a discretization method (our 2nd layer). The final step (corresponding to the *Classifier* on Fig. 1) is a predictor composed of an averaging of PETs' predictions.

[2]http://explo.cs.ucl.ac.uk/.

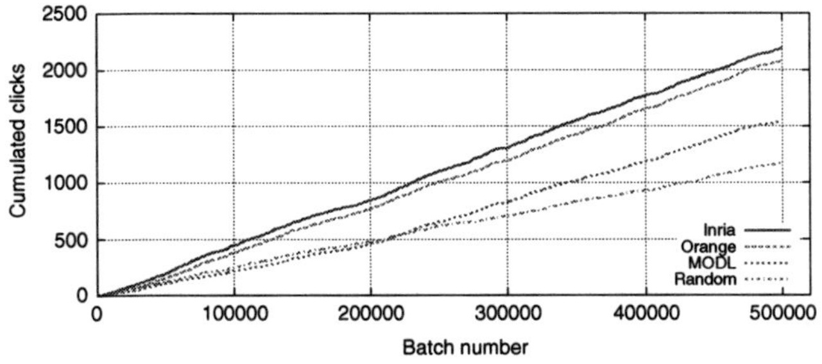

Fig. 3 ICML Exploration and Exploitation challenge 2011 results

Figure 3 shows results on this challenge: Inria (ranked 1st), our submission with PETs (ranked 2nd) and a random predictor. Our approach was competitive and provides good density estimations for building online tree classifiers.

5 Future Works

5.1 Extension to Nominal Features

The work presented before only addresses the discretization for numerical features. Many classification problems contain nominal features. Moreover the MODL approach for grouping modalities (Boullé 2005) is a competitive method to find groups on nominal features. The simplest summary for nominal features just keeps counts—it requires low memory and processing time and is practicable if the number of different values is low. Unfortunately the number of nominal values can be large, for example: *client ids, cookies, city names, etc*. As we want to bound memory we can only afford to focus on frequent values. This can be done using a hash function: nominal values are hashed and put into a fixed number of buckets in which classes counts are stored. In order to reduce errors several hashing functions may be combined as proposed in the count-min sketch algorithm (Cormode and Muthukrishnan 2005).

5.2 Dialogs Between Two Layers

A dialog between the two layers to control the number of tuples in the first layer could be beneficial to share memory between different features summaries. In a classification problem with many features, some of them may need a very fine discretization and some may not need it. As the second layer is non parametric and supervised it can inform the first layer if it needs more or less bins.

5.3 Online Trees

The stream mining community often uses trees to build online and incrementally classifiers. The most known ones are the Hoeffding trees proposed in VFDT (Domingos and Hulten 2000). In those trees, leaves keep statistics on the data. A leaf is transformed into a node using a split criterion (usually entropy or Gini index). As a split is a definitive decision the Hoeffding bound is used to set the confidence (δ) in the split. Another parameter (τ) is used to break ties between two attributes with similar criterion to avoid late splits.

Our two layers discretization can be used as summaries in the tree leaves. Moreover the second layer of our discretization method (MODL method applied on the first layer summary) gives cut points for a feature with a quality index. This quality index allows selecting the feature on which to split and the cut points where to split. If a feature is not considered as informative its index equals zero. The tree can expend by splitting a leaf on the feature having the greatest non-null index as it is sure that this feature is informative: the Hoeffding bound is not anymore needed. With this MODL criterion there is no need to have the two previous parameters δ and τ to build online trees.

6 Conclusion

The first experiments validate that with large data sets and bounded memory, our two layers discretization has a strong interest. Our approach uses order statistics on both levels and can be set up to use a fixed memory size or to stay beyond a given error. We used Greenwald and Khanna quantiles summary for the first layer and MODL discretization for the second layer as they are known to be amongst the most competitive methods to build quantiles summary and perform supervised discretization. Classifiers assuming features independence can be easily built on the summary as shown on the first experiment with a naive Bayes classifier. Some other classifiers such as online trees can also take advantage of our method.

References

Ben-Haim, Y., & Tom-Tov, E. (2010). A streaming parallel decision tree algorithm. *Journal of Machine Learning*, *11*, 849–872.

Boullé, M. (2005). A Bayes optimal approach for partitioning the values of categorical attributes. *Journal of Machine Learning Research*, *6*(04), 1431–1452.

Boullé, M. (2006). MODL: A Bayes optimal discretization method for continuous attributes. *Machine Learning*, *65*(1), 131–165.

Cormode, G., & Muthukrishnan, S. (2005). An improved data stream summary: the count-min sketch and its applications. *Journal of Algorithms*, *55*(1), 58–75.

Domingos, P., & Hulten, G. (2000). Mining high-speed data streams. In *Proceedings of the sixth ACM SIGKDD international conference on Knowledge discovery and data mining* (pp. 71–80). New York, NY: ACM.

Dougherty, J., Kohavi, R., & Sahami, M. (1995). Supervised and unsupervised discretization of continuous features. In *Proceedings of the twelfth international conference on machine learning* (pp. 194–202). San Francisco: Morgan Kaufmann.

Fayyad, U., & Irani, K. (1993). Multi-interval discretization of continuous-valued attributes for classification learning. In *Proceedings of the international joint conference on uncertainty in AI* (pp. 1022–1027).

Gama, J., & Pinto, C. (2006). Discretization from data streams: applications to histograms and data mining. In *Proceedings of the 2006 ACM symposium on applied computing* (pp. 662–667).

Greenwald, M., & Khanna, S. (2001). Aproximate medians and other quantiles in one pass and with limited memory. *ACM SIGMOD Record*, *27*(2), 426–435.

Manku, G. S., Rajagopalan, S., & Lindsay, B. G. (1998). *Lecture Notes in Computer Science Volume 5012*, New York.

Pfahringer, B., Holmes, G., & Kirkby, R. (2008). Handling numeric attributes in hoeffding trees. *Advances in Knowledge Discovery and Data Mining*, 296–307.

Provost, F., & Domingos, P. (2003). Tree induction for probability-based ranking. *Machine Learning*, *52*(3), 199–215.

Vitter, J. S. (1985). Random sampling with a reservoir. *ACM Transactions on Mathematical Software*, *11*(1), 37–57.

Interpreting Error Measurement: A Case Study Based on Rasch Tree Approach

Annalina Sarra, Lara Fontanella, Tonio Di Battista, and Riccardo Di Nisio

Abstract This paper describes the appropriateness of Differential Item Functioning (DIF) analysis performed via mixed-effects Rasch models. Groups of subjects with homogeneous Rasch item parameters are found automatically by a model-based partitioning (Rasch tree model). The unifying framework offers the advantage of including the terminal nodes of Rasch tree in the multilevel formulation of Rasch models. In such a way we are able to handle different measurement issues. The approach is illustrated with a cross-national survey on attitude towards female stereotypes. Evidence of groups DIF was detected and presented as well as the estimates of model parameters.

1 Introduction

The Rasch model is the simplest Item Response Theory (IRT) model, widely used especially in social science. It is concerned with the measurement of a latent construct referring to a person characteristic (such as attitude, ability, skill), assessed indirectly by a group of items. It has been demonstrated that IRT models can be regarded as mixed-effects models (see for example De Boeck 2008). The multilevel formulation of IRT models constitutes an interesting way to carry out item response data analysis, allowing to handle the measurement issues regarding the latent variable and take into proper account error measurements related to limited number of items, unreliability of the measurement instrument and the stochastic nature of human response behaviour. In the measurement process, Differential Item Functioning (DIF) has been long recognised as a potential source of bias. DIF is the statistical term traditionally used to describe a dependence of item response

A. Sarra (✉) · L. Fontanella · T. Di Battista · R. Di Nisio
University G. D'Annunzio, Chieti-Pescara, Italy
e-mail: asarra@dmqte.unich.it

P. Giudici et al. (eds.), *Statistical Models for Data Analysis*, Studies in Classification, Data Analysis, and Knowledge Organization, DOI 10.1007/978-3-319-00032-9_37,

and group membership after conditioning on the latent trait of interest. In this respect, it is worth noting that an item might show DIF but not considered biased if the difference exists because groups display actual differences in the underlying latent construct measured by the item. In that case it is more convenient to speak about item impact rather than item bias. A large number of statistical methods for detecting DIF require post-hoc sample split analyses and involve the comparison of parameter estimates between one group, labelled the reference group, and the other, labelled focal group. Accordingly, only those groups that are explicitly proposed by the researcher are tested for DIF. In this study we explore and illustrate how DIF analysis can be integrated into Rasch-mixed effects models, in which both the person and item parameters can be treated as random. The proposed approach is adopted with regard to a questionnaire on attitude towards female stereotypes. More specifically, we first employ a new methodology introduced by Strobl et al. (2010) for detecting DIF in Rasch models. This procedure, termed Rasch tree (RT) method, is based on a recursive partitioning of the sample and has the valuable advantage over traditional methods that groups exhibiting DIF are found automatically, from combination of person covariates, and statistical influence is also tested. As a second step, the terminal nodes produced in the tree structured partition of the covariate space of our sample, identifying groups of subjects with homogenous Rasch item parameters, are considered in the random item formulation of mixed-effects Rasch models. To enhance the recognition of female stereotype statements, different mixed-effects Rasch models have been fitted and compared. The remainder of the paper is organised as follows. Sections 2 and 3 briefly summary the underlying statistical framework followed in this study: i.e. the Rasch tree approach and the mixed-effects Rasch models, respectively. The reviewed methodology is then applied in our research, aiming to investigate gender stereotypes, which is presented in Sect. 4 along with the main results. Finally, conclusions are given in Sect. 5.

2 The Rasch Tree Approach

In this section we give some methodological details on the Rasch tree (RT) approach. RT is a model-based partitioning (see Zeiles et al. 2008) aiming at identifying groups of subjects with homogenous Rasch item parameters. Instabilities in the model parameters are found by using structural change tests (Zeiles and Hornik 2007). The RT approach verifies the null hypothesis that one-joint Rasch model can properly hold for the full sample of subjects. In other terms, the null hypothesis of parameter stability is tested against the alternative hypothesis of a structural break. In order to identify DIF in the Rasch model the method produces a tree-structured partition of the covariate space. The Rasch tree algorithm is essentially a four steps procedure. In the first step a Rasch model is fitted to all subjects in the current sample. The second step consists in assessing the stability of the Rasch model with respect to each available covariate. If significant instability is detected, the sample is splitted along the covariate with the strongest instability, leading to the individuation of a cutpoint and to the estimation of two separate Rasch models.

Splitting continues until a stop criterion is reached; that means there is no significant DIF or the subsample is too small. A formal description of the statistical inference framework of this procedure can be found in Strobl et al. (2010).

3 Mixed Effects Rasch Models

As known, the Rasch model provides a theoretical background to assess the consistency between a latent trait of interest and the specific responses to a set of items.

In order to include random effects, the classical formulation of Rasch models can be extended in a number of ways, leading to the so-called mixed-effects Rasch models. We deal with the Rasch model for binary responses. Let $\eta_{pi} = \theta_p + \beta_i$, where η_{pi} is the logit of the probability of a 1-response, $log[P(Y_{pi} = 1|\theta_p, \beta_i)/P(Y_{pi} = 0|\theta_p, \beta_i)]$, defined as the simple sum of the "ability" of person (θ_p) and the "easiness" of item (β_i), and Y_{pi} is the response of person p $(p = 1, \ldots, P)$ to item i $(i = 1, \ldots, I)$. Unlike the classic Rasch model, the difficulties and abilities are not both fixed and unknown parameters. Given a completely crossed design, i.e. each subject responds to all items, we consider mixed effect models with crossed independent random effects for subjects and items (Baayen et al. 2008). Although it is common practice in Item Response Theory (IRT) to consider items as fixed and persons as random, De Boeck (2008) shows that random item parameters make sense theoretically for two reasons. The first reason is that items can be thought of as a sample drawn from the population of all possible items on the subject matter; the second reason is the uncertainty about the item parameters. Depending on whether the persons and the items are either treated as random or fixed, four different kinds of Rasch models can be defined: (1) the fixed person-fixed items Rasch model (FPFI Rasch), (2) the random person-fixed items Rasch model (RPFI Rasch), (3) the fixed persons-random items Rasch model (FPRI Rasch) and (4) the random person-random item Rasch model (RPRI Rasch) (De Boeck 2008).

Mixed effect modelling is a useful approach to carry out item response data analysis as it allows to consider the Rasch parameters as randomly varying parameters and accommodate nested structures found in the data (e.g. items nested within dimensions). Besides, through mixed-effects Rasch models one is able to handle all factors that potentially contribute to understanding the structure of the data and retain respondents with invariant extreme response patterns.

4 Data and Empirical Results

This section examines the appropriateness of DIF analysis performed via mixed-effects Rasch models for the measurement of female stereotypes. As it is well known, stereotype is a crystallization of a collective imaginary on a group of people, which is more or less susceptible to change over time and space and is assimilated through the socialization process, contributing to the construction of both individual and group identity. Hence, stereotypes occur when individuals are classified by others as having something in common because they are members of a particular group or category of people. A stereotype is a cultural construct which has no scientific basis and it be either positive or negative. The current research deals with the recognition of the main female stereotypes around the world and data arise from a cross-national survey carried out from July to August 2010. A first analysis of these data is provided by Bernabei et al. (2010). The questionnaire has been completed by a sample of 2002 respondents consisting in majority of women (85%); the educational level is high for both gender, whereas age ranges from 16 to 84 years with an average of 38 years. In this study, we focus on the stereotypes related to two different dimensions: *temperament and emotional sphere* and *body and sexuality*. Each dimension includes 8 statements. In the first dimension we grouped the following items: "Women's behaviour is more guided by their emotions", "Women tend to be more unstable emotionally", "Women have to be protected as they are the "weak sex", "Women are more talkative even nagging", "Women are manipulative", "Women are easily more jealous than men", "Women spend more money than men", "Women are always late". In the other dimension we included statements containing beliefs about the body and sexuality sphere: "Women have to take care of their personal hygiene more than men", "Women who don't take care of themselves and their bodies are masculine", "Women feel less sexual pleasure than men", "Women tend to have fewer sexual partners in their lifetime", "Women cannot imagine sex without love", "Women are more affected by cultural models imposed by the media", "Women build their image and identity on the basis of fashion more than men", "The bi-sexuality in women is a more cultural condition than in men". Initial items, rated on a Likert scale, have been dichotomised in *agree* and *not agree*. So, we assess the item responses with respect to some sample covariates. The RT procedure offers the possibility to promptly generate a graph (the Rasch tree) and helps with visualizing which groups are affected by DIF with respect to which items. As displayed in Fig. 1, it is the combination of sex, age and country of origin that determine which items differ according to the observed person groups, whereas the variable education was not selecting for splitting. It is worth noting that with standard approaches, this pattern could only be detected if the interaction terms were explicitly included in the models or respective groups were explicitly pre-specified. The results were obtained using the R package psychotree (Zeiles et al. 2010).

Exploring the Rasch tree from the top to bottom we find different item parameters for males and females. Within the group of males, the Rasch tree indicates DIF in the variable country of origin, differentiating estimates of the item difficulty (in our context item intensity) for males coming from Africa (AF) and Asia (AS) from those of males belonging to the other continents (Europe (E), North-America (NTA), South-America (SA) and Oceania (O)). Moreover, it appears that for the female group the Rasch tree has a split in the variable country of origin as well as in the covariate referring to the age, exhibiting DIF between females up to the age of 23 and females over the age of 23. In this last group we observe a further

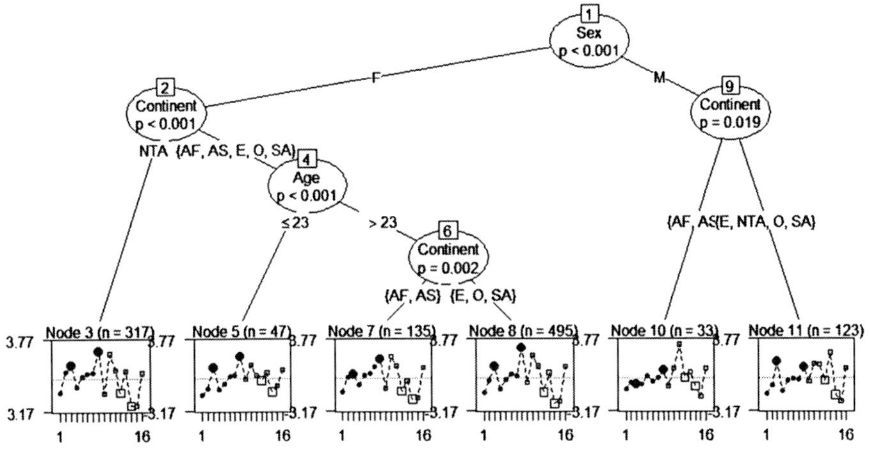

Fig. 1 Rasch tree for female stereotypes. In the terminal nodes, estimates of the item difficulty are displayed for each items

partition: significant instability of item parameter is detected in two subsamples. One subsample is constituted by females coming from Africa (AF) and Asia (AS) while in the other there are females coming from Europe (E), Oceania (O) and South-America (SA). In each terminal node of the tree, the item parameter estimates for the 16 items are displayed: the circle indicates stereotypes linked to temperament and emotional sphere and the square those related to body and sexuality. In this context the item parameter can be thought of as a crystallization of the stereotype such that a larger value correspond to a lower degree of acceptance. From Fig. 1 we notice that in general stereotypes are less crystallized for women. More specifically the items showing particularly marked DIF, (highlighted by a large symbol in Fig. 1) are: *Women have to be protected as they are the "weak sex"*, *Women are always late*, *Women cannot imagine sex without love*, *Women feel less sexual than men*. It is important to point out that we have also checked if the observed pattern of responses to items conforms to the Rasch model expectations by means of Andersen's LR-test (Andersen 1973). The fit analysis results, not displayed here, revealed the feasibility of the Rasch model for all nodes.

The complex interaction structure detected in the six terminal nodes of the Rasch tree is embedded in the RPRI formulation of the mixed effect Rasch model. As shown in Table 1, we specify 8 models that differ with respect to the inclusion of item and/or person partitions as item or person property covariates (De Boeck et al. 2011). The models have been estimated with lme4 package in R (Doran et al. 2007), using restricted maximum likelihood estimation. The simplest model (m_0) is the RPRI model with homoscedastic random effects for subjects, $\theta_p \sim N(0, \sigma^2_\theta)$, and items, $\beta_i \sim N(0, \sigma^2_\beta)$. Defining, in model m_1, a different intercept for each dimension ($k = 1, 2$) does not lead to a significant improvement of the goodness of fit (see Table 1), and this is also true if we consider model m_2 with heteroscedastic

Table 1 RPRI Rasch models comparison

	Model specification	Df	AIC	BIC	logLik	Chisq	Chi Df	Pr(>Chisq)
m_0	$\eta_{pi} = \gamma_0 + \theta_p + \beta_i$	3	17,011	17,035	−8502.5			
m_1	$\eta_{pi} = \gamma_k + \theta_p + \beta_i$	4	17,013	17,045	−8502.3	0.3	1	0.560
m_2	$\eta_{pi} = \gamma_k + \theta_p + \beta_{ik}$	6	17,015	17,063	−8501.7	1.3	2	0.524
m_3	$\eta_{pi} = \gamma_j + \theta_p + \beta_i$	8	16,941	17,005	−8462.4	78.6	2	< 2.2e−16
m_5	$\eta_{pi} = \gamma_{jk} + \theta_p + \beta_i$	14	16,828	16,940	−8400.0	124.9	6	< 2.2e−16
m_4	$\eta_{pi} = \gamma_j + \theta_{pj} + \beta_i$	28	16,966	17,190	−8455.1	0.0	14	1
m_7	$\eta_{pi} = \gamma_{jk} + \theta_p + \beta_{ij}$	34	16,768	17,040	−8350.2	209.9	6	< 2.2e−16
m_6	$\eta_{pi} = \gamma_{jk} + \theta_{pj} + \beta_{ik}$	36	16,862	17,149	−8395.0	0.0	2	1
m_8	$\eta_{pi} = \gamma_{jk} + \theta_{pj} + \beta_{ij}$	54	16,801	17,232	−8346.4	97.2	18	7.15E−13

Table 2 Fixed effects for model m_7

	Body and sexuality				*Emotion and temperament*			
Node	Estimate	Std. Error	z value	Pr(> |z|)	Estimate	Std. Error	z value	Pr(> |z|)
3	−2.046	0.496	−4.123	3.74E−05	−2.965	0.498	−5.952	2.65E−09
5	−1.349	0.379	−3.557	3.76E−04	−1.054	0.377	−2.796	0.005
7	−1.472	0.446	−3.299	9.70E−04	−2.079	0.448	−4.644	3.42E−06
8	−2.183	0.492	−4.437	9.10E−06	−2.551	0.492	−5.181	2.21E−07
10	−1.108	0.382	−2.906	3.67E−03	−0.653	0.376	−1.736	0.083
11	−1.911	0.441	−4.334	1.46E−05	−1.570	0.439	−3.579	3.44E−04

item random effects, $\boldsymbol{\beta}_i = (\beta_{i1}, \beta_{i2})' \sim N(\mathbf{0}, \boldsymbol{\Sigma}_\beta)$. The inclusion, in model m_3, of a different intercept for each of the 6 detected nodes ($j = 1, \ldots, 6$) leads to a better fit with respect to the previous models, not improved if we consider heteroscedastic person random effects, $\boldsymbol{\theta}_p = (\theta_{p1}, \ldots, \theta_{p6})' \sim N(\mathbf{0}, \boldsymbol{\Sigma}_\theta)$, as in m_4. Given the interaction of person and item covariates, we can specify a different intercept for each dimension in each node, assuming the same variance (m_5) or different variance (m_6) for both items and persons. The specification of a different intercept with respect to the person groups allow to consider a similar DIF for all the items (m_3, m_4) or for the items belonging to the same dimension (m_5, m_6); in addition we can assume item parameter heteroscedasticity depending on the node, $\boldsymbol{\beta}_i = (\beta_{i1}, \ldots, \beta_{i6})' \sim N(\mathbf{0}, \boldsymbol{\Sigma}_\beta)$, as in m_7 and m_8. The inspection of goodness-of-fit indexes (AIC, BIC and logLik values), listed in Table 1, suggests that we can choose model m_7 for which we have the smallest AIC and also the BIC is low, while the loglikelihood is the second highest.

For the selected model the fixed effect estimates are statistically significant for all the levels of the interaction dimension-node (see results in Table 2).

Results in Table 2 can be inferred considering that the plus sign in $\eta_{pi} = \gamma_{jk} + \theta_p + \beta_{ij}$ implies that the $\gamma_{jk} + \beta_{ij}$ should be interpreted as item easiness instead of item difficulty. Accordingly, the smaller are the values of the estimates the lower is the level of crystallization for the surveyed stereotypes. An interesting consequence of treating Rasch parameters as random is that the variance component and intraclass correlation can be determined. Table 3 displays the person variance

Table 3 Random effects for model m_7

Groups	Name	Variance	Std.Dev.	Corr				
Subject	(Intercept)	1.37	1.17					
Item	NODO3	1.91	1.38					
	NODO5	0.81	0.90	0.94				
	NODO7	1.46	1.21	0.97	0.91			
	NODO8	1.90	1.38	0.98	0.98	0.97		
	NODO10	0.67	0.82	0.91	0.73	0.91	0.86	
	NODO11	1.40	1.18	0.95	0.91	0.84	0.93	0.83

component and the variance of the item parameters for each node. Furthermore the intraclass correlation coefficients are shown.

Examining the group variance component we find out a greater variability mainly in nodes 3 and 8. As for the correlation structure we notice that all the nodes are highly associated with the other groups.

5 Concluding Remarks

The focus of this paper was to consider and illustrate how differential item analysis can be performed via Rasch mixed-effects models. Building on idea put forward in Strobl et al. (2010), we consider in this study an alternative methodology, named Rasch tree, to detect DIF that does not require post-hoc split analysis. This procedure has many attractive properties. Basically, it yields interpretable results of DIF in a quick and straightforward fashion by means of a visual representation (the Rasch-tree graph). Another notable feature of this approach is the natural treatment of categorical and numeric covariates. The latter do not need to be discretized in advance. In general, covariates which correspond to a significant change in the model parameters and interactions between them are automatically selected in a data-driven way, resulting in the terminal nodes of the Rasch tree graph. The subsequent innovative step of the integrated framework considered in this paper was to include the complex interaction structure detected in the terminal nodes of RT in the random formulation of mixed-effects Rasch models. They are a rather new approach in the domain of IRT. To assume the item difficulties are random variables may seem controversial because in most IRT models they are taken to be fixed, while the person parameters are regularly considered to be a random sample from a population. Here, we referred to a so-called crossed random effect model, defined by supposing that both item and person parameters are random effects. We believe that defining the difficulty of the items as random is an efficient and realistic way to model the variation in item difficulty. In addition, our formulation of these models takes into account the possibility that items difficulties differ across groups, previously identified by the RT approach. The appropriateness and usefulness of the proposed approach is demonstrated in interpreting the results of a female stereotypes survey. Data, arising from a cross-national survey aimed at enhancing the recognition of female stereotypes statements, have been analyzed by fitting different Rasch-mixed effects models. The empirical findings lead to the selection of *RPRI* model where a different intercept for each dimension in each node is assumed as well as potential item parameter heteroskedasticity with respect groups exhibiting DIF is accounted. The set of estimated parameters for that model includes the coefficients for the fixed effects on the one hand and the standard deviations and correlations for the random effects on the other hand. Previously, following the different branches of the Rasch tree, we were able to identify groups of people with certain characteristics affected by DIF and items displaying marked instability across classes of persons. In sum, from the application of this unified framework to stereotype data set, it may be concluded that this strategy is promising and useful to solve various kinds of issues, related to the measurement process. A limitation of our case study is worth noting. The application handles dichotomous response data. However, transforming the initial polytomous data to binary responses, by collapsing response categories to enforce dichotomous outcomes, leads to a loss of information contained in the data. Future work will aim at expanding application of the proposed approach for items with more than two response categories.

References

Andersen, E. B. (1973). A goodness of fit test for the Rasch model. *Psychometrika, 38*, 123–140.

Baayen, R. H., Davidson D. J., & Bates, D. M. (2008). Mixed-effects modeling with crossed random effects for subject and items. *Journal of Memory and Language, 59*, 390–412.

Bernabei, D., Di Zio, S., Fontanella, L., & Maretti, M. (2010). Attitude towards women stereotypes: the partial credit model for questionnaire validation. In Proceedings of *MTISD 2010, Methods, Models and Information Technologies for Decision Support Systems.*

De Boeck, P. (2008). Random item IRT models. *Psychometrika, 73*(4), 533–559.

De Boeck, P., Bakker, M., Zwitser, R., Nivard, M., Hofman, A., Tuerlinckx, F., & Partchev, I. (2011). The estimation of item response models with the lmer function from the lme4 package in R, **73**(4). *Journal of Statistical Software, 39*(12), 1–28.

Doran, H., Bates, D., Bliese, P., & Dowling M. (2007). Estimating the multilevel rasch model: with the lme4 package. *Journal of Statistical Software, 20*(2), 1–18.

Strobl, C., Kopf, J., & Zeiles, A. (2010). A new method for detecting differential item functioning in the rasch model. Technical Report 92, Department of Statistics, Ludwig-Maximilians Universitt Mnchen. http://epub.ub.uni-muenchen.de/1195/.

Zeiles, A., & Hornik, K. (2007). Generalised M-fluctuation tests for parameters instability. *Statistica Neerlandica, 61*(4), 488–508.

Zeiles, A., Hothor, T., & Hornik, K. (2008). Model-based recursive partitioning. *Journal of Computational and Graphical Statistics, 17*(2), 492–514.

Zeiles, A., Strobl, C., Wickelmaier, F., & Kopf, J. (2010). Psychotree: recursive partitioning based on psycometric models. R package version 0.11-1 htpp://CRAN.R-project.org/package= psychotree.

Importance Sampling: A Variance Reduction Method for Credit Risk Models

Gabriella Schoier and Federico Marsich

Abstract The problem of the asymmetric behaviour and fat tails of portfolios of credit risky corporate assets such as bonds has become very important, not only because of the impact of both defaults end migration from one rating class to another. This paper discusses the use of different copulas for credit risk management. Usual Monte Carlo (MC) techniques are compared with a variable reduction method i.e. Importance Sampling (IS) in order to reduce the variability of the estimators of the tails of the Profit & Loss distribution of a portfolio of bonds. This provides speed up for computing economic capital in the rare event quantile of the loss distribution that must be held in reserve by a lending institution for solvency. An application to a simulated portfolio of bonds ends the paper.

1 Introduction

In recent years, along with an ever increasing number of financial products together with the globalization and the financial environment, investors and financial organs faced comparatively formerly an ever increasing risk associated with their asset allocation or their investment strategies. Tail events that occur rarely but whose occurrence results in catastrophic losses are increasingly important in managerial decision making (Morgan 1997). Investors holding a portfolio of assets closely monitor and control the value-at-risk (a specified tail-percentile of the loss distribution) of their investment portfolio. Insurance companies hold huge reserves to

G. Schoier (✉)
Dipartimento di Scienze Economiche Aziendali Matematiche e Statistiche, Universitá di Trieste, Piazzale Europa 1, 34127 Trieste, Italia
e-mail: gabriella.schoier@econ.units.it

F. Marsich
Assicurazioni Generali via Macchiavelli 4, 34100 trieste, Italia
e-mail: federico.marsich@generali.com

P. Giudici et al. (eds.), *Statistical Models for Data Analysis*, Studies in Classification, Data Analysis, and Knowledge Organization, DOI 10.1007/978-3-319-00032-9_38,

protect against the possibility of rare but catastrophic losses. Different models have been proposed to consider changes correlated across the portfolio, i.e. the possibility of migration from one rating class to another and the default of each issuer.

In this paper we consider the structural model approach first proposed by Merton (1974) the basic idea of which is that, over a single time horizon, the firm's asset returns, for a portfolio of corporate credits are drawn from a multivariate Gaussian copula (Morokoff 2004). The distribution of equity asset returns are compared with those obtained when sampling from a t-copula (Kole et al. 2007). As marginal distributions we have considered the Gaussian, the Student-t and the generalized hyperbolic. We introduce the generalized hyperbolic distribution as it describes well both fat tails and skewness.

Moreover we introduce another joint distribution, the generalized hyperbolic, which can be used to model the distribution of equity asset returns. We model both the marginal and the joint distribution with the generalized hyperbolic.

In order to reduce variability of the estimators we use Importance Sampling (Glasserman and Li 2005; Kole et al. 2007) in the Gaussian, Student-t and generalized hyperbolic framework.

An application to a simulated portfolio of bonds ends the paper.

2 Credit Risk Portfolio Models

Events that can occur to a credit risky asset are t migration from one rating class to another and the possibility of default. Migration implies a change in the bond's price, while default implies a loss of all, or a considerable part, of the invested capital. Different approaches have been proposed to model the joint evolution of these events across the portfolio. Changes in credit quality are defined as the possibility of migration from one rating class to another or of default. Our starting point is the structural model approach first proposed by Merton (1974). Merton's model is based on the idea of treating the default of an issuer using the Black and Scholes model for options evaluation. In this case the fundamental stochastic variable is the underlying asset value of the corporation. When the asset value of a firm falls below a certain point derived from its outstanding liabilities it defaults. The credit quality of the firm and its bonds and loans value increase as the distance between the firm's asset value and its default point increases. The underlying stochastic process is assumed to follow a geometric Brownian motion.

In Merton's model the change in the asset's value is used to explain default-non default events. In this case it is sufficient to consider only one limit threshold: the default one. However, as our objective is to model not only the risk of default but also that of migration, we have considered a more complex structure. Merton's model has been extended assuming that not only default, but also the issuer's rating is a function of the total value of its assets at the end of time horizon.

Given the transition probabilities it is possible to evaluate the thresholds limits of the asset value for the different rating classes. In practice the issuer's asset values are not observable, but one can demonstrate that a proxy variable the equity of the issuers can be used (Hull 2006, 2009). A portfolio is composed of a great number of bonds, so it is necessary to specify not only the marginal probability of transition but also the joint structure by means of a correlation matrix. In practice it is impossible to evaluate all possible correlations; a solution is given by considering the correlations between a fixed number of equity indices (Morgan 1997). The strength of the tie between an issuer and his index is given by the coefficient R^2 evaluated for each issuer on the base of the linear regression between the historical series of returns of the price of the equity and those of the indices.

The asset returns Z_i for issuer $i = 1, \ldots, N$ are calculated according to

$$Z_i = R_{i,j} X_j + (\sqrt{1 - R_{i,j}^2}),\ \varepsilon_i \ \ i = 1, \ldots, N, j = 1, \ldots, n, \tag{1}$$

where $R_{i,j}^2$ is equal to the R^2 of the linear regression between the returns of the ith issuer and its referring index, X_j is the return of the jth index, ε_i is the noise factor distributed according to a $N(0, 1)$ distribution.

For simplicity it is assumed that the Loss Given Default (LGD) values of defaulted loans are independent and identically distributed according to a Beta distribution. Moreover it is assumed that the random LGD values are independent of the random asset returns. Finally, we assume that each issuer has only one bond issued.

The last step in our modelling procedure is to sample the tails of the associated multivariate distribution using Importance Sampling (Lemieux 2009). The choice of an Importance Sampling distribution from which to sample is very difficult (Sak et al. 2007; Sun and Hong 2010). A possible solution could consist of sampling observations from a multivariate distribution with a wider correlation matrix. A way to obtain this is to orthogonalise the covariance matrix of the asset returns and work with the eigenvector decomposition (Golub and Van Loan 1996). The eigenvector direction corresponding to the largest eigenvalue is exactly the single dimension that has the largest impact on the portfolio.

Let P be the correlation matrix between the standardized asset returns, Q be the orthogonal matrix whose columns are the orthonormal eigenvectors of P, and Λ be the diagonal matrix of eigenvalues sorted such that $\lambda_1 \geq \lambda_2 \geq \ldots \geq \lambda_n \geq 0$. Its spectral decomposition can be represented as

$$P = Q \Lambda Q', \tag{2}$$

Let q_1 be the first column of Q, corresponding to the largest eigenvalue. By multiplying q_1 by a scale factor η greater than 1 we can obtain $\tilde{P}$, that is the covariance matrix which results from scaling up the largest eigenvalue by the scale factor η.

In this orthogonal framework, we can scale up the variance in one coordinate direction (corresponding to the largest eigenvalue) independently of the other dimensions. The sampled scenarios will present an higher number of defaults with respect to a classic Monte Carlo simulation.

The simulations run with Importance Sampling would have the same structure of the Monte Carlo, except for the wider correlation matrix.

To obtain a point estimate of VaR at a fixed percentile (α, usually 99%, 99.5%), Importance Sampling weights are cumulated until the distribution of the percentile estimator's is achieved. The probability to have loss greater than VaR_α is given by $\beta\beta = 1 - \alpha$.

In the following we will define as σ_{MC} the standard error of the Monte Carlo estimator while that of Importance Sampling will be defined as σ_{IS}.

If we are interested in interval estimates at a certain confidence level we can use a normal approximation of the percentile estimator's distribution.

Importance Sampling (Glasserman and Li 2005; Lemieux 2009) is used in the Gaussian, Student-t and generalized hyperbolic framework. We consider a normal copula with normal marginals and a Student-t copula with Student-t and hyperbolic marginals. We introduce the generalized hyperbolic distribution as it describes well both fat tails and skewness.

We compared the results using a Gaussian copula with those obtained by sampling from a t-copula and hyperbolic copula. We do not consider Archimedean Gumbel copulas as goodness-of-fit-tests applied by Kole et al. (2007) demonstrate that a t-copula is preferred. As marginal distributions we used the Gaussian, the Student-t and the generalized hyperbolic.

3 The Algorithm and the Application

3.1 The Algorithm

Given the historical series of equities and indices the *Importance Sampling Credit Portfolio Model (ISCPM) Algorithm* comprises the following steps:

THE ISCPM Algorithm
Step 1. Evaluation of the correlation matrix between indices.
Step 2. Stress the correlation matrix using eigenvalue decomposition.
Step 3. Evaluation of the R square coefficients between indices and equities.
Step 4. Monte Carlo Simulation via Importance Sampling to obtain N one year issuers return scenarios (each scenario having n issuers) and N weights. Simulations of indices returns by using the stressed correlation matrix evaluated in step 2.
Step 5. Issuer rating calculation using the given transition matrices.
Step 6. Evaluation of P & L distribution.
Step 7. the VaR can be evaluated by using the Importance Sampling weights.

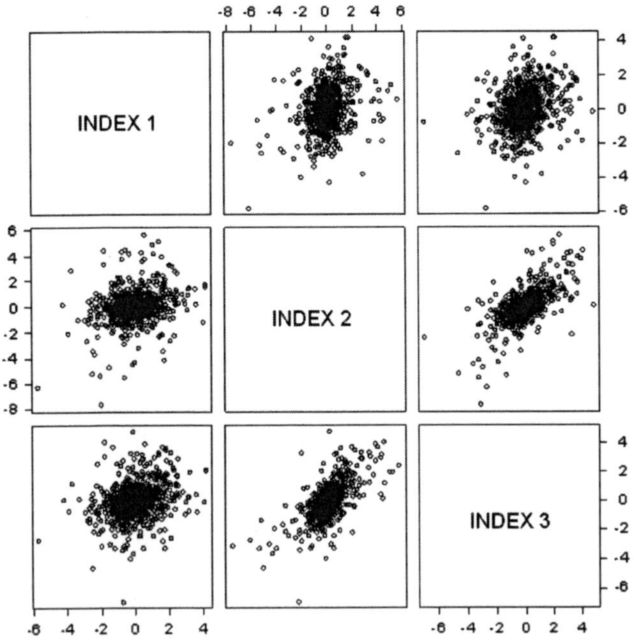

Fig. 1 Scatter plot of the indices

Remark 1. The *ISCPM* algorithm differs from the structure of a classical *Monte Carlo Credit Portfolio Model Algorithm* (*MCCPM*) in steps 4 and 7.

Remark 2. We used R language as our working environment and in particular some relevant R functions which have been implemented recently in order to work with the univariate Generalized Hyperbolic distribution and its special cases. A random generator from the *Rmetrics* package *f Basics* is used for generalized inverse Gaussian distributed random variates.

3.2 *The Application*

In order to test our *ISCPM* algorithm we consider a portfolio formed by three indices and 35 bonds with nominal value equal to one hundred dollars. Scatter plot of the indices returns are reported in Fig. 1.

The input data of the algorithm presented in the previous subsection are the prices of the bonds in all possible rating classes at the end of the temporal horizon, the matrix of correlation between the indices, the R-square of each issuer with the referred index, the bond's prices at the beginning and at the end (in all the possible rating classes) of the considered period, the number of simulations N,

Table 1 VaR estimators efficiency

N	η	$\frac{\sigma_{MC}^2}{\sigma_{IS}^2}$
50,000	2	9.67
50,000	2	3.92

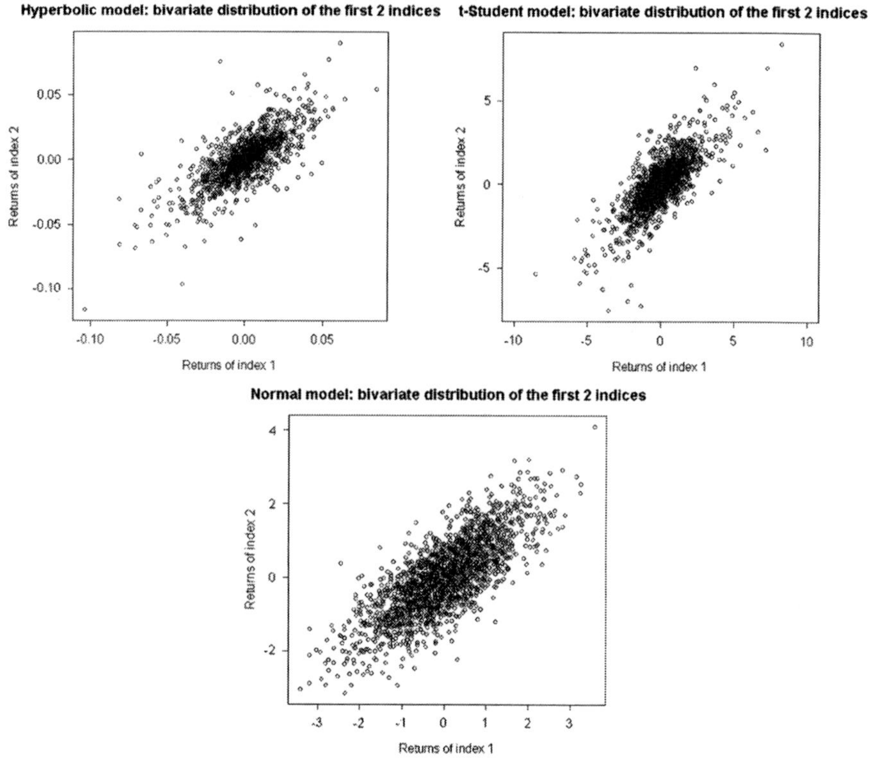

Fig. 2 Bivariate normal, Student-t and hyperbolic bivariate distributions

the confidence level (99.5%) and the transition matrix from which the threshold for the standardized asset returns are evaluated.

Both the *ISCPM* and the *MCCPM* algorithm have been implemented with the software *R*; the results show that the same accuracy is obtained with ten times fewer simulations using the Gaussian copula. Even if we use a t-copula the same accuracy is obtained with less simulations but, obviously, it depends on d (the degrees of freedom) in this paper we consider the proposal by Sak et al. (2007) and take $d = 5$ (Table 1), where η is a scale parameter.

At this point we considered the performance obtained by using Gaussian, Student-t and hyperbolic marginal distribution function, a graphical representation of which can be seen in Fig. 2

Table 2 Efficiency of VaR estimators with different marginal distributions

marginal distribution	99.5 estimate	$\frac{\sigma^2_{MC}}{\sigma^2_{IS}}$
normal	−7.97	5.48
Student-t	−8.96	4.96
hyperbolic	−12.16	3.02

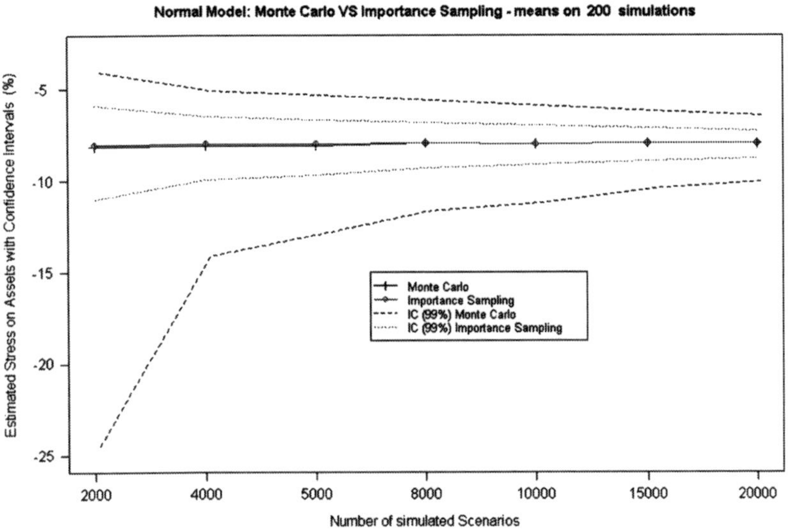

Fig. 3 Monte Carlo vs Importance Sampling Normal copula

As one can see from Table 2 generalized hyperbolic function approximate better than the Gaussian and the Student-t distributions the kernel density estimation of the distributions of the returns. The efficiency is better in the normal case, but a good improvement is obtained also for the other two models. We have to note that the normal VaR 99.5% is lower in absolute value than the VaR obtained from the other two models. This is due to the fact that the normal distribution does not present fat tails and therefore the risk factors are not well described. This fact may cause the underestimation of the portfolio's risk. The other two distributions describe better the risk leading to higher stress values.

We evaluate the VaR quantiles for this portfolio both with the traditional Monte Carlo Credit Portfolio Model algorithm (MCCPM) and the Importance Sampling Credit Portfolio Model algorithm (ISCPM) comparing the estimators' performances. To do this three copulas are considered: the Gaussian copula, the t-copula and the generalized hyperbolic copula with different marginals distributions (i.e. Student-t and hyperbolic).

In order to analyze the convergence of both estimators with increasing numbers of simulations we estimate the percentage VaR.

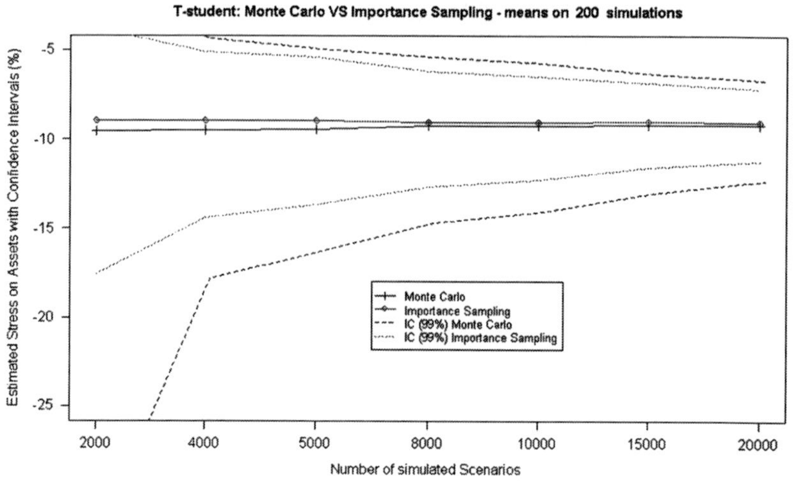

Fig. 4 Monte Carlo vs Importance Sampling Student-t copula

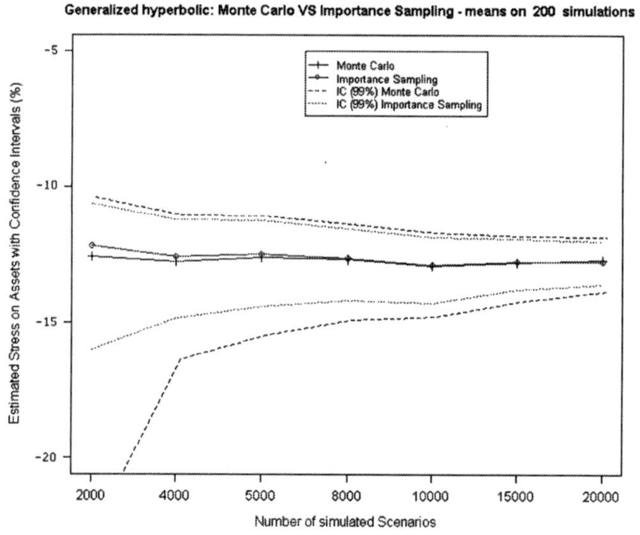

Fig. 5 MC vs Importance Sampling generalized hyperbolic copula

In Figs. 3, 4 and 5 the results are shown for the case of the Gaussian, Student-t and the hyperbolic copulas.

In order to check the behaviour of the model we performed a cross-validation analysis with a time frame that ranges from 01/01/2001 to 31/12/2007 instead that from the full (01/01/2001—07/20/2011)data set. This has been done to test the behaviour of the VaR before and after the 2008 crisis. The results are in line with our expectations: all the three VaRs are smaller before the crisis (see Table 2) than after (normal −6.98, Student-t −8.09, hyperbolic −13.64). This effect is due to the higher volatility and the increasing correlation levels in the more recent period. With higher level of correlation the (positive and) negative returns tend to occur at the same time, creating a higher number of defaults than in a non-stressed correlation period. This behaviour shows the sensitivity of the model with respect to the market situation: a feature which is highly desirable in some practical applications.

4 Conclusions

In this paper we considered the structural model approach first proposed by Merton (1974), the basic idea of which is that, over a single time horizon, the firm's asset returns, for a portfolio of corporate credits, are drawn from a multivariate Gaussian copula. The results are compared with those obtained when sampling from a t-copula (Kole et al. 2007).

In order to reduce the variability of the estimators we use Importance Sampling (Glasserman and Li 2005; Kole et al. 2007) for the Gaussian and Student-t with generalized hyperbolic marginal distributions.

We do not consider Archimedean Gumbel copulas as goodness-of-fit-tests applied by Kole et al. (2007) demonstrate that a t-copula is preferred. As marginal distributions we have considered the gaussian, the Student-t and the generalized hyperbolic. We introduce the generalized hyperbolic distribution as, in this context, it describes well both fat tails and skewness.

The present an application to a simulated portfolio of bonds. The results show the idea that the Student-t with generalized hyperbolic marginals model is more flexible than the classical gaussian copula. Moreover, by means of a Student-t copula, the risk can be described better because the tails are sampled more accurately.

We have successfully applied Importance Sampling to normal, Student-t and generalized hyperbolic copulas. Importance Sampling can reduce considerably the computational time needed to perform the calculations with Monte Carlo methods. We have seen that the efficiency for the t-copula and the hyperbolic copula is less than that in the gaussian framework. Despite this fact, Importance Sampling can reduce by half the computation time.

References

Glasserman, P., & Li, J. (2005). Importance sampling for portfolio credit risk. *Management Science*, *51*, 1463–1656.

Golub, G. H., & Van Loan, C. (1996). *Matrix computations* (3rd edn.). Baltimore: The John Hopkins University Press.

Hull, J. C. (2009). *Risk management and financial institutions* (2nd edn.). London: Prentice Hall.

Hull, J. C. (2006). *Options, futures and other derivatives* (6th edn.). London: Prentice Hall.

Kole, E., Koedijk, K., & Verbeek, M. (2007). Selecting copulas for risk management. *Journal of Banking & Finance, 51*, 2405–2423.
Lemieux, C. (2009). *Monte Carlo and Quasi-Monte Carlo sampling*. Heidelberg: Springer.
Merton, R. C. (1974). On the pricing of corporate debt: the risk structure of interest rates. *The Journal of Finance, 29*, 449–470.
Morgan, J. P. (1997). CreditMetricsTM, Technical Document.J.P. Morgan Co. Incorporated. Available http:///www.ma.hw.ac.uk/~mcneil/F79CR/CMTD1.pdf. Cited 16 Oct 2010.
Morokoff, J. W. (2004). An importance sampling method for portfolio credit risky assets. In R. G. Ingalls, M. D. Rossetti, J. S. Smith, & B. A. Peters (Eds.) *Proceedings of the 2004 Winter simulation conference* (pp. 1668–1676). Heidelberg: Springer. Available http://www.informs-org/wsc04papers/221.pdf. Cited 16 Oct 2010.
Sak, H., Hoermann, W., & Leyod, J. (2007). Efficient risk simulations for linear asset portfolios in the t-copula model. *European Journal of Operational Research, 202*, 802–809.
Sun, L., & Hong, L. J. (2010). Asymptotic representations for importance-sampling estimators of value-at-risk and conditional value-at-risk. *Operations Research Letters, 38*, 246–251.

A MCMC Approach for Learning the Structure of Gaussian Acyclic Directed Mixed Graphs

Ricardo Silva

Abstract Graphical models are widely used to encode conditional independence constraints and causal assumptions, the directed acyclic graph (DAG) being one of the most common families of models. However, DAGs are not closed under marginalization: that is, if a distribution is Markov with respect to a DAG, several of its marginals might not be representable with another DAG unless one discards some of the structural independencies. Acyclic directed mixed graphs (ADMGs) generalize DAGs so that closure under marginalization is possible. In a previous work, we showed how to perform Bayesian inference to infer the posterior distribution of the parameters of a given Gaussian ADMG model, where the graph is fixed. In this paper, we extend this procedure to allow for priors over graph structures.

1 Acyclic Directed Mixed Graph Models

Directed acyclic graphs (DAGs) provide a practical language to encode conditional independence constraints (see, e.g., Lauritzen 1996). However, such a family is not *closed under marginalization*. As an illustration of this concept, consider the following DAG:

$$Y_1 \rightarrow Y_2 \leftarrow X \rightarrow Y_3 \leftarrow Y_4$$

This model entails several conditional independencies. For instance, it encodes constraints such as $Y_2 \perp\!\!\!\perp Y_4$, as well as $Y_2 \not\perp\!\!\!\perp Y_4 \mid Y_3$ and $Y_2 \perp\!\!\!\perp Y_4 \mid \{Y_3, X\}$. Directed graphical models are non-monotonic independence models, in the sense

R. Silva (✉)
University College London, Gower Street, London WC1E 6BT, UK
e-mail: ricardo@stats.ucl.ac.uk

P. Giudici et al. (eds.), *Statistical Models for Data Analysis*, Studies in Classification, Data Analysis, and Knowledge Organization, DOI 10.1007/978-3-319-00032-9_39,

that conditioning on extra variables can destroy and re-create independencies, as the sequence $\{\emptyset, \{Y_3\}, \{Y_3, X\}\}$ has demonstrated.

If X is a latent variable which we are not interested in estimating, there might be no need to model explicitly its relationship to the observed variables $\{Y_1, Y_2, Y_3, Y_4\}$—a task which would require extra and perhaps undesirable assumptions.

However, marginalizing X results in a model that cannot be represented as a DAG structure without removing some of the known independence constraints. Since any constraint that conditions on X has to be dropped in the marginal for $\{Y_1, Y_2, Y_3, Y_4\}$ (for instance, $Y_2 \perp\!\!\!\perp Y_4 \mid \{Y_3, X\}$), we are forced to include extra edges in the DAG representation of the remaining variables. One possibility is $\{Y_1 \rightarrow Y_2 \leftarrow Y_3 \leftarrow Y_4, Y_4 \rightarrow Y_2\}$, where the extra edge $Y_4 \rightarrow Y_2$ is necessary to avoid constraints that we know should not hold, such as $Y_2 \perp\!\!\!\perp Y_4 \mid Y_3$. However, with that we lose the power to express known constraints such as $Y_2 \perp\!\!\!\perp Y_4$.

Acyclic directed mixed graphs (ADMGs) were introduced in order to provide independence models that result from marginalizing a DAG. ADMGs are *mixed* in the sense they contain more than one type of edge. In this case, *bi-directed* edges are also present. They are *acyclic* in the sense that there is no directed cycle composed of directed edges only. In principle, it is possible for two vertices to be linked by both a directed and a bi-directed edge. Moreover, let $sp(i)$ denote the "spouses" of Y_i in the graph (i.e., those Y_j such that $Y_i \leftrightarrow Y_j$ exists) and define $nsp(i)$ to be the non-spouses (Y_i is neither a spouse nor a non-spouse of itself).

In our example, the corresponding ADMG could be

$$Y_1 \rightarrow Y_2 \leftrightarrow Y_3 \leftarrow Y_4$$

Independences can be read off an ADMG using a criterion analogous to d-separation. More than one Markov equivalent ADMG can exist as the result of marginalizing a DAG (or marginalizing another ADMG). Moreover, other types of (non-independence) constraints can also result from an ADMG formulation if one allows two edges between two vertices. A detailed account of such independence models and a Gaussian parameterization are described at length by Richardson (2003), Richardson and Spirtes (2002). Generalizations are discussed by Sadeghi and Lauritzen (2012). An algorithm for maximum likelihood estimation in Gaussian ADMGs was introduced by Drton and Richardson (2004). A Bayesian method for estimating parameters of Gaussian ADMGs was introduced by Silva and Ghahramani (2009). In this paper, we extend Silva and Ghahramani (2009) by allowing the ADMG structure to be estimated from data, besides the parameters. Section 2 reviews the Bayesian formulation of the problem while Sect. 3 describes a sampler for inferring structure. A simple demonstration is given in Sect. 4.

2 A Review of the Gaussian Parametrization and Priors

Given a ADMG $\mathcal{G}$ and a p-dimensional distribution $\mathcal{P}$, each random variable in the distribution corresponds to a vertex in the graph. Let Y_i be a vertex with parents $Y_{i[1]}, Y_{i[2]}, \ldots, Y_{i[M_i]}$, $1 \leq i \leq p$, $1 \leq i[j] \leq p$, $1 \leq j \leq M_i$. We define a set of parameters $\{\lambda_{ij}\}$ according to the regression equation $Y_i = \sum_{j=1}^{M_i} \lambda_{ij} Y_{i[j]} + \epsilon_i$, where each error term ϵ_i is distributed as a zero mean Gaussian. Therefore, given the covariance matrix $\mathbf{V}$ of the error terms, we have a fully specified zero-mean Gaussian distribution. The parameterization of $\mathbf{V}$ is given by a sparse positive definite matrix: if there is no bi-directed edge $Y_i \leftrightarrow Y_j$, then we define $(\mathbf{V})_{ij} \equiv v_{ij} \equiv 0$. The remaining entries are free parameters within the space of (sparse) positive definite matrices. Priors for such models were described by Silva and Ghahramani (2009). Priors for each λ_{ij} are defined as independent zero-mean Gaussians, which in our experiments were given a prior variance of 3. The prior for $\mathbf{V}$ is given by

$$\pi_{\mathcal{G}}(\mathbf{V}) \propto |\mathbf{V}|^{-(\delta+2p)/2} \exp\left\{-\frac{1}{2} tr(\mathbf{V}^{-1}\mathbf{U})\right\} \tag{1}$$

for $\mathbf{V} \in M^+(\mathcal{G})$, the cone of positive definite matrices where $v_{ij} \equiv 0$ if there is no edge $Y_i \leftrightarrow Y_j$ in $\mathcal{G}$. This is called a *$\mathcal{G}$-inverse Wishart prior*. In general, there is no closed-form expression for the normalizing constant of this density function. Draws from the posterior distribution for parameters can be generated by a Gibbs sampler scheme as introduced by Silva and Ghahramani (2009).

3 A Sampler for Bi-directed Structures

In this section we consider the case where a given p-dimensional observed vector $\mathbf{Y}$ is generated according to a Gaussian ADMG model without directed edges. In this special case, the corresponding graph is called a *bi-directed graph*. That is, $\mathbf{Y}$ follows a zero-mean multivariate Gaussian with sparse covariance matrix $\mathbf{V}$. Conditional on a bi-directed graph $\mathcal{G}$, $\mathbf{V}$ is given the $\mathcal{G}$-inverse Wishart prior (1). For each pair of variables (Y_i, Y_j), $i < j$, we define a Bernoulli random variable z_{ij}, where $z_{ij} = 1$ if and only if there is an edge $Y_i \leftrightarrow Y_j$ in $\mathcal{G}$. Vector $\mathbf{z}$ denotes the vector obtained by stacking all z_{ij} variables. The prior for $\mathcal{G}$ is defined as the product of priors for each z_{ij}, $i < j$ (z_{ij} is also defined for $i > j$ and equal to z_{ji}), where $p(z_{ij} = 1) \equiv \eta_i \eta_j$, with each $\eta_i \sim Uniform(0, 1)$ a priori.

For a general ADMG with directed and bi-directed edges, directed edge coefficients can be sampled conditioned on $\mathbf{V}$ using a variety of off-the-shelf methods (e.g., using spike-and-slab priors corresponding to parameters associated with directed edges). Conditioned on edge coefficients $\{\lambda_{ij}\}$, the joint residual vector given by entries $Y_i - \sum_{j=1}^{M_i} \lambda_{ij} Y_{i[j]}$ follows a Gaussian bi-directed graph model, where sampling requires novel methods. Therefore, for simplicity of exposition, we describe only the sampler for the bi-directed structure. We present an (approximate) Gibbs sampler to generate posterior samples for $\mathbf{z}$ given a dataset $\mathscr{D} = \{\mathbf{Y}^{(1)}, \mathbf{Y}^{(2)}, \ldots, \mathbf{Y}^{(N)}\}$.

Let $\mathbf{V}_{\backslash i,\backslash i}$ the submatrix of $\mathbf{V}$ obtained by dropping its i-th column and row. Let $\mathbf{z}_{\backslash ij}$ be the set of edge indicator variables $\mathbf{z}$ without indicator z_{ij} (or z_{ji}). The sampler iterates over each vertex Y_i and performs the following:

1. For each $j \in \{1, 2, \ldots, p\} \backslash \{i\}$, we sample z_{ij} given $\mathbf{V}_{\backslash i,\backslash i}$ and $\mathbf{z}_{\backslash ij}$.
2. We sample the i-th row/column entries of $\mathbf{V}$, $\{v_{i1}, v_{i2}, \ldots, v_{ip}\}$ given $\mathbf{V}_{\backslash i,\backslash i}$ and $\mathbf{z}$.

The second step above is parameter sampling for sparse covariance matrices, described in detail by Silva and Ghahramani (2009). In the remainder of this section, we focus on the step of structure sampling. The conditional distribution for z_{ij} is given by

$$p(z_{ij} \mid \mathbf{V}_{\backslash i,\backslash i}, \mathbf{z}_{\backslash ij}, \mathscr{D}) \propto p(\mathscr{D} \mid \mathbf{V}_{\backslash i,\backslash i}, \mathbf{z}) \times p(\mathbf{V}_{\backslash i,\backslash i} \mid \mathbf{z}) \times p(z_{ij} \mid \mathbf{z}_{\backslash ij}) \quad (2)$$

One difficulty is introduced by the factor $p(\mathbf{V}_{\backslash i,\backslash i} \mid \mathbf{z})$, which is the marginal of a $\mathscr{G}$-inverse Wishart and where in general $p(\mathbf{V}_{\backslash i,\backslash i} \mid \mathbf{z}_{\backslash ij}, z_{ij} = 1) \neq p(\mathbf{V}_{\backslash i,\backslash i} \mid \mathbf{z}_{\backslash ij}, z_{ij} = 0)$. Computing this factor is expensive. However, in 1,000 preliminary runs with $p = 10$, $\delta = 1$ and $\mathbf{U}$ as the identity matrix, we found that errors introduced by the approximation

$$p(\mathbf{V}_{\backslash i,\backslash i} \mid \mathbf{z}_{\backslash ij}, z_{ij} = 1) \approx p(\mathbf{V}_{\backslash i,\backslash i} \mid \mathbf{z}_{\backslash ij}, z_{ij} = 0) \quad (3)$$

are minimal. No convergence problems for the Markov chains could be detected either. We adopt this approximation due to the large computational savings it brings, and as such the factor $p(\mathbf{V}_{\backslash i,\backslash i} \mid \mathbf{z})$ will be dropped without further consideration.

The first factor in (2) can be rewritten by completing and integrating away the remaining non-zero entries of $\mathbf{V}$, which we denote here by $\mathbf{V}_{i\cdot}$:

$$p(\mathscr{D} \mid \mathbf{V}_{\backslash i,\backslash i}, \mathbf{z}) = \int \prod_{d=1}^{N} p(\mathbf{Y}^{(d)} \mid \mathbf{V}) p(\mathbf{V}_{i\cdot} \mid \mathbf{V}_{\backslash i,\backslash i}, \mathbf{z}) \prod_{j \in E(\mathbf{z},i)} dv_{ij} \quad (4)$$

where $E(\mathbf{z}, i)$ is the set of indices for the spouses of Y_i in $\mathscr{G}$ (as given by $\mathbf{z}$), including Y_i itself. By definition, $v_{ij} = 0$ if Y_j is not a spouse of Y_i in $\mathscr{G}$.

In order to solve this integral, we appeal to some of the main results of Silva and Ghahramani (2009). Let $\mathscr{B}_i$ be a $1 \times (p-1)$ vector and γ_i a positive scalar such that

$$\mathbf{V}_{i,\backslash i} = \mathscr{B}_i \mathbf{V}_{\backslash i,\backslash i}, \quad v_{ii} = \gamma_i + \mathscr{B}_i \mathbf{V}_{\backslash i,\backslash i} \mathscr{B}_i^T \quad (5)$$

where $\mathbf{V}_{i,\backslash i}$ is the i-th row of $\mathbf{V}$ after removing entry v_{ii}. We define $\mathcal{B}_{sp(i)}$ and $\mathcal{B}_{nsp(i)}$ to be the subvectors of $\mathcal{B}_i$ that match the corresponding rows of $\mathbf{V}_{\backslash i,\backslash i}$. The "non-spouse" entries are not free parameters when considering the structural zeroes of $\mathbf{V}$.

By our definition of $\mathcal{B}_i$, we have that $\mathcal{B}_i \mathbf{V}_{\backslash i,nsp(i)}$ gives the covariance between Y_i and its non-spouses (where $\mathbf{V}_{\backslash i,nsp(i)}$ is the corresponding submatrix of $\mathbf{V}$). By assumption these covariances are zero, that is $\mathcal{B}_i \mathbf{V}_{\backslash i,nsp(i)} = 0$. It follows that

$$\mathcal{B}_{sp(i)} \mathbf{V}_{sp(i),nsp(i)} + \mathcal{B}_{nsp(i)} \mathbf{V}_{nsp(i),nsp(i)} = 0 \Rightarrow \mathcal{B}_{nsp(i)} = -\mathcal{B}_{sp(i)} \mathbf{V}_{sp(i),nsp(i)} \mathbf{V}^{-1}_{nsp(i),nsp(i)} \tag{6}$$

As in the unconstrained case, the mapping between the non-zero entries of $\mathbf{V}_{i\cdot}$ and $\{\mathcal{B}_{sp(i)}, \gamma_i\}$ is one-to-one. Following Silva and Ghahramani (2009), conditional density $p(\mathbf{V}_{i\cdot} \mid \mathbf{V}_{\backslash i,\backslash i}\mathbf{z})$ can be rewritten as:

$$p(\mathbf{V}_{i\cdot} \mid \mathbf{V}_{\backslash i,\backslash i}, \mathbf{z}) = p_{\mathcal{B}}(\mathcal{B}_{sp(i)} \mid \gamma_i, \mathbf{V}_{\backslash i,\backslash i}; \mu_{\mathcal{B}}, \gamma_i \mathbf{K}_{\mathcal{B}}) p_{\gamma_i}(\gamma_i \mid \mathbf{V}_{\backslash i,\backslash i}; \alpha_i, \beta_i) \tag{7}$$

where $p_{\mathcal{B}}(\cdot; \mu_{\mathcal{B}}, \gamma_i \mathbf{K}_{\mathcal{B}})$ is a Gaussian density function with mean $\mu_{\mathcal{B}}$ and covariance matrix $\gamma_i \mathbf{K}_{\mathcal{B}}$, and function $p_{\gamma_i}(\cdot; \alpha_i, \beta_i)$ is an inverse gamma density function with parameters α_i and β_i. Parameters $\{\mu_{\mathcal{B}}, \mathbf{K}_{\mathcal{B}}, \alpha_i, \beta_i\}$ are described in the appendix. Moreover, as a function of $\{z_{ij}, \mathcal{B}_{sp(i)}, \gamma_i, \mathbf{V}_{\backslash i,\backslash i}\}$, we can rewrite $p(\mathbf{Y}^{(d)} \mid \mathbf{V})$ as:

$$p(\mathbf{Y}^{(d)} \mid \mathbf{V}) \propto \gamma_i^{-1/2} \exp\left\{-\frac{1}{2\gamma_i}\left(Y_i^{(d)} - \mathcal{B}_{sp(i)} \mathbf{H}^{(d)}_{sp(i)}\right)^2\right\} \tag{8}$$

where $\mathbf{H}^{(d)}_{sp(i)}$ are the residuals of the regression of the spouses of Y_i on its non-spouses for datapoint d, as given by $\mathbf{V}_{\backslash i,\backslash i}$. That is

$$\mathbf{H}^{(d)}_{sp(i)} \equiv \mathbf{Y}^{(d)}_{sp(i)} - \mathbf{V}_{sp(i),nsp(i)} \mathbf{V}^{-1}_{nsp(i),nsp(i)} \mathbf{Y}^{(d)}_{nsp(i)} \tag{9}$$

Combining (7) and (8) allows us to rewrite (4) as

$$p(\mathcal{D} \mid \mathbf{V}_{\backslash i,\backslash i}, \mathbf{z}) \propto |\mathbf{K}_{\mathcal{B}}|^{-\frac{1}{2}} |\mathbf{T}|^{-\frac{1}{2}} \frac{\beta_i^{\alpha_i}}{\Gamma(\alpha_i)} \frac{\Gamma(\alpha_i')}{\beta_i'^{\alpha_i'}} \tag{10}$$

where $\{\mathbf{T}, \alpha_i', \beta_i'\}$ are defined in the Appendix. Every term above depends on the value of z_{ij}. Finally, $p(z_{ij} = 1 \mid \mathbf{V}_{\backslash i,\backslash i}, \mathbf{z}_{\backslash ij}, \mathcal{D}) \propto p(\mathcal{D} \mid \mathbf{V}_{\backslash i,\backslash i}, \mathbf{z}_{\backslash ij}, z_{ij} = 1)\eta_i \eta_j$ and $p(z_{ij} = 0 \mid \mathbf{V}_{\backslash i,\backslash i}, \mathbf{z}_{\backslash ij}, \mathcal{D}) \propto p(\mathcal{D} \mid \mathbf{V}_{\backslash i,\backslash i}, \mathbf{z}_{\backslash ij}, z_{ij} = 0)(1 - \eta_i \eta_j)$.

After resampling z_{ij} for all $1 \leq j \leq p, j \neq i$, we resample the corresponding non-zero covariances, as described at the beginning of this section, and iterate, alternating with steps to sample latent variables, regression coefficients and hyperparameters η_i as necessary.

4 Illustration: Learning Measurement Error Structure

One application of the methodology is learning the structure of measurement error for a latent variable model. Consider, for illustration purposes, the DAG in Fig. 1. Assume the goal is to learn from observed measurements $Y_1, Y_2, \ldots, Y_8$ what values the corresponding latent variables X_1 and X_2 should take (more precisely, to calculate functionals of the conditional distribution of $\{X_1, X_2\}$ given $\mathbf{Y}$). Other sources of variability explain the marginal distribution of $\mathbf{Y}$, but they are not of interest. In this example, X_3 and X_4 are the spurious sources. Not including them in the model introduces bias. Sometimes background knowledge is useful to provide which observed variable measures which target latent variable (e.g., Y_1 should be a child of X_1 but not of X_2). The literature in structural equation models and factor analysis (Bollen 1989; Bartholomew et al. 2011) provides some examples where observed variables are designed so that latent concepts of interest are measured (up to some measurement error). Background knowledge about other hidden common causes of the observed variables is less clear, though.

In this section, we provide a simple illustration on how to combine background knowledge about measurement with an adaptive methods that generates extra conditional dependencies among observed variables. Consider a more complex synthetic model given by a latent variable ADMG with four latent variables and 12 observed variables. Each observed variable has a single latent parent: the first three have X_1 as a common parent, the next three have X_2, and so on. The covariance matrix of the latent variables was sampled from an inverse Wishart distribution. Bi-directed edges among indicators were generated randomly with probability 0.2. To ensure identifiability, we pick 2 out the each 3 children of each latent variable and enforce that no bi-directed edges should exist within this set of 8 indicators. More flexible combinations can be enforced in the future using the results of Grzebyk et al. (2004).

The goal is: given knowledge about which directed edges exist and do not exist, learn the bi-directed structure. The algorithm in the previous section is used to sample error covariance matrices among observed variables (non-zero error covariances between a latent variable and an observed variable are prohibited for simplicity, and the covariance matrix among latent variables has no independence constraints). This is done as part of a Gibbs sampling procedure where the values of the latent variables are also sampled so that the procedure in Sect. 3 can be used without modification as if all variables were observed.

Figure 2 summarizes the analysis of the error covariance matrix and its corresponding bi-directed structure using a sample size of 2000 (and a Markov chain with 5,000 iterations). A bi-directed structure estimate is generated using the posterior samples. In this case, instead of using the most common structure as the estimator, we use a thresholding mechanism. Edges $Y_i \leftrightarrow Y_j$ such that the posterior expected value of the corresponding z_{ij} is greater than 0.5 are kept, while the others are estimated to be non-existent. A thresholding estimator for the structure is a practical alternative to choosing the most probable graph: a difficult task for Markov

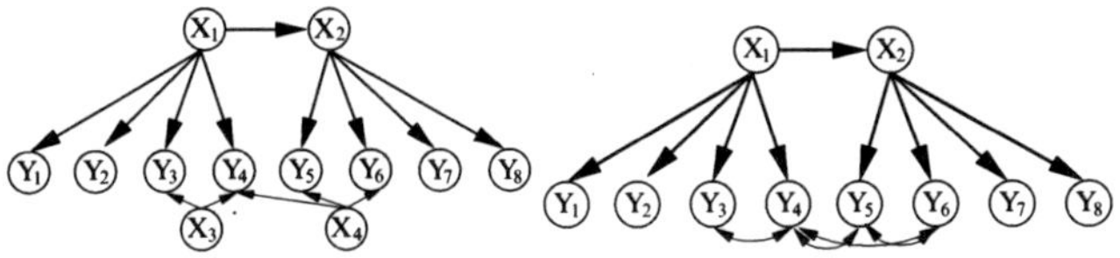

Fig. 1 In the *left*, a model where latent variables X_3 and X_4 provide an extra source of dependence for some of the observed variables that is not accounted by the target latent variables X_1 and X_2. In the *right*, a graphical representation of the marginal dependencies after marginalizing away some (X_3 and X_4) but not all of the latent variables

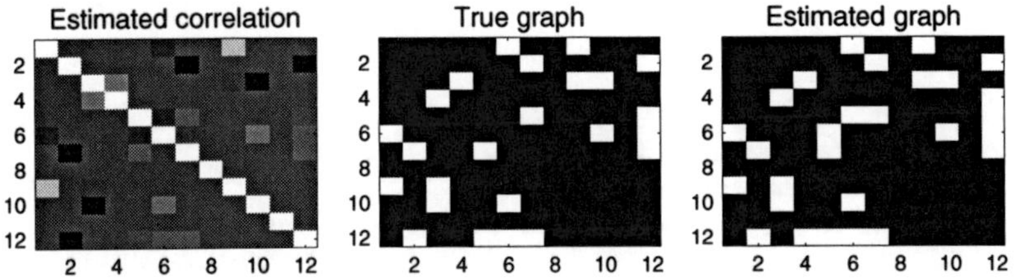

Fig. 2 In the *left*, the estimated error correlation matrix as given by the expected value of the marginal (hence, not sparse) posterior distribution of the rescaled error covariance $\mathbf{V}$. *Black dots* mean correlation of -1, *white dots* mean correlation of 1. In the *right*, the estimator of the structure (edge appears if its posterior probability is greater than 0.5). The procedure added two spurious edges, but the corresponding estimated correlations are still close to zero

chain Monte Carlo in discrete structures. An analysis of thresholding mechanisms is provided in other contexts by Barbieri and Berger (2004) and Carvalho and Polson (2010). However, since the estimated graph might not have occurred at any point during sampling, further parameter sampling conditioned on this graph will be necessary in order to obtain as estimator for the covariance matrix with structural zeroes matching the missing edges.

We also found that the choice of prior $p(z_{ij} = 1) \equiv \eta_i \eta_j$ to be particularly important. An alternative prior $p(z_{ij} = 1) = 0.5$ resulted in graphs with considerably more edges than the true one. A more extended discussion on how to enforce sparsity by priors over graphical structures is presented by Jones et al. (2005). An important line of future work will consist on designing and evaluating priors for mixed graph structures.

Appendix

We describe the parameters referred to in the sampler of Sect. 3. The full derivation is based on previous results described by Silva and Ghahramani (2009). Let **HH** be the statistic $\sum_{n=1}^{d} \mathbf{H}_{sp(i)}^{(d)} \mathbf{H}_{sp(i)}^{(d)\,T}$. Likewise, let $\mathbf{YH} \equiv \sum_{n=1}^{d} Y_i^{(d)} \mathbf{H}_{sp(i)}^{(d)}$ and $\mathbf{YY} \equiv \sum_{n=1}^{d} Y_i^{(d)2}$. Recall that the hyperparameters for the $\mathcal{G}$-inverse Wishart are δ and

U, as given by Eq. (1) and as such we are computing a "conditional normalizing constant" for the posterior of **V** integrating over only *one* of the row/columns of **V**.

First, let

$$\begin{aligned}\mathbf{A}_i &\equiv \mathbf{V}_{sp(i),nsp(i)}\mathbf{V}^{-1}_{nsp(i),nsp(i)}\\ \mathbf{M}_i &\equiv (\mathbf{U}_{\backslash i,\backslash i})^{-1}\mathbf{U}_{\backslash i,i}\\ \mathbf{m}_i &\equiv (\mathbf{U}_{ss} - \mathbf{A}_i\mathbf{U}_{ns})\mathbf{M}_{sp(i)} + (\mathbf{U}_{sn} - \mathbf{A}_i\mathbf{U}_{nn})\mathbf{M}_{nsp(i)}\\ \\ \mathbf{K}^{-1}_{\mathscr{B}} &\equiv \mathbf{U}_{ss} - \mathbf{A}_i\mathbf{U}_{ns} - \mathbf{U}_{sn}\mathbf{A}_i^T + \mathbf{A}_i\mathbf{U}_{nn}\mathbf{A}_i^T\\ \mu_{\mathscr{B}} &\equiv \mathbf{K}_{\mathscr{B}}\mathbf{m}_i\end{aligned} \tag{11}$$

where

$$\begin{bmatrix}\mathbf{U}_{ss} & \mathbf{U}_{sn}\\ \mathbf{U}_{ns} & \mathbf{U}_{nn}\end{bmatrix} \equiv \begin{bmatrix}\mathbf{U}_{sp(i),sp(i)} & \mathbf{U}_{sp(i),nsp(i)}\\ \mathbf{U}_{nsp(i),sp(i)} & \mathbf{U}_{nsp(i),nsp(i)}\end{bmatrix} \tag{12}$$

Moreover, let

$$\begin{aligned}\mathscr{U}_i &\equiv \mathbf{M}_i^T\mathbf{U}_{\backslash i,\backslash i}\mathbf{M}_i - \mathbf{m}_i^T\mathbf{K}_i\mathbf{m}_i\\ u_{ii.\backslash i} &\equiv \mathbf{U}_{ii} - \mathbf{U}_{i,\backslash i}(\mathbf{U}_{\backslash i,\backslash i})^{-1}\mathbf{U}_{\backslash i,i}\\ \\ \alpha_i &\equiv (\delta + p - 1 + \#nsp(i))\,/2\\ \beta_i &\equiv \left(u_{ii.\backslash i} + \mathscr{U}_i\right)/2\\ \mathbf{T} &\equiv \mathbf{K}^{-1}_{\mathscr{B}} + \mathbf{HH}\\ \mathbf{q} &\equiv \mathbf{YH} + \mathbf{K}^{-1}_{\mathscr{B}}\mu_{\mathscr{B}}\end{aligned} \tag{13}$$

where $\#nsp(i)$ is the number of non-spouses of Y_i (i.e., $(p-1) - \sum_{j=1}^{p} z_{ij}$).

Finally,

$$\begin{aligned}\alpha_i' &\equiv \frac{N}{2} + \alpha_i,\\ \beta_i' &\equiv \frac{\mathbf{YY} + \mu_{\mathscr{B}}^T\mathbf{K}^{-1}_{\mathscr{B}}\mu_{\mathscr{B}} - \mathbf{q}^T\mathbf{T}^{-1}\mathbf{q}}{2} + \beta_i\end{aligned} \tag{14}$$

Notice that each calculation of $\mathbf{A}_i$ (and related products) takes $\mathscr{O}(p^3)$ steps (assuming the number of non-spouses is $\mathscr{O}(p)$ and the number of spouses is $\mathscr{O}(1)$, which will be the case in sparse graphs). For each vertex Y_i, an iteration could take $\mathscr{O}(p^4)$ steps, and a full sweep would take prohibitive $\mathscr{O}(p^5)$ steps. In order to scale this procedure up, some tricks can be employed. For instance, when iterating over each candidate spouse for a fixed Y_i, the number of spouses increases or decreases by 1: this means fast matrix update schemes can be implemented to obtain a new $\mathbf{A}_i$ from its current value. However, even in this case the cost would still be $\mathscr{O}(p^4)$. More speed-ups follow from solving for $\mathbf{V}_{sp(i),nsp(i)}\mathbf{V}^{-1}_{nsp(i),nsp(i)}$ using sparse matrix representations, which should cost less than $\mathscr{O}(p^3)$ (but for small to moderate p, sparse matrix inversion might be slower than dense matrix inversion). Moreover, one might not try to evaluate all pairs $Y_i \leftrightarrow Y_j$ if some pre-screening is done by looking only at pairs where the magnitude of corresponding correlation sampled in the last step lies within some interval.

References

Barbieri, M. M., & Berger, J. (2004). Optimal predictive model selection. *The Annals of Statistics, 32*, 870–897.

Bartholomew, D., Knott, M. & Moustaki, I. (2011). *Latent variable models and factor analysis: a unified approach*. Wiley Series in Probability and Statistics.

Bollen, K. (1989). *Structural equations with latent variables*. New York: Wiley.

Carvalho, C., & Polson, N. (2010). The horseshoe estimator for sparse signals. *Biometrika, 97*, 465–480.

Drton, M., & Richardson, T. (2004). Iterative conditional fitting for Gaussian ancestral graph models. In *Proceedings of the 20th conference on uncertainty in artificial intelligence*, (pp. 130–137). AUAI Press, Arlington, Virginia.

Grzebyk, M., Wild, P., & Chouaniere, D. (2004). On identification of multi-factor models with correlated residuals. *Biometrika, 91*, 141–151.

Jones, B., Carvalho, C., Dobra, A., Hans, C., Carter, C., & West, M. (2005). Experiments in stochastic computation for high-dimensional graphical models. *Statistical Science, 20*, 388–400.

Lauritzen, S. (1996). *Graphical models*. Oxford: Oxford University Press.

Richardson, T. (2003). Markov properties for acyclic directed mixed graphs. *Scandinavian Journal of Statistics, 30*, 145–157.

Richardson, T., & Spirtes, P. (2002). Ancestral graph Markov models. *Annals of Statistics, 30*, 962–1030.

Sadeghi, K., & Lauritzen, S. (2012). Markov properties for mixed graphs. 247 arXiv:1109.5909v4.

Silva, R., & Ghahramani, Z. (2009). The hidden life of latent variables: Bayesian learning with mixed graph models. *Journal of Machine Learning Research, 10*, 1187–1238.

Symbolic Cluster Representations for SVM in Credit Client Classification Tasks

Ralf Stecking and Klaus B. Schebesch

Abstract Credit client scoring on medium sized data sets can be accomplished by means of Support Vector Machines (SVM), a powerful and robust machine learning method. However, real life credit client data sets are usually huge, containing up to hundred thousands of records, with good credit clients vastly outnumbering the defaulting ones. Such data pose severe computational barriers for SVM and other kernel methods, especially if all pairwise data point similarities are requested. Hence, methods which avoid extensive training on the complete data are in high demand. A possible solution may be a combined cluster and classification approach. Computationally efficient clustering can compress information from the large data set in a robust way, especially in conjunction with a symbolic cluster representation. Credit client data clustered with this procedure will be used in order to estimate classification models.

1 Introduction

Clustering the training data to deal with large and often imbalanced data has been examined several times. Appropriate cluster procedures include K-means (Japkowicz 2002), Kernel K-means (Yuan et al. 2006), Micro Cluster Trees (Wang et al. 2008; Yu et al. 2003), Rocchio Score based clustering (Shih et al. 2003) and constrained clustering (Evgeniou and Pontil 2002). The large original data set then usually is replaced by cluster representatives and given to a training algorithm.

R. Stecking (✉)
Department of Economics, Carl von Ossietzky University Oldenburg, D-26111 Oldenburg, Germany
e-mail: ralf.w.stecking@uni-oldenburg.de

K.B. Schebesch
Faculty of Economics, Vasile Goldiş Western University Arad, Romania
e-mail: kbschebesch@uvvg.ro

P. Giudici et al. (eds.), *Statistical Models for Data Analysis*, Studies in Classification, Data Analysis, and Knowledge Organization, DOI 10.1007/978-3-319-00032-9_40,

Different aspects of combining clustering with SVM are treated in the more recent literature: Li et al. (2007) introduce a Support Cluster Machine (SCM) as a general extension of the SVM with the radial basis function (RBF) kernel, where cluster size and cluster covariance information is incorporated into the kernel function. Evgeniou and Pontil (2002) propose a special clustering algorithm for binary class data, that tends to produce large clusters of training examples which are far away from the boundary between the two classes and small clusters near the boundary. Yuan et al. (2006) concentrate on large *imbalanced* data sets. They partition the examples from the bigger negative class into disjoint clusters and train an initial SVM with RBF kernel using positive examples and cluster representatives of the negative examples. Yu et al. (2003) construct two micro-cluster trees from positive and negative training examples respectively and train a linear SVM on the centroids of the root nodes. Subsequently, training examples near the class boundary are identified and split up into finer levels using the tree structure. Wang et al. (2008) generalize this approach for solving nonlinear classification problems. The main results of these former approaches are (*i*) combining clustering with SVM can be used for very large data sets with up to millions of training examples, (*ii*) they lead to reduced training time and (*iii*) the classification results for all these techniques are *comparable* but usually *slightly worse* than for models that, if possible, are trained on the full data set.

While various *clustering procedures* have been examined in the past, much less is known about the right *cluster representations*. The most common way is to use cluster centers, a practice that will become inappropriate as soon as categorical variables are concerned. Moreover, mean values in general seem to be too limited to adequately express characteristics of quantitative variables with special distributions.

The outline of this paper is as follows: in Sect. 2 we describe our credit client data set, in Sect. 3 the symbolic representation of the data clusters. Section 4 details the classification method we use. In Sect. 5 we discuss the experimental setup including model validation, followed by a presentation of the results of our work in Sect. 6. Finally, we present our conclusions in Sect. 7.

2 Credit Client Data Set Description

The data set used consists of 139,951 clients for a building and loan credit. There are twelve variables per client of which eight are categorical (with two up to five categories) and four are quantitative. Due to binary coding of categorical variables each single credit client is represented by a 25-dimensional input pattern. Input variables include personal attributes like *client's profession*, loan related attributes like *repayment rate* and object related attributes like *house type*. The default rate within one year is 2.6% (3,692 out of 139,951). Imbalanced data sets usually call for some special treatment like under- or oversampling, one class learning, cost-sensitive learning, weighting, boosting and many more (Weiss 2004). In this work, an alternative way of dealing with large and imbalanced data sets will be shown: First, the data set is divided into "good" and "bad" credit clients. Subsequently, unsupervised k-means clustering (MacQueen 1967) is used, partitioning the large data set into equal numbers of clusters from "good" and "bad" classes respectively, while preserving the class labels. The advantage of a preprocessing type cluster analysis is massive down-sizing of the large data set of N cases into a much smaller set of $n \ll N$ clusters. These cluster solutions may include a weighting scheme, e.g. using a balanced number of "good" and "bad" credit client clusters.

Information from *data clusters* are organized in the following way: For each of the labeled clusters the relative frequencies of all outcomes per *categorical variable* are recorded. *Quantitative variables* are divided into four equally sized intervals, with quartiles of the full variable range as interval borders. Relative frequencies for these intervals are also recorded. A data cluster finally is represented by 43 inputs between zero and one.

3 Symbolic Representation of Data Clusters

In contrast to "classical" data analysis where elements are individuals with single variable outcomes (*"first degree objects"*), in symbolic data analysis elements, respectively *objects* in general, are *classes of individuals* or *categories of variables* that usually will be described by more than one outcome per variable (*"second degree objects"*) (Bock and Diday 2000). Clusters of credit clients can be seen as such symbolic objects with e.g. an interval representation of the amount of credit that was given to the cluster members. However, a special data description is needed to represent the variable outcomes of a symbolic object. A complete overview of symbolic variable types can be found in Billard and Diday (2006). In the present work the clusters (the symbolic objects) are described by *modal variables* where categories or intervals (of categorical or quantitative) variables appear with a given probability. Each symbolic variable is represented by a vector of its outcomes and the respective probabilities, e.g. $X_{li} = \{\eta_{i1}, p_{i1}; \ldots; \eta_{is}, p_{is}\}$ where outcome η_{ik} (of cluster i relating to variable X_l with $k = 1, \ldots, s$ outcomes per variable and $l = 1, \ldots, m$ as the number of variables) is given with a probability p_{ik} and $\sum_{k=1}^{s} p_{ik} = 1$. Outcomes η_{ik} may either be categories or (non-overlapping) intervals. Therefore, categorical as well as quantitative variables are represented by a vector of *standardized* values between zero and one.

Distances between two clusters, that are coded as symbolic objects, are computed over all $l = 1, \ldots, m$ symbolic variables. The *squared Euclidian distance* between two cluster u and v then reads (Billard and Diday 2006)

$$d_{uv}^2 = \frac{1}{m} \sum_{l=1}^{m} \sum_{k=1}^{s} w_l \frac{(p_{ukl} - p_{vkl})^2}{\sum_{i=1}^{n} p_{ikl}}$$

with m as the number of variables, s as the number of categories and n as the number of clusters. Within distance d_{uv}^2 the *squared sums of deviations* are weighted by the *sums of probabilities* per category. The distance function may also include a weighting scheme $\{w_l > 0\}$ for the set of $l = 1, \ldots, m$ variables.

4 Classification Methods Used

The first step in establishing our classification method is to partition the large data set of N credit clients into a much smaller set of n clusters. We use clustering in order to find a shorter (compressed) description of the original training set and not in order to discover self contained clusters structures in the data, hence we do not pursue the standard goal of unsupervised clustering. For simplicity, we also stick to non-overlapping clusterings. A clustering procedure, which solves such a problem in the sense of grouping similar points into non-overlapping clusters and which does not require the computation of all mutual client distances is the *k-means* algorithm (MacQueen 1967), readily available in many popular statistical computer packages like for instance R-Cran and SPSS.

Subsequently, we are given a set of $n \ll N$ training examples $\{X_i, y_i\}$, $i = 1, \ldots, n$, with $X_i = \{X_{1i}, \ldots, X_{mi}\}$ as the symbolic description of the m input variables for each cluster i and y_i as the associated labels $y_i \in \{-1, 1\}$ for "good" and "bad" credit client cluster, respectively. A classification model $s(x)$ now predicts the label y of a new credit applicant described by feature vector x. Note that $s(x)$ is a separating function for both credit client classes and that the forecasting model is a binary classification model. Support Vector Machines (SVM) are such binary classification models, which produce a forecasting rule, i.e. a separating function of the type

$$y^{pred} = \mathbf{sign}(s(x)) = \mathbf{sign}\Big(\sum_{i=1}^{n} y_i \alpha_i^* k(X_i, x) + b^*\Big),$$

with parameters $0 \leq \alpha_i^* \leq C$ and b^* the result of the SVM optimization. Here X_i refers to the symbolic description of cluster i, while x is the feature vector of a new client with as yet unknown y. Support vectors are those training examples i from $\{1, \ldots, n\}$ for which $\alpha_i^* > 0$. They are located near the class boundaries. For given kernel type and hyperparameters support vectors generate the optimal SVM separating function between classes. The first user selected hyperparameter $C > 0$ controls the amount of misclassification (or "softness") of the SVM model, i.e. C is an upper bound for the dual variables $\alpha_i^* > 0$ of the SVM optimization problem. Hence, C implicitly selects the effective functional form of $s(x)$. Note that $\alpha_i^* > 0$ actively contribute to the separating function by invoking the ith training example via a user defined kernel $k(.,.)$, which in most cases is selected to be a semi-positive, symmetric and monotonous function of the distance between pairs of clients (Schölkopf and Smola 2002). A very general nonlinear kernel function to be used in later sections is the RBF kernel defined by $k(u, v) = \exp\left(-g \cdot d_{uv}^2\right)$, with d_{uv}^2 being the (e.g. Euclidean) symbolic distance between clusters u and v of Sect. 3, and with $g = \frac{1}{\sigma^2} > 0$ being the second hyperparameter of the SVM model.

5 Experimental Setup

Evaluating the comparative expected classification performance of models trained on cluster representations and to ease their comparison with "conventional" models trained on the full data, we have to use a procedure, which is statistically safe and which also enables a fair comparison of the models from structurally different model classes. Therefore, we use a tenfold out of sample validation procedure. To this end we randomly subdivide the data into $\frac{2}{3}N$ examples used for training while the remaining $\frac{1}{3}N$ examples are held out for validation. This is repeated ten times generating ten different training-validation sets. The expected classification performance of a model is compared on these respective ten validation sets, returning thus the average, the best and the worst measured classification performance computed for that model and for a chosen error function.

In order to further describe the experimental setup, the model classes have to be considered separately. For overall model estimation, we can (*i*) train a classification model directly on the full training data (One-stage) or (*ii*) use a clustering procedure on the full training data set and subsequently train a classification model on the cluster centers (Two-stages). For our large credit client data set we narrow down the choice of alternative clustering procedures to that of k-means clustering. Such algorithms do not require the beforehand computation of the complete distance matrix between pairs of training points and are therefore efficient even for very large data sets. Our two-stage experiments then proceed as follows:

Repeat the steps 1–6 ten times:

1. Divide credit client data set randomly into *training* ($N_T = 93,301$) and *validation* ($N_V = 46,650$) set with $N_T : N_V = 2 : 1$.
2. Split training data set into "good" and "bad" credit clients.
3. K-means cluster analysis: extract $\frac{n}{2}$ clusters from "good" and "bad" classes respectively and preserve labels.
4. Train SVM with small data set of n *symbolic cluster descriptions.*
5. Use SVM classification function to predict credit client default on the validation set.
6. Calculate ROC curve and area under ROC curve *(AUC)*. Return to step 1.

For benchmark purposes we also train a *linear SVM* on the *full data set*, following the above procedure but omitting steps 2 and 3 (One-stage model). Subsequently, the linear SVM is trained on the whole set of $N_T = 93,301$ data points, respectively.

6 Classification Results

Applying the validation procedure outlined in Sect. 5, equal numbers of clusters are chosen to under-sample the much bigger class of non defaulting credit clients as one possible solution to imbalanced class problems (Weiss 2004). The number of clusters necessary to best represent the respective training sets must be determined experimentally: Too few large clusters may not include important characteristics of the data source, whereas too many small clusters may inadequately highlight noise in the data. Therefore, an ascending list of cluster numbers is tested, starting with $n = 50$ up to $n = 1,000$ clusters. Each cluster is represented by a symbolic description of modal variables and then given as input to SVM with RBF kernel. For this type of SVM at first the width parameter σ of the kernel function $k(u, v) = \exp(-g \cdot d_{uv}^2)$, where $g = \frac{1}{\sigma^2}$, is initialized to the mean Euclidian distance of pairwise comparisons of all input examples. Grid search optimization of width parameter σ and model capacity control parameter C, which is the bound of the dual variables α of the SVM (used in Sect. 4), then leads to values of $\sigma = 2.58$ and $C = 15$.

In order to predict the behavior of unknown examples each credit client of the validation set is assigned to a category or, in case of quantitative variables, an interval with a probability of either zero or one. In this way data descriptions for individuals match with the format of those of the data clusters defined in Sect. 3. Consequently, future credit client behavior can be predicted by a classification function that was previously estimated on a training set containing cluster representatives.

The extremely unequal class sizes of our validation sets make standard measures of accuracy, like e.g. the misclassification rate, impractical. They must be replaced by a more complex measure. A good candidate is the receiver operating characteristics *(ROC)* curve (Hanley and McNeil 1982) which offsets true and false positive rates of a classification model using model predictions against reference output (i.e. the true labels of the data set). The area under curve *(AUC)* then denotes the area between the abscissa and the *ROC* curve as a measure for the cut-off independent classification power of a decision function. Note, that the *AUC* for "pure chance" models i.e. models with no predictive power, is 0.5.

Table 1 shows the mean, standard deviation, minimum and maximum of the *AUC* computed for our ten randomly selected validation sets. The smallest models with just 50 training examples lead to an average *AUC* of 0.692 over the validation sets. We observe the highest classification accuracy for SVM models that were trained on 100, 200 and 400 cluster representations. Hereafter, the mean *AUC* decreases slowly. Furthermore, a linear SVM could be trained on the full training sets with 93,301 observations each. The mean *AUC* for this benchmark model is 0.702, which is slightly, but not significantly, better than the cluster based results we obtained from the reduced training sets.

An open question for cluster based approaches is, which number of clusters to choose as model input. Experimentally, we observe a connection between these numbers of clusters and C, which is the upper bound of the dual variables α, that controls for the accuracy of the SVM model. Figure 1 shows, that smaller numbers

Table 1 Area under curve (AUC) statistics computed for ten randomly selected validation sets with $N = 46,650$ each. SVM Models with RBF kernel are trained on 50–1,000 cluster representations

	SVM RBF ($C = 15$, $\sigma = 2.6$) AUC (Validation Set, $N = 46,650$)			
No. of Clusters	Mean	Std. Dev.	Minimum	Maximum
50	0.692	0.009	0.674	0.699
100	0.700	0.006	0.689	0.709
200	0.700	0.005	0.692	0.707
300	0.697	0.005	0.685	0.704
400	0.700	0.006	0.694	0.711
500	0.697	0.008	0.684	0.711
600	0.696	0.006	0.688	0.704
700	0.696	0.004	0.688	0.703
800	0.692	0.005	0.682	0.700
900	0.690	0.005	0.682	0.696
1,000	0.686	0.005	0.676	0.693
Linear SVM on full Set	0.702	0.005	0.695	0.707

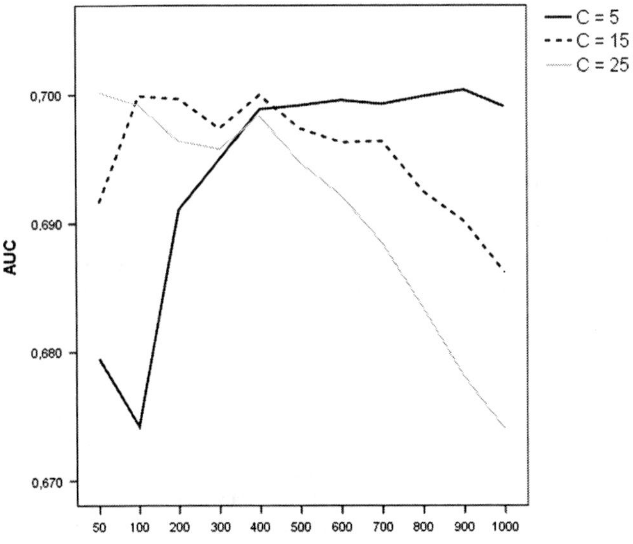

Fig. 1 Mean *AUC* for SVM models trained on $n = 50$–$1,000$ cluster representations respectively, with three different SVM capacity control parameters C

of clusters lead to higher AUC, when higher values of C are considered, i.e. smaller models need to more closely fit the data and vice versa. In more detail, small models with $n = 50$ achieve maximum AUC for $C = 25$ (grey line), medium sized models with $n = 100$ up to $n = 400$ do profit from a medium $C = 15$

(dashed line) and larger models with $n = 500$ up to $n = 1{,}000$ are better for small $C = 5$ (black line). With growing numbers of clusters the optimal C decreases: obviously larger models are prone to overfitting. In summary, there appears a trade-off between model *accuracy* and model *capacity*, which can be accounted for by a proper combination of SVM hyperparameter C and the number of clusters n.

7 Conclusions

In general terms, our exploratory experimental approach indicates that cluster based SVM models used on very large credit client data sets are successful in describing and predicting credit client defaulting behavior. Furthermore, symbolic coding of clusters is introduced, which enables representation of more complex cluster information. This leads to competitive classification performance with regard to *ROC* properties of small classification models, i.e. models trained on a small number of symbolic cluster representatives.

References

Billard, L., & Diday, E. (2006). *Symbolic data analysis*. New York: Wiley.

Bock, H. -H., & Diday, E. (2000). *Analysis of symbolic data: exploratory methods for extracting statistical information from complex data*. Berlin: Springer.

Evgeniou, T., & Pontil, M. (2002). Support vector machines with clustering for training with very large datasets. *Lectures Notes in Artificial Intelligence, 2308*, 346–354.

Hanley, A., & McNeil, B. (1982). The meaning and use of the area under a receiver operating characteristics (ROC) curve. *Diagnostic Radiology, 143*, 29–36.

Japkowicz, N. (2002). Supervised learning with unsupervised output separation. In *Proceedings of the 6th International Conference on Artificial Intelligence and Soft Computing* (pp. 321–325).

Li, B., Chi, M., Fan, J., & Xue, X. (2007). Support cluster machine. In *Proceedings of the 24th international conference on machine learning* (pp. 505–512).

MacQueen, J. B. (1967). Some methods for classification and analysis of multivariate observations. In *Proceedings of the fifth symposium on math, statistics and probability* (pp. 281–297).

Schölkopf, B., & Smola, A. (2002). *Learning with kernels*. Cambridge: MIT Press.

Shih, L., Rennie, J. D. M., Chang, Y. H., & Karger, D. R. (2003). Text bundling: Statistics-based data reduction. In *Twentieth international conference on machine learning* (pp. 696–703).

Wang, Y., Zhang, X., Wang, S., & Lai, K. K. (2008). Nonlinear clustering–based support vector machine for large data sets. *Optimization Methods & Software – Mathematical Programming in Data Mining and Machine Learning, 23*(4), 533–549.

Weiss, G. M. (2004). Mining with rarity: A unifying framework. *SIGKDD Explorations, 6*(1), 7–19.

Yu, H., Yang, J., & Han, J. (2003). Classifying large data sets using SVM with hierarchical clusters. In *Ninth ACM SIGKDD international conference on knowledge discovery and data mining* (pp. 306–315).

Yuan, J., Li, J., & Zhang, B. (2006). Learning concepts from large scale imbalanced data sets using support cluster machines. In *Proceedings of the ACM international conference on multimedia* (pp. 441–450).

A Further Proposal to Perform Multiple Imputation on a Bunch of Polytomous Items Based on Latent Class Analysis

Isabella Sulis

Abstract This work advances an imputation procedure for categorical scales which relays on the results of Latent Class Analysis and Multiple Imputation Analysis. The procedure allows us to use the information stored in the joint multivariate structure of the data set and to take into account the uncertainty related to the true unobserved values. The accuracy of the results is validated in the Item Response Models framework by assessing the accuracy in estimation of key parameters in a data set in which observations are simulated Missing at Random. The sensitivity of the multiple imputation methods is assessed with respect to the following factors: the number of latent classes set up in the Latent Class Model and the rate of missing observations in each variable. The relative accuracy in estimation is assessed with respect to the Multiple Imputation By Chained Equation missing data handling method for categorical variables.

1 Introduction

Missing data is a problem in the analysis of questionnaires to evaluate services quality or to measure people abilities, attitudes or skills, where bunches of multi-item Likert scales are used to collect information on several aspects of the phenomena under investigation. The solution of proceeding with an imputation method before to analyze the data with standard statistical tools is often adopted in order to avoid reduction in the sample size and, depending on which mechanism is generating missing observations, bias or loss in efficiency, or both, in the estimation of parameters.

I. Sulis (✉)
Viale S. Ignazio 78, Dipartimento di Scienze Sociali e delle Istituzioni, Università di Cagliari, Cagliari, Italy
e-mail: isulis@unica.it

P. Giudici et al. (eds.), *Statistical Models for Data Analysis*, Studies in Classification, Data Analysis, and Knowledge Organization, DOI 10.1007/978-3-319-00032-9_41,

The aim of this work is to advance a model-based approach to recover for missingness which specifically takes into account both the information provided by the joint multivariate structure of the data and the uncertainty related to the true value of unobserved unities. *Multiple Imputation Analysis (MIA)* (Rubin 1987) consists of imputing more than one value for each unobserved value by drawing observations from a plausible distribution. *Latent Class Analysis (LCA)* for polythomous items is a multivariate modelling approach which allows us to identify latent classes of observations from a multi-way table of categorical variables and to highlight the profiles of the classes in terms of the matrix of probabilities that respondents in the same class have to provide an answer in each category of each item. Units are classified into clusters based upon membership probabilities (posterior probabilities) estimated directly from units response pattern to the items of the questionnaire. In this article we advance a further proposal to perform a MIA in order to recover for missing responses based on the results of Latent Class Analysis (Vermunt et al. 2008), that will be denoted from here on **MILCA**. The approach allows us to predict plausible values for missing responses when missing data structure is complex and all variables in the data set are affected by missingness using the whole information available on the interconnection across categorical items. The **MILCA** approach replaces missing values for a categorical item with random draws from a *multinomial* distribution whit vector of parameters equal to the *estimated item response probability conditional upon the latent class membership* of the unit (Linzer and Lewis 2011). The procedure has been implemented in the miLCApol function written in the R language. miLCApol uses the poLCA function (Linzer and Lewis 2011) to apply LCA to a set of polythomous items and to estimate the key parameters from which the random draws are made.

The structure of this work is as follows. In Sect. 2 possible implications related to the missing data generating process are discussed and the philosophy underlying multiple imputation analysis is presented. Section 3 is devoted to the description of **MILCA** procedure. In Sect. 4 a simulation study in Item Response Theory framework is presented. **MILCA** has been validated according to the estimation accuracy (EA) criterium using a real complete data set in which observations are set *Missing at Random* (MAR) (Rubin 1987). Further a validation analysis aims to compare the relative EA of **MILCA** with respect to another widely adopted MI method for categorical data: Multiple Imputation by Chain Equation (**MICE**) (van Buuren and Groothuis-Oudshoorn 2011). An analysis of the sensitivity of the results of **MILCA** to the choice of the number of latent classes is carried out. Lastly, in Sect. 4 the code to implement **MILCA** in the R environment is provided.

2 Implications of Missingness and MIA Approach

The way to handle missing information is affected by the missing data generating process. Rubin (1987) defines the probability distribution of the data $\boldsymbol{Y}$ as the joint probability of the observed $\boldsymbol{Y}_o$ and missing responses $\boldsymbol{Y}_m$, and specifies a matrix of missingness $\boldsymbol{R}$, where R_{ij} is equal to 1 if the related value in $\boldsymbol{Y}$ $\{Y_{ij}\}$ is missing, 0 if it is observed. From the analysis of the conditional distribution of $\boldsymbol{R}$ given $\boldsymbol{Y}$, he identifies three probabilistic mechanisms which govern the missing data process. Specifically, if the distribution of observing a pattern of missing values is independent on what it is observed and unobserved in the data matrix $P(\boldsymbol{R}|\boldsymbol{Y}_o, \boldsymbol{Y}_m) = P(\boldsymbol{R})$, missing observations are said *Missing Completely at Random* (MCAR). The missing data mechanism is considered ignorable (Little and Rubin 2002); a *Complete Case Analysis* does not bias the final results since missing data are considered a random sample from $\boldsymbol{Y}$. However, if the size of the data set is strongly affected, the loss of sample size may cause a loss of efficiency (Little and Rubin 2002).

A weaker condition considers unobserved units MAR. The observations are MAR when $P(\boldsymbol{R}|\boldsymbol{Y}_o, \boldsymbol{Y}_m) = P(\boldsymbol{R}|\boldsymbol{Y}_o)$. After conditioning upon $\boldsymbol{Y}$ the distribution of missing data is the same among the cases in which $\boldsymbol{Y}$ is observed or unobserved. This implies that the missing process does not depend on $\boldsymbol{Y}_m$ and missing observations are predictable using $\boldsymbol{Y}_o$. Finally, the missing process is said to be *Not Missing at Random* (NMAR) if the probability to observe a missing value depends also on unobserved responses. In this last case missing responses are not predictable conditional upon what it is observed in the data set. Thus whenever MCAR and MAR conditions hold, the solution to replace missing observations with imputed values before to proceed with further analysis is recommended.

MIA (Rubin 1987) is a method introduced by Rubin in 1987 (Rubin 1987) which allow us to overcome the uncertainty related to the unknown values of the observations when the missing data are imputed with plausible values. It consists of imputing more than one value for each missing observation drawing M ($m = 1, \ldots, M$) values from a plausible distribution. The imputed and observed values are jointed and M completed data sets are built up. Each data set is analyzed separately, using standard statistical analysis, and then estimated parameters and their standard errors are combined together using formula provided by Rubin (1987) to obtain an overall inferential statement. If $\hat{\theta}_m$ is an estimate for a scalar parameter θ in data set m and $\sqrt{V_m}$ is the related standard error, the final estimate for θ is the mean of $\hat{\theta}_m$ taken over M data sets

$$\bar{\theta} = M^{-1} \sum_{m=1}^{M} \hat{\theta}_m.$$

The total uncertainty related to the parameter, namely $T = W + (1 + M^{-1})B$, is a combination of the *within* and *between* imputation data sets variance. The *within variance*—$W = M^{-1} \sum_{m=1}^{M} V_m$—is considered the variance we would observe if there were not missing information in the data set, while the *between variance*—$B = (M-1)^{-1} \sum_{m=1}^{M} (\hat{\theta}_m - \bar{\theta})^2$—is the component which takes into account the uncertainty on the true value of θ due to unobserved units. Since the overall inferential uncertainty is applied only to the missing part of the data set, it is possible to achieve a satisfactory level of accuracy in the estimation of key parameters also with a relative low number of draws (Rubin 1987; Schafer 1997).

3 A Further Proposal to Perform MIA Using LCA

LCA assumes that any dependency across responses provided to manifest categorical indicators is explained "*by a single unobserved 'latent' categorical variable*" (Linzer and Lewis 2011) z which has R categories $(z_1, \ldots, z_R)$. Let us denote with $\boldsymbol{y}_i$ a vector which contains the responses of unit i to the J $(j = 1, \ldots, J)$ categorical indicators and with y_{ijk} the indicator variable which assumes value 1 if observation i $(i = 1, \ldots, n)$ selects category k $(k = 1, \ldots, K$ the categories) of item j, the joint probability density function of $\boldsymbol{y}_i$ is specified as

$$P(\boldsymbol{y}_i \,|\, \boldsymbol{p}\,, \boldsymbol{\pi}) = \sum_{r=1}^{R} p_r \prod_{j=1}^{J} \prod_{k=1}^{K} (\pi_{rjk})^{y_{ijk}}. \tag{1}$$

π_{rjk} denotes the probability that an observation in latent class r provides the k outcome to item j and p_r is the probability to belong to each of the r classes. The poLCA package in R-language (Linzer and Lewis 2011) uses iteratively the EM algorithm until convergence is reached: (i) in the expectation step the posterior class membership $\hat{P}(r|\boldsymbol{y}_i)$ of each unit i is calculated using Bayes' formula (in the first iteration two arbitrary values of $\hat{\pi}_{rjk}$ and $\hat{p}_r$ are plugged in the $\hat{P}(r|\boldsymbol{y}_i)$ formula); (ii) in the maximization step new values of the key parameters are estimated

$$\hat{\pi}_{rjk}^{new} = \frac{\sum_{i=1}^{n} y_{ijk} \hat{P}(r|\boldsymbol{y}_i)}{\sum_{i=1}^{n} \hat{P}(r|\boldsymbol{y}_i)} \quad \hat{p}_r^{new} = \frac{1}{n} \sum_{i=1}^{n} \hat{P}(r|\boldsymbol{y}_i); \tag{2}$$

(iii) the new values are set as new parameters in (i).

The **LC MI** method proposed by Vermunt et al. (2008) to impute missing observations using LCA consists of 4 phases (for details see Vermunt et al. 2008):

1. M complete data sets $\{\boldsymbol{Y}^1, \ldots, \boldsymbol{Y}^M\}$ from the incomplete data matrix $\boldsymbol{Y}$ are generated using M bootstrap samples. For any of the M data sets.
2. LCA is applied and LCA key parameters ($\hat{\boldsymbol{\pi}}^m$ and $\hat{\boldsymbol{p}}_r^m$) are estimated.
3. Person are randomly assigned to one of the R latent classes using posterior subject's latent class membership probabilities $\hat{P}(r|\boldsymbol{y}_{i.obs})^m$.
4. Missing observations for unit i are generated by sampling units from $\hat{P}(\boldsymbol{y}_{i.mis}|r)^m$. Thus, for each item j imputed values are drawn from a univariate multinomial distribution.

The **MILCA** procedure to multiple imputes unobserved values using the results of the LCA requires that all items are measured on the same categorical scale (the same number of categories). It follows the following steps:

1. Missing observations are classified in a new category, labeled as "$K + 1$", that will be treated as a known modality in the LCA estimation process.
2. The LCA is applied and on the basis of the posterior estimates $\hat{P}(r|\boldsymbol{y}_i)$ each unit i is classified in one of the R LCs by modal assignment.
3. For each unit i a missing value in item j (for $j = 1, \ldots, J$) is replaced by randomly generating a draw from a multinomial distribution with vector of

parameters equal to *the estimated vector of item response probabilities* of the class to which the unit has been classified in step 2: $\hat{\boldsymbol{\pi}}_{jr}(\hat{\pi}_{jr1}, \ldots, \hat{\pi}_{jr(K+1)})$.
4. If the randomly generated value is equal to the $(K + 1)$ category, the generated value is rejected, otherwise it is imputed.
5. If the draw generated in step 4 is rejected, the procedure is iterated until a valid draw is generated.
6. M values are imputed for each unobserved value in the data set.

By replacing $\boldsymbol{Y_m}$ in the $\boldsymbol{Y}$ matrix with the imputed values generated in the m-th draw $\boldsymbol{Y_i}^m$, M complete data sets are generated. The number of latent classes R has to be assessed in advance, before to apply the **MILCA** procedure, by recoding all missing responses in each of the J items in the new category "$K + 1$". The LCA analysis will be carried out on the bunch of items measured on a scale with $K + 1$ categories. The suitable number of latent classes will be selected according to the Bayesian Information Criterium (BIC) or other index of fit based on the Akaike's information criterium (AIC) (Vermunt et al. 2008). However since the aim of LCA analysis is predictive, the sensitivity of the accuracy of the results to the choice of the number of LCs is specifically analyzed in Sect. 4. Multiple imputed data sets are analyzed with standard statistical tools and results are summarized using formula for MI analysis provided by Rubin (1987).

4 A Simulation Study to Validate the Imputation Procedure

A simulation study on a real complete data set has been carried on to assess the degree of accuracy of the **MILCA** procedure. The same simulation scheme and data set adopted by Sulis and Porcu (2008) to validate the MISR procedure has been used. The data set is composed by 8 categorical items ($L_1 - L_8$) measured on a four-category Likert scale which contains information on students' evaluation of several aspects of university courses. Observations have been simulated MAR (using function miss.AR Sulis and Porcu 2008) according to two covariates (X_1 and X_2) measured on ordered categorical scales (4 categories). For details on the simulation scheme see Sulis and Porcu (2008). Several data sets with MAR units have been generating by allowing the rate of missingness to vary between 5% to 20%. The number of imputed data sets in all simulation studies has been set equal to $M = 5$. The data sets have been imputed using **MILCA** and **MICE** procedures. **MICE** adopts Gibbs Sampling to generate multiple imputations for incomplete data (van Buuren and Groothuis-Oudshoorn 2011). Specifically, the **polyreg** function has been adopted to impute missing values for item j as a function of the others $J - 1$ items (for details see Sulis and Porcu 2008). The EA has been assessed by comparing the estimates of the location and discrimination parameters of a Graded Response Model (GRM) in the real data set with the one obtained using data imputed with **MILCA**. The GRM (Rizopoulos 2006) specifies the

$$logit\,(P(Y_{ij} \leq k)) = \lambda_j(\theta_i - \beta_{jk}) \tag{3}$$

Table 1 Parameter estimates and standard errors (in brackets)

Item	(a) Real parameters (RP) (se) β^*_{i1}		β^*_{i2}		β^*_{i3}		λ_i	
L_1	−1.271	(0.14)	−0.344	(0.13)	0.911	(0.39)	2.643	(0.12)
L_2	−1.987	(0.17)	−1.048	(0.17)	0.278	(0.14)	2.258	(0.10)
L_3	−2.327	(0.18)	−1.423	(0.17)	−0.089	(0.16)	1.983	(0.10)
L_4	−2.115	(0.16)	−0.982	(0.15)	0.324	(0.14)	2.000	(0.09)
L_5	−1.411	(0.18)	−0.617	(0.18)	0.495	(0.17)	3.092	(0.15)
L_6	−1.076	(0.09)	0.165	(0.06)	1.496	(0.55)	1.749	(0.08)
L_7	−2.904	(0.14)	−2.282	(0.14)	−0.853	(0.13)	1.295	(0.08)
L_8	−1.383	(0.20)	−0.606	(0.19)	0.662	(0.28)	3.272	(0.16)

	Miss. observations 20% (b) RP/MILCA				(c) RP /MICE			
	β^*_{i1}	β^*_{i2}	β^*_{i3}	λ_i	β^*_{i1}	β^*_{i2}	β^*_{i3}	λ_i
L_1	0.964	0.884	0.994	1.048	0.960	0.886	1.011	1.021
L_2	0.992	0.992	1.039	1.033	1.001	0.997	1.035	0.998
L_3	0.973	0.970	0.832	1.058	0.990	0.979	0.866	1.028
L_4	0.987	0.988	1.001	1.033	0.987	1.007	1.031	1.011
L_5	0.985	0.921	1.016	1.072	0.964	0.948	1.016	1.018
L_6	1.009	0.956	1.029	0.988	0.993	0.983	1.010	0.992
L_7	0.922	0.954	0.961	1.078	0.928	0.955	0.932	1.068
L_8	0.977	0.937	0.963	1.077	0.940	0.922	0.990	1.046
Geom. mean	0.987				0.983			

in terms of a discrimination parameter λ_j, a location parameter β_{jk} and a person parameter θ_i. Simulation results reveal a high accuracy in the estimates of the parameters obtained by adopting **MILCA** with 5 LCs when the rate of missingness is severe (20% in each item): panel (b) and (c) in Table 1 show the ratio between the estimates of the parameters in the real data set and in data sets imputed using MILCA (b) and MICE (c). The geometric mean of the ratios of the parameters has been calculated (see last row Table 1 Panel b and c) to summarize single results and compare the two MI methods. Table 2 shows the values of Mean Square Errors (MSE) of the estimates of the coefficients parameters for data sets with rate of missingness from 5% to 20% in each item.

In order to compare the overall EA of the two MI procedures **MILCA** and **MICE** the sum of the MSE (O_{mse} index) of the parameters has been calculated for each simulation scheme (see Table 2 O_{mse}). The **MILCA** procedure seems to have an overall EA very close to the one reached by **MICE**. The sensitivity of the **MILCA** procedure to the choice of number of the LCs has been assessed and results are listed in Table 3. The value of the O_{mse} index has bee calculated by allowing the rate of missingness (5%,. . .,20%) and the number of LCs in the **MILCA** procedure to vary from 3 to 6. From the results arise a good performance of the **MILCA** procedure in terms of EA even when the rate of missingness is severe. The results

Table 2 MSE of coefficient parameters: MILCA with 5 LCs, MICE

	MSE MILCA															
	(a) Miss. observations 5%				(b) Miss. observations 10%				(c) Miss. observations 15%				(d) Miss. observations 20%			
	β^*_{j1}	β^*_{j2}	β^*_{j3}	λ_j	β^*_{j1}	β^*_{j2}	β^*_{j3}	λ_j	β^*_{j1}	β^*_{j2}	β^*_{j3}	λ_j	β^*_{j1}	β^*_{j2}	β^*_{j3}	λ_j
L_1	0.019	0.016	0.153	0.017	0.019	0.016	0.135	0.022	0.022	0.018	0.142	0.019	0.022	0.018	0.121	0.032
L_2	0.030	0.028	0.021	0.012	0.033	0.031	0.021	0.015	0.041	0.032	0.021	0.022	0.030	0.027	0.020	0.017
L_3	0.031	0.030	0.025	0.012	0.032	0.030	0.025	0.011	0.037	0.032	0.025	0.015	0.042	0.034	0.025	0.027
L_4	0.026	0.024	0.019	0.009	0.028	0.024	0.018	0.010	0.026	0.024	0.019	0.012	0.028	0.023	0.018	0.014
L_5	0.033	0.030	0.031	0.022	0.032	0.030	0.031	0.030	0.035	0.031	0.030	0.031	0.031	0.031	0.024	0.064
L_6	0.008	0.006	0.333	0.007	0.009	0.005	0.325	0.007	0.010	0.007	0.352	0.009	0.008	0.005	0.289	0.009
L_7	0.026	0.023	0.017	0.009	0.030	0.025	0.018	0.009	0.057	0.033	0.022	0.013	0.088	0.037	0.020	0.016
L_8	0.039	0.037	0.090	0.026	0.039	0.036	0.086	0.031	0.041	0.037	0.091	0.030	0.036	0.034	0.069	0.084
$O_{mse} = \sum MSE$	1.205				1.210				1.331				1.343			
	MSE MICE															
	(e) Miss. observations 5%				(f) Miss. observations 10%				(g) Miss. observations 15%				(h) Miss. observations 20%			
	β^*_{j1}	β^*_{j2}	β^*_{j3}	λ_j	β^*_{j1}	β^*_{j2}	β^*_{j3}	λ_i	β^*_{j1}	β^*_{j2}	β^*_{j3}	λ_j	β^*_{j1}	β^*_{j2}	β^*_{j3}	λ_i
L_1	0.020	0.016	0.156	0.016	0.020	0.016	0.140	0.022	0.026	0.020	0.110	0.017	0.025	0.020	0.125	0.026
L_2	0.031	0.029	0.021	0.012	0.031	0.029	0.021	0.012	0.033	0.028	0.021	0.019	0.033	0.029	0.021	0.016
L_3	0.032	0.030	0.025	0.012	0.033	0.031	0.026	0.012	0.035	0.032	0.026	0.015	0.036	0.031	0.025	0.014
L_4	0.027	0.024	0.019	0.010	0.025	0.023	0.018	0.011	0.027	0.025	0.020	0.011	0.030	0.025	0.019	0.010
L_5	0.033	0.030	0.028	0.026	0.035	0.033	0.030	0.025	0.041	0.036	0.029	0.035	0.038	0.035	0.028	0.038
L_6	0.008	0.005	0.321	0.007	0.009	0.005	0.310	0.007	0.008	0.005	0.324	0.008	0.008	0.005	0.302	0.008
L_7	0.023	0.021	0.017	0.008	0.037	0.029	0.019	0.013	0.067	0.037	0.025	0.014	0.080	0.037	0.024	0.015
L_8	0.039	0.037	0.082	0.029	0.039	0.036	0.082	0.040	0.047	0.039	0.062	0.057	0.049	0.040	0.069	0.052
$O_{mse} = \sum MSE$	1.195				1.216				1.298				1.312			

Table 3 Overall measure of EA in data sets imputed with MILCA: O_{mse} index

N. LCs	% missingness in the data sets				N LCs	% missingness in the data sets			
MILCA	5%	10%	15%	20%	MILCA	5%	10%	15%	20%
6 CLs	1.203	1.228	1.328	1.464	4 CLs	1.192	1.187	1.315	1.487
5 CLs	1.206	1.210	1.331	1.343	3 CLs	1.166	1.236	1.577	1.500

of the imputation procedure seem not to be strongly affected by the choice of the number of latent classes until the rate of missingness is under 20%. A good accuracy is reached also when the model is underfitted.

Appendix: miLCApol Function Written in the R Language

Description: Function to implement the MILCA procedure
Use: miLCApol(item, m, K, cl, rep, fs)
Arguments:

item: A data frame containing the J categorical variables (the same specified in **fs** formula) all measured on a categorical scale with $K-1$ categories. The categorical variables in **item** must be coded with consecutive values from 1 to $K-1$. All missing values should be coded with NA (see poLCA manual Linzer and Lewis (2011) for details)
fs: A formula expression which uses as responses the items contained in the data frame **item** e.g. $fs < -cbind(Y_1, \ldots, Y_J) \sim 1$ (see poLCA manual Linzer and Lewis (2011) for details)
m: The number of M randomly imputed data sets
K: The number of categories of the items plus 1
class: The number of latent classes (see poLCA manual Linzer and Lewis (2011) for details)
rep: The number of times the poLCA procedure has to be iterated in order to avoid local maxima (see poLCA manual)

Function

```
miLCApol<-function(m,K, cl, rep, fs, item){
 replacemiss<-function(item){
itemp<-matrix(NA,nrow(item), ncol(item))
for(i in 1:ncol(item)){
itemp[,i]<-ifelse(is.na(item[,i]),K,item[,i])    }
return(itemp)    }
 itempr<-replacemiss(item)
library(poLCA)
itempr<-as.data.frame(itempr)
dimnames(itempr)<-dimnames(item)
##see poLCA manual to specify further options  in poLCA
msim<-poLCA(fs,nclass=cl, itempr, nrep=rep ,na.rm=FALSE)
pr<-msim$probs
classm<-msim$predclass
n<-nrow(itempr)
R<-length(table(classm))
J<-ncol(itempr)
p<-array(NA,c(J,K, R))
for(r in 1:R){
for(j in 1:J){
p[j,,r]<-pr[[j]][r,] }}
 impm<-array(NA, c(n,J,m))
 for(t in 1:m){
 for(i in 1:n){
 r<-classm[i]
 for(j in 1:J){impm[i,j,t]<- if(itempr[i,j]==K){
 cate<-rmultinom(1, 1, p[j,,r])
 for(k in 1:K){
```

```
cate[k]<-ifelse(cate[k]==1, k, cate[k])}
label<-sum(cate)
while(label>K-1){  cate<-rmultinom(1, 1, p[j,,r])
for(k in 1:K){
cate[k]<-ifelse(cate[k]==1, k, cate[k])}
label<-sum(cate) }
label }
else(itempr[i,j])}}}
return(impm) }
```

References

Linzer, D. A., & Lewis, J. B. (2011). poLCA: An R package for polytomous variable latent class analysis. *Journal of Statistical Software, 42*(10), 1–29. http://www.jstatsoft.org/v42/i10/.

Little, R. J. A., & Rubin, D. B. (2002). *Statistical analysis with missing data* (2nd edn.). New York: Wiley.

Rizopoulos, D. (2006). *ltm: latent trait models under IRT*. R package version 0.5–0.

Rubin, D. B. (1987). *Multiple imputation for nonresponse in surveys*. New York: Wiley.

Schafer, J. (1997). *Analysis of incomplete multivariate data*. Boca Raton, FL: Chapman and Hall.

Sulis, I., & Porcu, M. (2008). Assessing the effectiveness of a stochastic regression imputation method for ordered categorical data. Working paper. *Quaderni di Ricerca CRENoS, 4*. http://crenos.unica.it/crenos/it/node/269.

van Buuren, S., & Groothuis-Oudshoorn, K. (2011). Mice: Multivariate imputation by chained equations in R. *Journal of Statistical Software, 45*(3), 1–67. http://www.jstatsoft.org/v45/i03/.

Vermunt, J. K., Van Ginkel, J. R., Van der Ark, L. A., & Sijtsma, K. (2008). Multiple imputation of categorical data using latent class analysis. *Sociological Methodology, 33*, 269–297.

A New Distance Function for Prototype Based Clustering Algorithms in High Dimensional Spaces

Roland Winkler, Frank Klawonn, and Rudolf Kruse

Abstract High dimensional data analysis poses some interesting and counter intuitive problems. One of this problems is, that some clustering algorithms do not work or work only very poorly if the dimensionality of the feature space is high. The reason for this is an effect called distance concentration. In this paper, we show that the effect can be countered for prototype based clustering algorithms by using a clever alteration of the distance function. We show the success of this process by applying (but not restricting) it on FCM. A useful side effect is, that our method can also be used to estimate the number of clusters in a data set.

1 Introduction

The curse of dimensionality for clustering can be best described by means of distance concentration. Beyer et al. (1999), introduced the effect of distance concentration for nearest neighbour queries. They showed that a nearest neighbour query is not meaningful if the relative variance of distances to other data objects converges to 0. In other words: the difference between the nearest and furthest data object becomes negligible with increasing dimensionality. Durrant and Kabán (2008) expanded the argumentation by showing that the implication in Bayer et al.'s paper is indeed an equivalence. Since clustering is the task to find meaningful

R. Winkler (✉)
German Aerospace Center Braunschweig, Germany
e-mail: roland.winkler@gmail.com

F. Klawonn
Ostfalia, University of Applied Sciences, Germany
e-mail: f.klawonn@ostfalia.de

R. Kruse
Otto-von-Guericke University Magdeburg, Germany
e-mail: kruse@iws.cs.uni-magdeburg.de

P. Giudici et al. (eds.), *Statistical Models for Data Analysis*, Studies in Classification, Data Analysis, and Knowledge Organization, DOI 10.1007/978-3-319-00032-9_42,

structure solely by analysing the spacial distribution of data objects, the results of Beyer et al. and Durrant and Kabán are relevant for all clustering algorithms in high-dimensional feature spaces. Distance concentration is especially a problem if relations of distances are analysed as it is the case for FCM and other prototype based clustering algorithms.

In this paper, we present a distance function that counters the effect of distance concentration. Our approach does not only counter the effect of distance concentration, it also presents a solution for the problem of finding the correct number of clusters which is a specific problem for prototype based clustering algorithms.

The paper is structured as follows. In the next section, the effect of distance concentration is defined. In Sect. 3 the new distance function is presented and used to specify the clustering algorithm. We apply the proposed algorithm and several others on a data set of aircraft movements in Sect. 4. Finally, this paper ends with the conclusions and references in Sect. 5.

2 Distance Concentration and FCM Type Clustering Algorithms

Let $X \subset \mathbb{R}^m$ be a finite set of m-dimensional real data objects, i.i.d. sampled from some unknown probability distribution F_X in $\mathbb{R}^m$. Let $p > 0$ be a constant, $Q \in \mathbb{R}^m$ be an arbitrary sample point, $\|\cdot\| : \mathbb{R}^m \longrightarrow \mathbb{R}$ a metric and $D_X^p(Q) = \{\|x - Q\|^p : x \in X\}$ be the set of distances, from the viewpoint of Q. Let $\overline{E}(D_X^p(Q))$ be the mean (sample expectation value) and $\overline{V}(D_X^p(Q))$ the sample variance of $D_X^p(Q)$. Then

$$\overline{RV}(D_X^p(Q)) = \frac{\overline{V}(D_X^p(Q))}{\overline{E}^2(D_X^p(Q))}$$

is called the sample relative variance of $D_X^p(Q)$.

Formally, distance concentration occurs if for a sequence of probability distributions F_m and resulting sequences of data sets X_m and query points Q_m holds:

$$\lim_{m \to \infty} \overline{RV}(D_{X_m}^p(Q_m)) = 0.$$

Or in other words, the relative variance of distances becomes negligible.

The occurrence of distance concentration depends on the norm $\|\cdot\|$ and the distribution F_X. Let here, $\|\cdot\|$ be one of the $\mathscr{L}_p$ norms with $p \geq 1$. A result from Hinneburg et al. (2000) shows that distance concentration can only occur for norms with $p > 1$, which means only the Manhattan distance $\mathscr{L}_1$ norm is stable. If a norm is unstable, distance concentration can occur for a wide range of data set distributions (Beyer et al. 1999). For example for an m-dimensional normal distribution with i.i.d. dimensions: $F_m = (\mathscr{N}(1,0), \ldots, \mathscr{N}(1,0))^{\perp}$ with

$\perp$ denoting the transposed vector and $\mathcal{N}(1,0)$ denoting the 1-dimensional standard normal distribution. Also more complex distributions like a uniform distribution on the hypercube surface (no pair of dimensions are independent given any subset of other dimensions) is suffering from distance concentration.

There are two problems with this probability theory result in clustering applications. First, no data set is really going to have an infinite number of features. Second, distance concentration might not occur for the data set it self as it is supposed to be clumped up into several clusters, otherwise clustering would not make any sense in the first place. However, even if the relative variance of distances of a given data set is not 0, clustering algorithms still have their problems because the probability distribution F_X is not known in advance and the clumping effect of clusters might be too weak for the algorithm to recognise. Especially for fuzzy prototype based clustering algorithms this is a problem because they tend to evaluate relative distances in order to assign fuzzy values.

Let $X = \{x_1, \ldots, x_n\} \subset \mathbb{R}^m$ be a m-dimensional data set with n data objects, $Y = \{y_1, \ldots, y_c\} \subset \mathbb{R}^m$ a set of c prototypes, $\| \cdot \| = \mathcal{L}_2$ the euclidean metric, $1 < \omega \in \mathbb{R}$ the fuzzifier and $U \in [0,1]^{c \times n}$ the membership matrix with $u_{ij} \in [0,1]$ as elements subjective to $1 = \sum_{i=1}^{c} u_{ij}$. The symbol $d_{ij} = \|y_i - x_j\|$ denotes the distance between a data object and a prototype with $\| \cdot \| = \mathcal{L}_2$ being the euclidean distance. The fuzzy c-means algorithm (Dunn 1973; Bezdek 1981) is defined by minimizing the objective function with Lagrange multipliers $\Lambda = \{\lambda_1, \ldots, \lambda_n\}$:

$$J_{\text{FCM}}(X, Y, U, \Lambda) = \sum_{i=1}^{c} \sum_{j=1}^{n} u_{ij}^{\omega} d_{ij}^{2} - \sum_{j=1}^{n} \lambda_j \left(\sum_{i=1}^{c} u_{ij} - 1 \right). \tag{1}$$

The objective function is minimized using the alternative optimization algorithm which iteratively optimizes the prototype locations Y and membership values U. The update equations for defining the next iteration $(t+1)$ from the current iteration (t) with the time variable $t \in \mathbb{N}$ are

$$u_{ij}^{(t+1)} = \frac{\left(\frac{1}{d_{ij}^{t}}\right)^{\frac{2}{\omega-1}}}{\sum\limits_{k=1}^{c} \left(\frac{1}{d_{ij}^{t}}\right)^{\frac{2}{\omega-1}}} \quad \text{and} \quad y_i^{t+1} = \frac{\sum\limits_{j=1}^{n} \left(u_{ij}^{t+1}\right)^{\omega} x_j}{\sum\limits_{j=1}^{n} \left(u_{ij}^{t+1}\right)^{\omega}}. \tag{2}$$

When FCM is applied on a high-dimensional data set, this update rule becomes problematic. It starts with the initialization, the initial positions $Y^0 = \{y_1^0, \ldots, y_c^0\}$ of prototypes must be somehow determined. A sample of prototype positions as a subset of the data set, $Y \subset X$, is usually not a good idea as this almost guaranties that not all clusters are found. Therefore, Y^0 is usually sampled from some distribution F_{init} of the feature space, for example a uniform distribution on the smallest data set enclosing hyperrectangle. From the view point of the data object ($Q = x_j$), according to the last section, all distances to the members of a sample of a probability distribution F_{init}, like Y^0, becomes equal. Formally, let $Q = x_j \in X$, for an $1 \leq j \leq n$, then

$$d_j^* = \overline{E}(D_Y(x_j)) \approx \|y - x_i\|, \forall y \in Y. \tag{3}$$

This has very bad implications on the performance of FCM. Especially because the distances to the prototypes w.r.t. a data object are not evaluated by their absolute value, but by their relative value to one another. Following from Eq. (3):

$$u_{ij} \approx \frac{\left(\frac{1}{d_j^*}\right)^{\frac{2}{\omega-1}}}{\sum_{k=1}^{c}\left(\frac{1}{d_j^*}\right)^{\frac{2}{\omega-1}}} = \frac{\left(\frac{1}{d_j^*}\right)^{\frac{2}{\omega-1}}}{c \cdot \left(\frac{1}{d_j^*}\right)^{\frac{2}{\omega-1}}} = \frac{1}{c}; \quad y_i \approx \frac{\sum_{j=1}^{n}\left(\frac{1}{c}\right)^{\omega} x_j}{\sum_{j=1}^{n}\left(\frac{1}{c}\right)^{\omega}} = \frac{\sum_{j=1}^{n}\left(\frac{1}{c}\right)^{\omega} x_j}{n \cdot \left(\frac{1}{c}\right)^{\omega}} = \frac{1}{n}\sum_{j=1}^{n} x_j.$$

All prototypes are updated to a position, close to the centre of gravity of the data set X (for experimental proof, see Sect. 4). Our previous work (Winkler et al. 2011) shows, that this can only be prevented by initializing the prototypes near the clusters in X which would increase the variance in $D_Y(x_j)$ for all data objects of a cluster with a prototype nearby. The probability that all or at least most prototypes are initialized near a cluster is almost 0 because the hypervolume of the space near the clusters is very small, compared to the complete relevant feature space. This means that either the distribution of data objects F_X has to be known in advance, which is usually not the case. Or another clustering algorithm must be used to determine the initial location of the prototypes, which would make the application of FCM unnecessary. Also the question is, if there is an other, reliable clustering algorithm for high-dimensional data.

It should be noted that the EM algorithm and other FCM related algorithms like noise clustering (Dave 1991) and in fact most prototype based fuzzy type algorithms are affected by the curse of dimensionality. Hierarchical clustering algorithms are usually also not a good choice in high dimensional spaces either, because the distances between clusters tend to be similar to the distances of data objects of one cluster. This prevents a "natural" choice of cutting the cluster hierarchy. Density based clustering algorithms like DBScan are difficult to adjust, because the correct parameter setting is very difficult to find. The difference between finding just one cluster or dividing the dataset in a very large number of tiny clusters is incredibly small.

3 Alternative Distance Functions

In the last section, we determined that it is almost impossible to initialize FCM in a high-dimensional space in such a way that prototypes find a cluster. The idea is, to adjust the distance function according to the new circumstances. Hsu and Chen (2009) proposed a new distance function (which is not a norm):

$$SDP(x,y) = \sum_{k=1}^{dim} \omega_k f_{s_{k1},s_{k2}} |x_k - y_k| \text{ and } f_{s_{k1},s_{k2}}(x) = \begin{cases} 0 & if \quad x < s_{k1} \\ x & if \; s_{k1} < x < s_{k2} \\ e^x & if \; s_{k2} < x \end{cases}$$

As this function is very useful and versatile, it contains many (3$\cdot$*dim*) parameters that are difficult to set. An other problem is, that the update equation for the prototypes in FCM must be solvable for the prototype location, which is not the case for most unusual norms as for example the SDP function. We propose an alternative distance function that is also useful for clustering purposes.

The reason FCM does not work very well is, that the distances have not enough contrast to be useful for assigning membership values. So the goal is to increase the contrast in distance values but leaving the update equation for the prototypes solvable for the prototype location. The *DCR* (**D**istance **C**oncentration **R**esistant) function is defined for a distance correction value $\delta \geq 0$:

$$DCR_\delta(x,y) = \|x-y\|^2 - \delta$$

This function is not a norm because its value can be less than 0. However, this function is very useful for replacing the distance function in FCM. With the parameter δ, it is possible to increase the contrast in distance values and $\nabla_y DCR_\delta(x,y) = \nabla_y \|x-y\|^2$ because δ is a constant value.

In the objective function of FCM, the distance function d_{ij} is replaced with $DCR_{ij} = DCR_{\delta_i}(x_j, y_i)$:

$$J_{\text{DCRFCM}}(X,Y,U,\Lambda) = \sum_{i=1}^{c}\sum_{j=1}^{n} u_{ij}^{\omega} DCR_{ij} - \sum_{j=1}^{n} \lambda_j \left(\sum_{i=1}^{c} u_{ij} - 1 \right). \quad (4)$$

With the parameters δ_i, $i = 1,\ldots,c$ it is possible to adjust the distance values in such a way, that the effect of distance concentration in high dimensions is nullified. The cleanest approach would be, to set $\delta_i = \min(D_X^2(y_i))$, because this way, all distances would remain positive or equal to 0. However, practical tests have shown that this is not enough, the prototypes would get stuck on randomly scattered noise data objects.

We use a more radical approach. For a parameter $\alpha \in \mathbb{R}$, $\alpha > 0$, set $\delta_i = \max\{0, \overline{E}(D_X^2(y_i)) - \alpha \cdot \overline{V}(D_X^2(y_i))\}$. So the distance reduction value δ_i is set to the mean of distances (from the point of view of the prototype), reduced by α times the sample variance of the distances. A value of $\alpha = 3$ is usually a good choice because the Cantelli inequality (one sided Chebyshev's inequality) guarantees that at most 10% of the data objects are closer to y_i than δ_i. That however implies that there might be negative distance values. That is not a problem for the objective function as its actual value is not important. For updating the membership values however, the condition of $u_{ij} \geq 0$ must be ensured using the Karush–Kuhn–Tucker multiplier. Because that is computationally difficult, the condition is satisfied manually: if $DCR_{ij} < 0$, the corresponding membership value is set to $u_{ij} = 1$. If there are

k prototypes with negative DCR value, the corresponding membership values are set to $\frac{1}{k}$.

If two or more prototypes are coming close to a cluster, they tend to move very close together due to the equal sharing of membership values of nearby data objects. This multiple representation of clusters can be resolved by simply removing all redundant prototypes. Therefore, the algorithm can end with less prototypes than it started, which means that it has to be initialised with an overestimation of prototypes. This also solves the problem of defining the number of clusters in a data set, which is often not known and hard to do, especially for high-dimensional data sets. Due to the overestimation of prototypes in the beginning, some prototypes end up covering very small cluster or random noise that created a denser area by chance. These prototypes usually represent only very small number of data objects which can easily be detected at the end of the update process. By removing all prototypes which have a sum of membership values below a predefined threshold: $\sum_{j=1}^{n} u_{ij} < \xi$ with threshold $0 < \xi \in \mathbb{R}$, these unnecessary prototypes are removed.

4 Application in S.O.D.A.

In this section we want to demonstrate the problems of FCM and similar algorithms as well as demonstrate the advantages of the proposed algorithm. We use two examples, one real world example in cooperation with Fraport AG and one artificial generated example to demonstrate that the problems are not induced by the specific data set. The Fraport AG develops an analysis tool called S.O.D.A. (Surveillance Data Analysis Tool) to analyse the movement patterns of aircraft on the airfield of the Frankfurt airport. The database contains approx. 700.000 aircraft tracks and the goal is to find groups of aircraft that move similar routes.

Due to the large number of tracks in the database and the complexity of comparing two tracks directly, we decided to simplify the task by transforming the data. A set of 457 reference points is added to the airport structure, for each track, the closest distance to each reference point is computed. To simplify the data further, the distance values are transformed using a simple, trapezoid fuzzy rule: let $d \in \mathbb{R}$ be the minimal distance of a reference point to an aircraft track, then $f(d) = 1$ for $d \leq a$, $f(d) = \frac{d-a}{b-a}$ for $a < d < b$ and $f(d) = 0$ if $d \geq b$ with $a = 25m$ and $b = 50m$. This rule simply states that for $f(d) = 1$, it is sure that the aircraft passed over this point, for $f(d) \in (0, 1)$ the case is unsure and for $f(d) = 0$ it is sure the aircraft did not pass over the reference point. Each fuzzified distance value corresponds to one dimension, the resulting dataset is therefore 457-dimensional. In Fig. 1 (leftmost subfigure), 10'000 transformed aircraft tracks are presented as grey points, projected on two of the 457 dimensions and with some jitter for demonstration purposes. In the second left subfigure of Fig. 1, an artificial dataset with 50 dimensions, 100 uniform distributed clusters which have in turn are sampled from a 100-dimensional normal distribution with the location of the cluster as expectation vector. Also the artificial data set contains 10% uniform distributed noise.

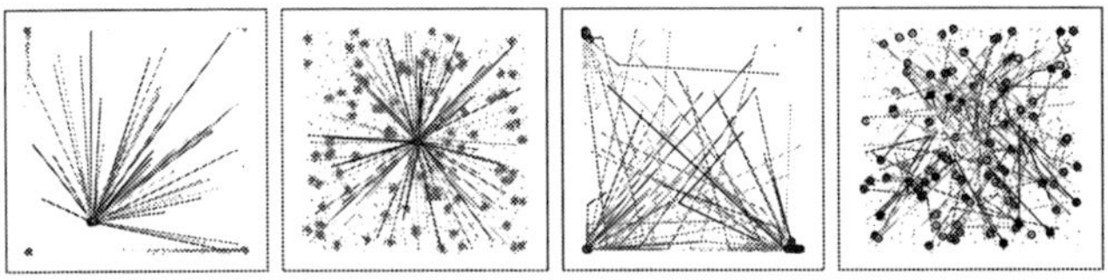

Fig. 1 High-dimensional data sets, projected on 2 dimensions

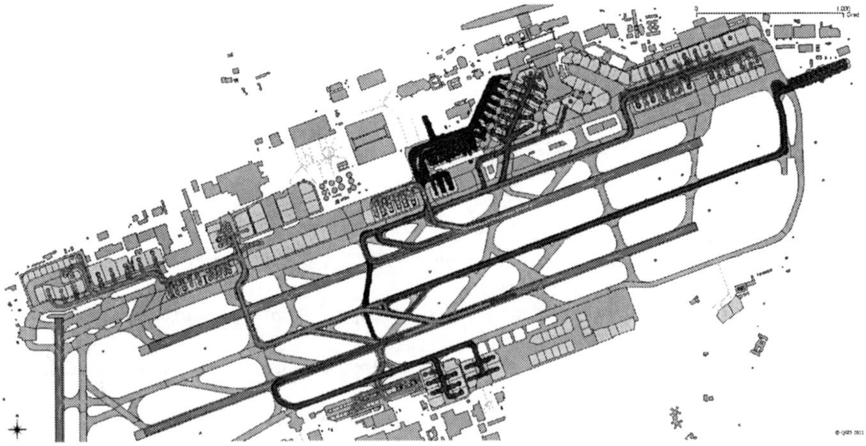

Fig. 2 8 Clusters in the S.O.D.A. dataset

The effect of FCM on this data set is demonstrated by the colourful circles which represent the prototypes. The lines represent the ways, the prototypes took from their initial position to their final position. It is clearly visible that FCM is not working in both cases. The colour of the data objects indicate that they are shared equally by all clusters as their colour indicate their cluster membership. The third and fourth subfigure of Fig. 1 present the same datasets, but instead of the euclidean distance, DCR is used. The algorithm was initialised with 200 prototypes in both cases. In the S.O.D.A. dataset, 62 clusters were found and on the artificial dataset, 99 out of 100 clusters which is almost perfect. In Fig. 2, 8 out of the 62 clusters of the S.O.D.A. dataset are presented. To reduce the overlapping effect of fuzzy clustering for this figure, only tracks with a membership value of at least 0.85 to their respective cluster are shown.

5 Conclusions

We presented a very simple alteration to the distance function that is very effective in countering effect of distance concentration on a prototype based clustering algorithm. The alteration provides additionally the chance of estimating the number of clusters in a data set by overestimating the number of prototypes needed and removing unnecessary ones. The process has been shown for FCM in particular but is not restricted to it, the distance function can also be useful for EM, NC and similar algorithms. To prove our point, we have applied the algorithm on aircraft movement data and on an artificial data set.

Acknowledgements We like to thank the FRAPORT AG for providing the data for scientific analysis, represented by Steffen Wendeberg, Thilo Schneider and Andreas Figur. We also would like to thank the engineers of DLR Braunschweig for setting up the database system, namely Hans Kawohl and his staff.

References

Beyer, K., Goldstein, J., Ramakrishnan, R., & Shaft, U. (1999). When is nearest neighbor meaningful? In *Database theory - ICDT'99*, vol.1540 of *Lecture notes in computer science* (pp. 217–235). Berlin/Heidelberg: Springer.

Bezdek, J. C. (1981). *Pattern recognition with fuzzy objective function algorithms*. New York: Plenum Press.

Dave, R. N. (1991). Characterization and detection of noise in clustering. *Pattern Recognition Letters, 12*(11), 657–664.

Dunn, J. C. (1973). A fuzzy relative of the isodata process and its use in detecting compact well-separated clusters. *Cybernetics and Systems: An International Journal, 3*(3), 32–57.

Durrant, R. J., & Kabán, A. (2008). When is 'nearest neighbour' meaningful: A converse theorem and implications. *Journal of Complexity, 25*(4), 385–397.

Hinneburg, A., Aggarwal, C. C., & Keim, D. A. (2000). What is the nearest neighbor in high dimensional spaces? In *VLDB '00: Proceedings of the 26th international conference on very large data bases* (pp. 506–515). San Francisco, CA: Morgan Kaufmann Publishers.

Hsu, C. -M., & Chen, M. -S. (2009). On the design and applicability of distance functions in high-dimensional data space. *IEEE Transactions on Knowledge and Data Engineering, 21*(4), 523–536.

Winkler, R., Klawonn, F., & Kruse, R. (2011). Fuzzy c-means in high dimensional spaces. *IJFSA, 1*(1), 1–16.

A Simplified Latent Variable Structural Equation Model with Observable Variables Assessed on Ordinal Scales

Angelo Zanella, Giuseppe Boari, Andrea Bonanomi, and Gabriele Cantaluppi

Abstract The communication is related to a wide empirical research promoted by the Università Cattolica del Sacro Cuore of Milan (UCSC) aimed at acquiring an insight into the real work possibilities of its graduates in the last seven years, as well as the appreciation and satisfaction of the firms which offered them a job position. The group of 1,264 firms which have a special connection with UCSC, regarding new job appointments, was considered and they were given a questionnaire, using web for sending and answering. The analysis of the 203 complete answers was conducted by having recourse to a structural equation model with latent variables.

1 Introduction

Nowadays several statistical studies are related to judgments concerning immaterial properties or conceptual aspects of an empirical situation, described by ordered qualitative attributes, giving rise to ordinal scales. Remember the evaluation of the Customer Satisfaction of a good or a service, of the efficacy of a public institution, like a University, or a bank/insurance service, etc.; correspondingly it is typical, with regard to an appropriate questionnaire, to ask respondents to express their judgment on a Likert scale, which is represented by a finite set, typically of a few integer numbers, for example $\{1, 2, 3, 4\}$, of which one has to be chosen by the respondent.

The problem of assessing the casual relationships presence among latent traits only by having recourse to qualitative ordered variables has been widely treated in literature, see for example Bollen 1989, Ch. 9, p. 433, Muthén (1984), Winship and Mare (1984), Jöreskog (2005). This because measurements of latent traits in the usual physical sense are impossible. Nevertheless, practitioners are not always

A. Zanella · G. Boari (✉) · A. Bonanomi · G. Cantaluppi
Dipartimento di Scienze statistiche, Università Cattolica del Sacro Cuore, Milano, Italy
e-mail: angelo.zanella@unicatt.it; giuseppe.boari@unicatt.it; andrea.bonanomi@unicatt.it; gabriele.cantaluppi@unicatt.it

P. Giudici et al. (eds.), *Statistical Models for Data Analysis*, Studies in Classification, Data Analysis, and Knowledge Organization, DOI 10.1007/978-3-319-00032-9_43,

aware that, for example, the numbers of a Likert scale only represent conventional ratings. Namely they cannot be treated like values of an interval or a ratio scale and, thus, current statistical techniques, like regression analysis, become in fact meaningless. It follows that their use requires a different data representation, which is based on a rather involved procedure. Since we are still dealing with a real case for which all observations are on ordinal scales, we considered it appropriate to take this opportunity to apply and discuss an approach, which is consistent with ordinal variables, as it is needed in the case at hand.

2 The Real Case Considered

Università Cattolica del Sacro Cuore of Milan (UCSC) has planned an extensive investigation on the work opportunities offered to its graduates, three-year degree and Master degree, by the Italian job market, in particular with regard to the province of Milan (Zanella et al. 2011).

As starting point the important reference (Chiandotto 2004) was considered looking for other aspects of particular relevance not examined there. In this regard the managing staff of UCSC is especially interested in assessing the satisfaction of companies with respect to the performance of personnel who graduated at UCSC in the last seven years in Economics, Political Sciences, Psychology, and Banking, Finance and Insurance.

For this purpose we used the list of 1,264 firms, which hold a special connection with UCSC in their activity of personnel recruitment, and whose e-mail addresses were available.

Regarding job performance's satisfaction of the employer we set up a questionnaire, to be filled only when graduated collaborators from UCSC of the specified type were present, composed of 15 items aimed at designing a construct with 4 so-called dimensions or implied concepts:

- "Employer's satisfaction" (later on latent variable η), with 2 observable ordinal indicators (Y_1: overall satisfaction, Y_2: a kind of loyalty inclination).
- "Employer's satisfaction regarding employee's personal technical preparation and potentialities" (later on latent variable ξ_1), with 6 observable ordinal indicators:
  - X_2 Ability to implement theoretical knowledge
  - X_5 Theoretical knowledge
  - X_6 Openness to novelty
  - X_{10} Problem solving
  - X_{11} Leadership
  - X_{13} Creativity
- "Employer's satisfaction regarding employee's personal behavior" (later on latent variable ξ_2), with 4 observable ordinal indicators:

  - X_1 Autonomous work
  - X_3 Work life balance
  - X_4 Working involvement
  - X_7 Timeliness

- "Employer's satisfaction regarding employee's capacity of socialization" (later on latent variable ξ_3), with 3 observable indicators:

  - X_8 Attitude to communicate with management
  - X_9 Attitude to communicate with colleagues
  - X_{12} Workgroup attitude

All indicators were measured on a 4 point Likert scale $\{1, 2, 3, 4\}$: each respondent, who received the questionnaire was required to answer by web and choose a number keeping in mind that the greater the number the greater is the satisfaction he intended to express according to some mental thresholds.

Correspondingly, the set of responses, expressed on a conventional scale, gives rise to a K-dimensional ($K = 15$) random categorical variable, say $X = (X_1, \ldots, X_K)'$ (where $X_1 = Y_1$, $X_2 = Y_2$ for simplicity), whose components may assume I ordered categories, denoted by the conventional integer values $i = 1, \ldots, I$.

Let $P(X_k = i) = p_{ki}$, with $\sum_{i=1}^{I} p_{ki} = 1, \forall k$, be the corresponding marginal probabilities and let

$$F_k(i) = \sum_{j \leq i} p_{kj} \tag{1}$$

be the cumulative probability of observing a conventional value x_k for X_k not larger than i. Furthermore assume that to each categorical variable X_k there corresponds an unobservable latent variable X_k^*, which is represented on an interval scale, with a continuous distribution function $\Phi_k(x_k^*)$. The distribution for the continuous K-dimensional latent random variable $(X_1^*, \ldots, X_K^*)$ is usually assumed to be multinormal. Each observed ordinal indicator $X_k, k = 1, \ldots, K$, is related to the corresponding latent continuous X_k^*, also called instrumental variable, by means of a non linear monotone function, see Bollen (1989):

$$X_k = \begin{cases} 1 & \text{if} \quad X_k^* \leq a_{k,1} \\ 2 & \text{if} \quad a_{k,1} < X_k^* \leq a_{k,2} \\ \vdots & \\ I_k - 1 & \text{if} \quad a_{k,I_k-2} < X_k^* \leq a_{k,I_k-1} \\ I_k & \text{if} \quad a_{k,I_k-1} < X_k^* \end{cases} \tag{2}$$

where $a_{k,1}, \ldots, a_{k,I_k-1}$ are marginal threshold values defined as $a_{k,i} = \Phi^{-1}(F_k(i))$, $i = 1, \ldots, I_k - 1$, being $\Phi(\cdot)$ the cumulative distribution function of a random variable, usually the standard Normal, Jöreskog (2005), and $I_k \leq I$ depending on

Table 1 Reliability analysis

Variable	Cronbach's α	Composite reliability	Comp. rel. 95% conf. interval
ξ_1	$\hat{\alpha} = 0.9167$	0.9260	(0.9124, 0.9395)
ξ_2	$\hat{\alpha} = 0.9146$	0.9215	(0.9066, 0.9364)
η	$\hat{\alpha} = 0.6440$	0.9012	(0.8533, 0.9490)

the categories effectively used by the respondents ($I_k = I = 4$ when each category has been chosen by at least one respondent).

For two generic ordinal categorical variables X_h and X_k, $h, k \in \{1, \ldots, K\}$, it is possible to evaluate the *polychoric* correlation, defined as the value of ρ maximizing the loglikelihood typically conditional on the univariate marginal threshold estimates $\sum_{i=1}^{I_h} \sum_{j=1}^{I_k} n_{ij} \ln(\pi_{ij})$, where n_{ij} is the number of observations for the categories ith of X_h and jth of X_k,

$$\pi_{ij} = \Phi_2(a_{h,i}, a_{k,j}) - \Phi_2(a_{h,i-1}, a_{k,j}) - \Phi_2(a_{h,i}, a_{k,j-1}) + \Phi_2(a_{h,i-1}, a_{k,j-1}),$$

being $\Phi_2(\cdot)$ the standard bivariate Normal distribution function with correlation ρ conditional on $a_{h,i}, a_{k,j}$, threshold values for X_h^* and X_k^*, respectively estimated by having recourse to the two marginal latent standard Normal variates according to the usual two step computation (Jöreskog 2005), $a_{k,0} = -\infty$ and $a_{k,I_k} = +\infty$.

Thus we can derive the polychoric correlation (and covariance, see Jöreskog 2005) matrix necessary for the parameter estimation of a structural model with latent variables.

The validity of the measurement model was first assessed by means of the reliability analysis, obtaining the Cronbach's Alpha index properly having recourse to the polychoric covariance matrix, $\mathbf{C}^*$, according to the following formula (Zanella and Cantaluppi 2006, p. 257):

$$\hat{\alpha} = \frac{q_j}{q_j - 1}\left(1 - \frac{\sum_{i=1}^{q_j} c_{ii}^*}{\sum_{i=1}^{q_j} \sum_{i'=1}^{q_j} c_{ii'}^*}\right)$$

where q_j is the number of manifest ordinal indicators linked to the latent variable ξ_j and $c_{ii'}^*$ is the polychoric covariance between the ith and i'th indicators of ξ_j that is the covariance between their corresponding instrumental continuous variables.

At this stage of the analysis only Cronbach's α coefficient can be computed, since the complete structural model has not been estimated yet. Cronbach's α represents a lower bound for reliability in presence of congeneric measures (Sijtsma 2009). After having defined relationships among constructs it will be possible by having recourse to a Confirmatory Factor Analysis to estimate composite reliability and its confidence intervals (Raykov 2002a,b; Green and Yang 2009), see Table 1.

The validity of the preliminary stated model (defined on 4 dimensions) was rejected by subsequent inferential analysis.

Table 2 Average variance extracted and shared variance estimates

Variable	Items	ξ_1	ξ_2	η
ξ_1	5	**0.6963**	0.5941	0.6833
ξ_2	4	0.7708	**0.7426**	0.6611
η	2	0.8266	0.8131	**0.6892**

Correlations below the diagonal, squared correlations above the diagonal, and AVE estimates are presented on the diagonal.

A discriminant validity analysis (Farrell 2010; Fornell and Larcker 1981) was first performed suggesting to remove indicator X_5 related to the latent construct ξ_1: employers do prefer employees able to implement theoretical knowledge. This operation improved discriminant validity.

From the statistical analysis for the resulting structural model it turned out that indicator ξ_3 had to be neglected since its effect on η in the structural equation model was not significant. Namely it showed a z-score ($\hat{\gamma}_3/\hat{\sigma}_{\hat{\gamma}_3} = 1.183$) while for $\hat{\gamma}_1$ and $\hat{\gamma}_2$ we obtained z-scores larger than 3. We have to remark that we got only a sample of 203 complete answers to the questionnaire i.e. a fraction of about 16% of the population of firms.

Table 2 reports discriminant validity results for the reduced model (without ξ_3). It was assessed by comparing the average variance extracted (AVE) of each construct with the shared variance between constructs. All shared variances are not larger than the AVEs and we can maintain that discriminant validity is sufficiently supported.

In conclusion we propose the following formulation of the model involving the latent variables η, ξ_1 and ξ_2 previously defined:

$$\eta = \gamma_1\xi_1 + \gamma_2\xi_2 + \zeta \tag{3}$$

$$X^*_{ji} = {}_X\lambda_{ji}\xi_j + \delta_{ji}, \quad j = 1, 2, \quad i = 1, 2, \ldots, q_j \tag{4}$$

$$Y^*_i = {}_Y\lambda_i\eta + \varepsilon_i, \quad i = 1, 2, \tag{5}$$

with $q_1 = 5$ and $q_2 = 4$, and where the structural error ζ and the measurement errors $\delta_{ji}, \varepsilon_i$ are assumed to be uncorrelated zero mean random variables.

Recall that estimation procedures for covariance based models (Bollen 1989) are based on the minimization of some distance measure, $d(\cdot)$, between the empirical covariance matrix $\mathbf{S}$ of the observed variables and its theoretical counterpart $\boldsymbol{\Sigma}(\boldsymbol{\theta})$ expressed as a function of the unknown theoretical parameters in the structural equation model:

$$\boldsymbol{\Sigma}(\boldsymbol{\theta}) = \begin{bmatrix} \boldsymbol{\Sigma}_{YY} & \boldsymbol{\Sigma}_{YX} \\ \boldsymbol{\Sigma}_{XY} & \boldsymbol{\Sigma}_{XX} \end{bmatrix} = \begin{bmatrix} \boldsymbol{\Lambda}_Y \boldsymbol{\Sigma}_{\eta\eta} \boldsymbol{\Lambda}'_Y + \boldsymbol{\Theta}_\varepsilon & \boldsymbol{\Lambda}_Y \boldsymbol{\Sigma}_{\eta\xi} \boldsymbol{\Lambda}'_X \\ \boldsymbol{\Lambda}_X \boldsymbol{\Sigma}_{\xi\eta} \boldsymbol{\Lambda}'_Y & \boldsymbol{\Lambda}_X \boldsymbol{\Phi} \boldsymbol{\Lambda}'_X + \boldsymbol{\Theta}_\delta \end{bmatrix}.$$

Since in the present case the manifest variables are of the ordinal type, the empirical covariance matrix is not defined. Following Jöreskog (2005, p. 23) we can have

Table 3 Parameter estimates

Structural Model
$\hat{\gamma}_1 = 0.49, \hat{\gamma}_2 = 0.43$
$\hat{\psi} = \hat{\sigma}^2_\zeta = 0.24, \hat{\phi}_{12} = \hat{\sigma}_{\xi_1\xi_2} = 0.77 \quad (\hat{\sigma}_{\xi_1\eta} = 0.83, \hat{\sigma}_{\xi_2\eta} = 0.81)$
Measurement Model
${}_X\hat{\lambda}_{1i} = \mathbf{0.96}, 0.78, 0.67, 0.82, 0.91$
${}_X\hat{\lambda}_{2i} = \mathbf{0.91}, 0.79, 0.85, \mathbf{0.90}$
${}_Y\hat{\lambda}_i = \mathbf{0.97}, 0.67$
$\hat{\sigma}^2_{\delta_{1i}} = \mathbf{0.09}, 0.39, 0.55, 0.32, 0.16$
$\hat{\sigma}^2_{\delta_{2i}} = \mathbf{0.18}, 0.38, 0.27, \mathbf{0.19}$
$\hat{\sigma}^2_{\varepsilon_i} = \mathbf{0.07}, 0.55$

recourse to the polychoric covariance matrix and estimate the unknown parameters in model (3)–(5) by using the following criterion:

$$\arg\min_{\boldsymbol{\theta}} d\left(\mathbf{C}^*, \boldsymbol{\Sigma}(\boldsymbol{\theta})\right).$$

where Y and X in $\boldsymbol{\Sigma}(\boldsymbol{\theta})$ are replaced by Y^* and X^* and the empirical covariance matrix $\mathbf{S}$ by the estimated polychoric matrix $\mathbf{C}^*$.

Estimates were obtained by using both the Lisrel 8.80 Student Edition (Jöreskog 2005)—by specifying the ordinal nature of the variables—and the R library `lavaan` (Rosseel 2012)—by providing the polychoric covariance matrix (Jöreskog 2005), since at present `lavaan` does not deal with ordinal manifest variables.

Table 3, see also Fig. 1, reports the parameter estimates for relationships (3), (4) and (5), involving standardized manifest and latent variables all having unit variance. Bold-faced types denote the most relevant estimates.

To assess the model fit, standard errors and goodness of fit statistics provided by the Lisrel software were considered, which are based on the asymptotic covariance matrix among ordinal variables (Jöreskog 1994).

With regard to the goodness of fit statistics the following positive results were obtained.

- If the model is correct the value of the Satorra–Bentler Scaled Chi-Square test must be not significant, in our case it is 45.6643 (p-value $= 0.28$), d.o.f. 41.
- The Root Mean Square Error of Approximation (RMSEA), a measure of the *discrepancy per degree of freedom* between the population covariance matrix and its estimate, is equal to 0.0237, which is lower than 0.05, and its 90% Confidence Interval is $(0, 0.0554)$ with the upper level lower than 0.08, giving evidence of a good model fit, Browne and Cudeck (1993), p. 144.
- The Expected Cross Validation Index (ECVI) is 0.4736, with a 90% Confidence Interval $(0.4505, 0.5763)$, and the values 0.6535 and 21.6767 respectively for the saturated and independence models indicate that the hypothesized model has a better predictive validity than the saturated one.

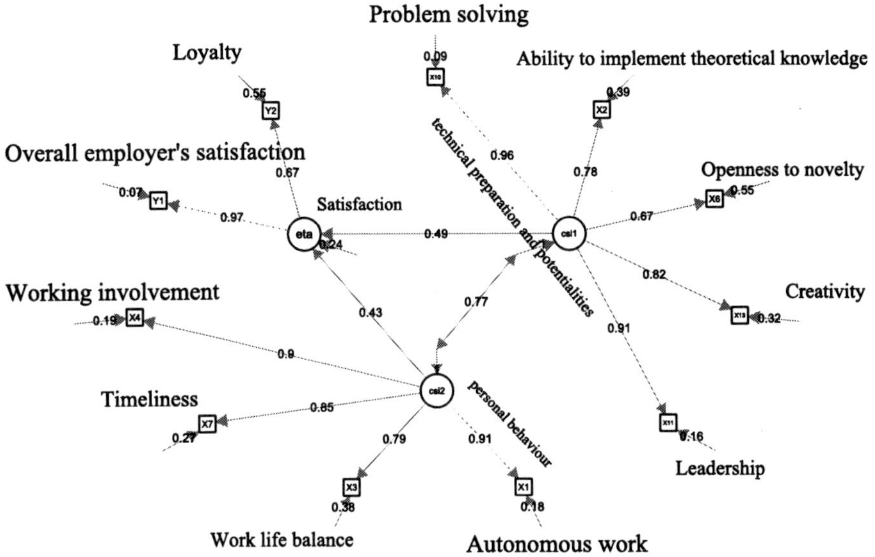

Fig. 1 Path model diagram with standardized estimates

- The Normed Fit Index (NFI) and Non-Normed Fit Index (NNFI) are 0.9895 and 0.9985 both very close to 1, indicating that the fit is very good.

3 Conclusion: Implications and Further Research

We are confident that the results obtained in this research may become useful for improving the University Courses contents and programs.

In particular, the final estimated model shows that η, expressing employer's satisfaction, depends more strongly on ξ_1 (representing satisfaction for employee's technical preparation and potentialities), than on ξ_2 (expressing satisfaction regarding personal behavior). Moreover ξ_1 reflects mostly on the indicators concerning problem solving ability and leadership, ξ_2 on propensity for self standing work and working involvement. Consequently Universities should give students the opportunity of better developing the characteristics best appreciated by the future employers and students must be aware of this chance. The study could be repeated in order to obtain more than 203 answers by the group of 1,264 firms, having a special relationship with UCSC for the recruiting of new graduated personnel, and representing the population to which the research was directed.

The questionnaire should be refined. With regard to variable Y_2, the present version of the questionnaire required that the firm responsible for communications with UCSC should indicate how much his firm might be ready to recommend to another company to give new appointments to UCSC graduates. This was first interpreted as an expression of the degree of loyalty of the firm–customer; but the question could also be perceived as controversial since one can be also inclined to keep for his own firm the possible advantages coming from a particular choice of graduated personnel. Perhaps the question should be better formulated. We found difficult to find an alternative observable indicator aimed at measuring overall satisfaction.

Also the questions referring to socialization should be expressed in a more intelligible form since it is astonishing that the respondents gave little importance to this aspect, which led to eliminate the corresponding construct described by ξ_3.

Then there are at least two methodological not trivial aspects which should still be delved into. The first concerns the use of the threshold method to obtain metric measurements from conventional ordinal rating scores by having recourse to polychoric correlations, which are based on the assumption that for any pair of observed variables there exists a latent bivariate normal distribution. This poses the problem of weakening the latter assumption by trying to consider some not symmetric bivariate distributions, which seem more suitable to comply with the skewness shown by the data distributions.

Finally, discriminant validity assessed by the method suggested in Fornell and Larcker (1981) could be further investigated as far as the theoretical justification of the inequality on which is based is concerned and a statistical test ensuring the significance of the differences between the average variance extracted by the observed indicators (AVE) and the squared correlations of the construct variables should be better considered.

References

Bollen, K. A. (1989). *Structural equations with latent variables*. New York: Wiley.

Browne, M. W., & Cudeck, R. (1993). Alternative ways of assessing model fit. In K. A. Bollen, & J. S. Long (Eds.) *Testing structural equation models* (pp. 136–162). Newbury Park: Sage.

Chiandotto, B. (2004). Sulla misura della qualità della formazione universitaria. *Studi e note di economia*, *3*, 27–61.

Farrell, A. M. (2010). Insufficient discriminant validity: A comment on Bove, Pervan, Beatty and Shiu (2009). *Journal of Business Research*, *63*, 324–327.

Fornell, C., & Larcker, D. F. (1981). Evaluating structural equation models with unobservable variables and measurement error. *Journal of Marketing Research*, *18*(1), 39–50.

Green, S. B., & Yang, Y. (2009). Reliability of summed item scores using structural equation modeling: an alternative to coefficient alpha. *Psychometrika*, *74*(1), 155–167.

Jöreskog, K. G. (2005). Structural equation modeling with ordinal variables using LISREL. http://www.ssicentral.com/lisrel/techdocs/ordinal.pdf. Cited 14 May 2012.

Jöreskog, K. G. (1994). On the estimation of polychoric correlations and their asymptotic covariance matrix. *Psychometrika*, *59*, 381–389.

Muthén, B. O. (1984). A general structural equation model with dichotomous, ordered, categorical, and continuous latent variable indicators. *Psychometrika*, *49*, 115–132.

Raykov, T. (2002). Scale reliability evaluation with LISREL 8.50. http://www.ssicentral.com/lisrel/techdocs/reliabil.pdf. Cited 14 May 2012.

Raykov, T. (2002). Analytic estimation of standard error and confidence interval for scale reliability. *Multivariate Behavioral Research, 37*(1), 89–103.

Rosseel, Y. (2012). lavaan: An R package for structural equation modeling. *Journal of Statistical Software, 48*(2), 1–36.

Sijtsma, K. (2009). On the use, the misuse, and the very limited usefulness of Cronbach's alpha. *Psychometrika, 74*(1), 107–120.

Winship, C., & Mare, R. D. (1984). Regression models with ordinal variables. *American Sociological Review, 49*, 512–525.

Zanella, A., Boari, G., Baldi, P. L., Cantaluppi, G., & Facchinetti, D. (2011). Soddisfazione delle aziende nei confronti dei collaboratori laureati all'Università Cattolica del Sacro Cuore di Milano. In B. V. Frosini (Ed.) *La valutazione della ricerca e la valutazione della didattica* (pp. 147–175). Milano: Vita e Pensiero.

Zanella, A., & Cantaluppi, G. (2006). Some remarks on a test for assessing whether Cronbach's coefficient α exceeds a given theoretical value. *Statistica Applicata, 18*, 251–275.

Optimal Decision Rules for Constrained Record Linkage: An Evolutionary Approach

Diego Zardetto and Monica Scannapieco

Abstract Record Linkage (RL) aims at identifying pairs of records coming from different sources and representing the same real-world entity. Probabilistic RL methods assume that the pairwise distances computed in the record-comparison process obey a well defined statistical model, and exploit the statistical inference machinery to draw conclusions on the unknown Match/Unmatch status of each pair. Once model parameters have been estimated, classical Decision Theory results (e.g. the MAP rule) can generally be used to obtain a probabilistic clustering of the pairs into Matches and Unmatches. Constrained RL tasks (arising whenever one knows in advance that either or both the data sets to be linked do not contain duplicates) represent a relevant exception. In this paper we propose an Evolutionary Algorithm to find optimal decision rules according to arbitrary objectives (e.g. Maximum complete-Likelihood) while fulfilling 1:1, 1:N and N:1 matching constraints. We also present some experiments on real-world constrained RL instances, showing the accuracy and efficiency of our approach.

1 The Decision Problem in Probabilistic Record Linkage

Record Linkage (RL) aims at identifying pairs of records coming from different sources and representing the same real-world entity; such pairs are named *Matches* according to a consolidated jargon.

At a very high level of abstraction, every RL workflow can be seen as the sequence of two fundamental processes: a *comparison* process followed by a *decision* process. The comparison process takes as input the data sets to be linked and performs distance (or, equivalently, similarity) measures on record pairs.

D. Zardetto (✉) · M. Scannapieco
Istat - Italian National Institute of Statistics, Rome, Italy
e-mail: zardetto@istat.it; scannapi@istat.it

P. Giudici et al. (eds.), *Statistical Models for Data Analysis*, Studies in Classification, Data Analysis, and Knowledge Organization, DOI 10.1007/978-3-319-00032-9_44,

The subsequent decision process takes such computed measures as input and, by applying to them a rule of some kind, eventually classifies each record pair as belonging to the class of Matches (M) or to the one of Unmatches (U).

Probabilistic RL methods assume that the pairwise distances computed in the comparison phase obey a well defined statistical model, and exploit the statistical inference machinery to draw conclusions on the unknown class-membership of each pair. In this framework the adoption of a *mixture model* (see e.g. McLachlan and Peel 2000) seems quite natural, as one can see the observed pairwise distances as arising from two genuinely distinct probability distributions, the one stemming from the M subpopulation and the other from the U one: $f(d\,;\,\Psi) = \pi_M f_M(d\,;\,\theta_M) + \pi_U f_U(d\,;\,\theta_U)$.

The mixture approach structures the decision phase of a RL process into two consecutive tasks. First, mixture parameters have to be estimated by *fitting* the model to the observed distance measures between pairs. Then, a probabilistic *clustering* of the pairs into M and U must be obtained by exploiting the fitted model.

The fitting step is crucial, as it implicitly determines the quality of the subsequent clustering results. Moreover, it represents a very hard task. This is mainly due to the *extremely imbalanced* nature of RL data, where M-pairs are always overwhelmed by U-pairs (real-world applications often exhibit Match Rates ranging from 10^{-3} up to 10^{-6}, see Table 1 for selected examples). Indeed, unless some smart countermeasure is adopted, whatever fitting algorithm (the EM Dempster et al. 1977 being no exception) would tend to tune *all* the model parameters so as to better describe some peculiar feature of the dominating U distance distribution, hence failing to detect properly the feeble signal arising from the extremely rare Matches.

In this paper we shall not discuss further the interesting fitting problem, but rather focus on the clustering task. Thus we assume that good estimates of mixture parameters $\hat{\Psi} = (\hat{\pi}_M, \hat{\theta}_M, \hat{\theta}_U)$ have been somehow computed (e.g. via Maximum Likelihood). We then incorporate the *latent* class-membership indicator z (with *true* value 1 if the pair is a Match and 0 otherwise) inside the original mixture model for the distance pdf trough a classical demarginalization argument, yielding the *complete* mixture density: $g(d, z;\ \Psi) = [\pi_M f_M(d\,;\,\theta_M)]^z\,[\pi_U f_U(d\,;\,\theta_U)]^{1-z}$. Correspondingly, the complete log-Likelihood reads:

$$\log \mathscr{L}^c(\mathbf{d}, \mathbf{z}\,;\,\Psi) = \sum_i \log[\pi_U f_U(d_i\,;\,\theta_U)] + \sum_i z_i \log\left[\frac{\pi_M f_M(d_i\,;\,\theta_M)}{\pi_U f_U(d_i\,;\,\theta_U)}\right] \tag{1}$$

The true value of z_i is obviously *unknown*: it will precisely represent the target of our inferences. Indeed, a *decision rule* that assigns each pair to a class is nothing but a rule to infer a value $\hat{z}_i$ for the hidden variable z_i. An *optimal* rule has moreover to work in such a way as to optimize some global objective function. A natural (yet not mandatory) choice is to select as objective function the complete log-Likelihood itself. We thus look for a classification vector $\hat{\mathbf{z}}$ that maximizes the complete data Likelihood under the fitted model, namely:

$$\hat{\mathbf{z}} = \underset{\mathbf{z}}{\operatorname{argmax}}\left[\log \mathscr{L}^c(\mathbf{d}, \mathbf{z}\,;\,\hat{\Psi})\right] \tag{2}$$

For *unconstrained* RL (also known as N:M[1] matching), the solution of (2) would follow easily from the structure of (1):

$$\hat{z}_i = \begin{cases} 1 \text{ if } & \hat{\tau}_i^M \geq \hat{\tau}_i^U \\ 0 \text{ otherwise} \end{cases} \tag{3}$$

with $\hat{\tau}_i^{M,U}$ representing the estimated *posterior probability* that the i-th pair belongs to class M and U respectively:

$$\hat{\tau}_i^C(d_i\, ;\, \hat{\Psi}) = \hat{\pi}_C f_C(d_i\, ;\, \hat{\theta}_C)/f(d_i\, ;\, \hat{\Psi}) \qquad C = \{M, U\} \tag{4}$$

Formula (3) is a classical Decision Theory result (see e.g. Duda et al. 2000), known as "Maximum a Posteriori (MAP) rule": it assigns each pair to the class to which the pair has the highest estimated posterior probability of belonging. We stress here that the easy, closed-form nature of the MAP solution (3) is a lucky byproduct of the *unconstrained* nature of the Maximum complete-Likelihood problem (2).

2 The Impact of Matching Constraints

The decision problem (2) becomes harder if *matching constraints* (i.e. 1:1, 1:N, N:1 linkage restrictions) are imposed: these arise whenever one knows in advance that both (1:1) or either (1:N or N:1) of the data sets to be linked do not contain *duplicates*.[2] Due to space limitations, in what follows we shall focus on 1:1 matching, being the extension to 1:N or N:1 straightforward. To express such constraints in a formal way, we switch to a more convenient matrix notation. We arrange the observed distance values d_i into a $n_{min} \times n_{max}$ matrix D, in such a way that element D_{ij} represents the distance between the i-th record of the smaller data set and the j-th record of the bigger data set. Accordingly, quantities depending on features of the generic i-th pair have to be replaced, inside all previous formulae, by the corresponding two-index quantities. The optimal decision problem with Maximum complete-Likelihood objective under 1:1 constraints now reads:

[1]The expression "N:M" means that each record of either data set can in principle match many records of the other, and viceversa.

[2]For *duplicates* we mean records that *(i)* correspond to the same real-world entity *and (ii)* belong to the *same* data set.

$$\hat{Z} = \underset{Z}{\text{argmax}} \Big[\log \mathscr{L}^c(D, Z \,;\, \hat{\Psi}) \Big] \tag{5}$$

subject to:

$$\sum_{j=1}^{n_{max}} Z_{ij} \leq 1 \quad \forall i \tag{6}$$

$$\sum_{i=1}^{n_{min}} Z_{ijs} \leq 1 \quad \forall j \tag{7}$$

with (6) and (7) obviously implying $\sum_{ij} Z_{ij} \leq n_{min}$.

Constraints (6) and (7) heavily affect the complexity of the problem: now, indeed, a decision taken on a pair *influences* decisions to be taken on other pairs. Moreover, clustering results based on the MAP rule will not, in general, fulfill 1:1 constraints.

Here we shall face the constrained decision problem by means of a purposefully designed Evolutionary Algorithm (EA). Before going into further details, we briefly argue why we chose an EA. A few papers from the RL literature (Jaro 1989; Winkler 1994) tackled the 1:1 problem by using Simplex-based algorithms. Indeed, since both the objective function and the constraints are linear in Z_{ij}, Eqs. (5)–(7) can be formulated as a Binary Linear Programming (BLP) problem. The main concern with this approach is tied to memory usage. If n denotes the size of the data sets to be matched (i.e. $n_{min} \simeq n_{max} \simeq n$), the number of unknowns and inequalities for the BLP problem grow like n^2 and n respectively, yielding a Simplex solver space complexity of $O(n^3)$ at least. The net result is that the BLP approach cannot be applied to real-world data sets, unless a very efficient previous blocking step has been performed. On the contrary, the size of the biggest data structure stored by our EA grows only linearly with n. Finally, we observe that our EA can be readily applied to clustering tasks that involve more complex objective functions than (1) (e.g. nonlinear Loss/Gain Functions, like the estimated F-measure), which definitely cannot be addressed by a Simplex-based algorithm.

3 An Evolutionary Algorithm for 1:1 Matching

The Evolutionary metaheuristic is so versatile that EAs can often be employed to find a satisfactory solution even for optimization problems for which no other solution strategy is known. Here, we assume a basic knowledge of EAs (referring the reader to Michalewicz (1996) for more advanced topics) and, due to space limitations, we restrict ourselves to a very concise outline of our clustering EA. In what follows we list the algorithm pseudocode and only sketch basic choices, parameters and operators.

EVOLUTIONARY ALGORITHM PSEUDOCODE

```
EA[n_ind, n_gen, p_muta, g_stall]
g ← 0
generate Initial Population[n_ind]
compute Fitness
while (¬ Termination Criterion[n_gen, g_stall]) do
   g ← g + 1
   apply Selection
   apply Reproduction + Repair
   apply Mutation[p_muta]
   compute Fitness
end while
return Best Fit individual found
```

Search Space. The EA search space is the set of all the $n_{min} \times n_{max}$ matrices Z with $\{0, 1\}$ elements that fulfill 1:1 constraints (6) and (7). It is a *huge* search space whose cardinality is given by $\mathcal{N}_Z = \sum_{k=0}^{n_{min}} \binom{n_{max}}{k}\binom{n_{min}}{k} k!$ (just to get an impression: if $n_{max} = n_{min} = 150$ then $\mathcal{N}_Z \simeq 1.28 \cdot 10^{272}$).

Representation. We encode a generic candidate solution Z (*phenotype*) by means of a vector ζ of length n_{min} (*genotype*). Elements of ζ (*alleles*) can be 0 or integers between 1 and n_{max}, namely $\zeta_k \in \{0, 1, \ldots, n_{max}\}$ with $k = 1, 2, \ldots, n_{min}$. The meaning of the alleles is easily understood. If $\zeta_k = 0$, then the candidate solution states that the k-th record of the smaller data set *does not match any record* of the bigger data set. If, on the contrary, $\zeta_k = j > 0$, then the candidate solution states that the k-th record of the smaller data set *does match* the j-th record of the bigger data set. Obviously a *legal* genotype, that is a genotype encoding a *feasible* candidate solution Z, is not allowed to contain *duplicated* alleles other than the 0 allele.

Fitness. The `Fitness` functions is obviously modeled on the objective function (1): $\text{Fitness}(\zeta) = \sum_{k:\zeta_k>0} \log\left(\hat{\tau}^M_{k\zeta_k} / \hat{\tau}^U_{k\zeta_k}\right)$, where uninfluential constant terms appearing in (1) have been dropped.

Constraints. Even though only legal individuals are generated in the initial population, some *illegal* genotype may arise during evolution, due to `Reproduction`. In order to maintain a population of feasible candidate solution, these illegal genotypes are *repaired* by means of a purposefully designed operator. The `Repair` operator, `Repair` : $\zeta \to \zeta^*$, acts as a *stochastic* function mapping a genotype, ζ, into a *randomly repaired* version of it, ζ^*. If the ζ individual is legal, `Repair` leaves it unchanged. If, instead, ζ is illegal, `Repair` works as follows. Suppose ζ has ρ groups of duplicated non-zero alleles, with multiplicities n_r where $r = 1, \ldots, \rho$. For each group r, `Repair` first randomly selects inside the group just a single allele to be left unchanged, then it substitutes all the remaining $n_r - 1$ duplicates with the 0 allele.

Initial Population. As the search space of our EA is so huge, generating a good initial population is crucial. It is apparent that creating n_{ind} random individuals by *uniformly* sampling the search space would be a very poor choice. On the contrary, our algorithm samples more heavily those regions of the search space that are believed to be "more promising" on the basis of the fitted posterior probabilities. This is accomplished by the following Monte Carlo (MC) technique. First, the following posterior probabilities are computed for each record k in the smaller data set:

$$p_k^0 = \Pr(Z_{ki} = 0\ \forall i) = \prod_i \hat{\tau}_{ki}^U \tag{8}$$

$$p_k^j = \Pr\Big((Z_{kj}=1) \text{ AND } (Z_{ki}=0\ \forall i \neq j)\Big) = \hat{\tau}_{kj}^M \prod_{i \neq j} \hat{\tau}_{ki}^U \tag{9}$$

where $j = 1, \ldots, n_{max}$. Value p_k^0 is the posterior probability that the k-th record of the smaller data set does not match any record of the bigger, while p_k^j gives the posterior probability that the k-th object matches *only* the j-th. The MC procedure generates *each* element ζ_k (for $k = 1, \ldots, n_{min}$) of *each* genotype ζ of the initial population by sampling an allele value from $\{0, 1, \ldots, n_{max}\}$ with probability proportional to (8) and (9), i.e. $\Pr(\zeta_k = 0) \propto p_k^0$ and $\Pr(\zeta_k = j > 0) \propto p_k^j$. These MC generated genotypes are eventually processed by the `Repair` operator, in order to warranty that the whole initial population is *legal*.

Selection. `Selection` is performed by means of a *rank-2 tournament*. Random pairs of individuals are formed. For each pair, the fitness of the individuals are compared. The fitter individual survives whereas the weaker dies and is dropped from the population, so as to make room for new individuals to be generated in the `Reproduction` phase. Notice that this `Selection` method is intrinsically *elitist*: the fittest individual of a generation surely survives and passes to the next generation.

Reproduction. `Reproduction` is performed as follows. Individuals that survived to `Selection` are randomly paired. Each pair generates two children. These children take the place of individuals that have been eliminated in the previous `Selection` phase. As a consequence the size of the population n_{ind} is kept *fixed* during the evolution. Children genotypes are obtained by merging those of the parents by means of *one-point crossover*. Call ζ^{p1} and ζ^{p2} the parents and ζ^{c1} and ζ^{c2} the children. A random cut point $cut \in \{1, \ldots, n_{min} - 1\}$ is selected for the parents genotypes. Hence both ζ^{p1} and ζ^{p2} are cut into a *left* portion and a *right* portion. The first child receives the left portion from the first parent and the right from the second, i.e. $\zeta^{c1} = (\zeta_1^{p1}, \ldots, \zeta_{cut}^{p1}, \zeta_{cut+1}^{p2}, \ldots, \zeta_{n_{min}}^{p2})$. The second child receives the left portion from the second parent and the right from the first, i.e. $\zeta^{c2} = (\zeta_1^{p2}, \ldots, \zeta_{cut}^{p2}, \zeta_{cut+1}^{p1}, \ldots, \zeta_{n_{min}}^{p1})$. As there is no warranty that the generated children are legal, they eventually undergo the `Repair` treatment before being plugged into the population.

Mutation. Each individual of the population has the same probability p_{muta} of undergoing `Mutation`. `Mutation` acts on a genotype ζ by affecting only a single allele. The outcome can be either that a nonzero allele is replaced by 0 (i.e. a declared Match is deleted from Z), or that a 0 allele is turned into a nonzero allele (i.e. a new declared Match is inserted into Z). The stochastic algorithm implementing `Mutation` exploits the MAP estimate of the number of Matches $\hat{n}_M = \sum_i \hat{z}_i^{MAP}$. A random integer $p \in \{0, 1, \ldots, n_{min}\}$ is drawn from a Binomial distribution with size n_{min} and success probability $\hat{n}_M / n_{min}$. If the genotype to mutate has more than p nonzero alleles, $\sum_k \text{sgn}(\zeta_k) > p$, then a random nonzero allele is replaced by 0. Otherwise, i.e. when $\sum_k \text{sgn}(\zeta_k) \leq p$, a random 0 allele is replaced by a nonzero one randomly selected from the set $\{1, \ldots, n_{max}\} \setminus \zeta$ (namely, by a new *legal* nonzero allele that did not already appear in ζ). Notice that, since the expected value of p is exactly $\hat{n}_M$, `Mutation` *tends on average* to delete declared Matches from candidate solution that contain *"too many"* of them, and conversely to add declared Matches to candidate solution that contain *"too few"* of them.

Termination Criterion. The `Termination Criterion` for the EA is two-fold. A first parameter, n_{gen}, controls the maximum number of generations that can be spent during evolution. If g denotes a generations counter, then the EA would stop as soon as $g > n_{\text{gen}}$. A second parameter, g_{stall}, gives the maximum number of generations that the EA is allowed to process without achieving a fitness improvement. If g' denotes the number of generations elapsed from the *last* fitness improvement, then the EA would stop as soon as $g' > g_{\text{stall}}$. The EA effectively stops as soon as either of the two conditions is verified.

Return Value. The return value of the EA is the genotype ζ^{Best} of the `Best Fit` individual found during evolution. This genotype is readily decoded into the corresponding phenotype matrix Z^{Best}, which in turn yields the *final* clustering result for the RL problem.

Space Complexity and Parameters Values. Storing a whole population of candidate solutions determines the EA memory overhead. As population size is kept fixed during evolution, if n denotes the size of the data sets to be matched then memory usage grows like $n_{\text{ind}} \cdot n$, i.e. almost linearly with n. Indeed, only a weak (less than linear) dependence of n_{ind} on n is expected. As an evidence, we stress that all the successful case studies listed in Sect. 4, despite their n values span over nearly an order of magnitude, have been carried out with the following default values for the EA parameters: $n_{\text{ind}} = 300$, $n_{\text{gen}} = 200$, $p_{\text{muta}} = 0.1$, $g_{\text{stall}} = 50$.

4 Experiments and Conclusions

We conclude by concisely illustrating some experimental tests of our EA clustering method on selected 1:1 RL instances. All the sources are publicly available and, with the only exception of CENS, involve real-world data. Table 1 reports basic information on such instances (see also Kopcke et al. 2010).

We treated all four RL instances as follows. First, we computed the pairwise distances d_i by applying uniformly the *Levenshtein* distance to the matching variables reported in Table 1 and averaging the obtained values. Then, we wrote a two-component mixture model and found ML estimates $\hat{\Psi}$ for the mixture parameters.

Table 1 Relevant features of selected 1:1 RL instances

RL Instance	Data Origin	Matching Variables	Data Nature	Pairs ($n_{min} \times n_{max}$)	Number of Matches	Match Rate
PARKS	SecondString[a]	`name`	Real	101,394 (258 × 393)	247	$2.4 \cdot 10^{-3}$
CENS	SecondString[a]	`surname, name, midinit, street, number`	Artificial	176,008 (392 × 449)	327	$1.9 \cdot 10^{-3}$
RESTAURANTS	Riddle[b]	`name, address, city, type`	Real	176,423 (331 × 533)	112	$6.3 \cdot 10^{-4}$
DBLP-ACM	Leipzig Univ[c]	`title, authors, year`	Real	6,001,104 (2,294×2,616)	2,224	$3.7 \cdot 10^{-4}$

[a] Available at: www.cs.utexas.edu/users/ml/riddle/data/secondstring.tar.gz
[b] Available at: www.cs.utexas.edu/users/ml/riddle/index.html
[c] Available at: dbs.uni-leipzig.de/en/research/projects/object_matching

Lastly, we exploited the fitted model to find optimal decision rules according to the traditional Maximum a Posteriori rule (MAP) and to our Evolutionary Algorithm (EA).[3] The results of our experiments are collectively shown in Table 2, with $\Delta_{\{\text{Prec, Rec, F}\}}$ expressing the percent performance gain (or loss) of our EA versus the traditional MAP solution, with respect to a given quality measure (for Recall, Precision and F-measure definitions and properties, see e.g. Christen and Goiser 2007).

A first look to the average F-measure gain (*+14.0%*) achieved by our EA immediately reveals the remarkable effectiveness of our proposal. Moreover, besides guaranteeing a full compliance with 1:1 matching constraints, our algorithm *always* produces RL decisions of higher overall accuracy. This result strongly supports the robustness of our methods. More importantly: the F-measure gain of our EA is obtained by *systematically increasing* Precision (*+24.6%* on average) while *nearly preserving* Recall (*−0.8%* on average). This proves that our EA is actually able to enforce 1:1 matching constraints almost without erroneously deleting any true Match. As a last remark, we stress that our EA was able to solve all the 1:1 problems listed above while running in an ordinary PC environment. On the contrary, handling the same problems via a Simplex solver required us to deploy the solver on a large memory server, without achieving any significant improvement in the quality of the results.[4]

[3] We performed 10 runs of our Evolutionary Algorithm on each instance, owing to its stochastic nature. Anyway, we found a negligible variability in the results.

[4] The Precision, Recall and F-measure increase (when present) turned out to be of *0.1%* at most.

Table 2 Precision, recall and F-measure results: MAP vs. EA

RL Instance	$Prec_{MAP}$	Rec_{MAP}	F_{MAP}	$Prec_{EA}$	Rec_{EA}	F_{EA}	$\Delta_{Prec}(\%)$	$\Delta_{Rec}(\%)$	$\Delta_{F}(\%)$
PARKS	0.712	0.960	0.817	0.971	0.960	0.965	+26.7%	+0.0%	+15.3%
CENS	0.634	0.997	0.775	1.000	0.994	0.997	+36.6%	−0.3%	+22.2%
RESTAURANTS	0.832	0.884	0.857	0.925	0.875	0.899	+10.0%	−1.0%	+4.7%
DBLP-ACM	0.751	0.987	0.853	0.987	0.970	0.978	+23.9%	−1.8%	+12.8%
Average	*0.732*	*0.957*	*0.826*	*0.971*	*0.950*	*0.960*	*+24.6%*	*−0.8%*	*+14.0%*

In this paper we proposed an Evolutionary Algorithm to find optimal decision rules for constrained Record Linkage. We also presented some experiments on real-world RL instances, showing the effectiveness of our approach. We designed our EA as a part of a new, comprehensive RL software that we are currently developing in Istat. At present, the system is undergoing beta testing and is planned to be released as a standard R package on CRAN (the Comprehensive R Archive Network).

References

Christen, P., & Goiser, K. (2007). Quality and complexity measures for data linkage and deduplication. *Quality Measures in Data Mining*, Springer Studies in Computational Intelligence, 2007, 43.

Dempster, A., Laird, N., & Rubin, D. (1977). Maximum-likelihood from incomplete data via the EM algorithm. *JRSS, Series B, 39*, 1.

Duda, R., Hart, P., & Stork D. (2000). *Pattern classification*. New York: Wiley.

Jaro, M. (1989). Advances in record-linkage methodology as applied to matching the 1985 census of tampa, Florida. *JASA*, 84, 406, 1989.

Kopcke, H., Thor, A., & Rahm, E. (2010). Evaluation of entity resolution approaches on real-world match problems. *PVLDB*, 3, 1, 2010.

McLachlan, G., & Peel, D. (2000). *Finite mixture models*. New York: Wiley.

Michalewicz, Z. (1996). *Genetic algorithms + data structures = evolution programs*. Berlin: Springer.

Winkler, W. (1994). Advanced methods for record linkage. In *Proc. of Survey Research Methods*, ASA, 1994.

On Matters of Invariance in Latent Variable Models: Reflections on the Concept, and its Relations in Classical and Item Response Theory

Bruno D. Zumbo

Abstract An overview is provided of the author's program of research on measurement invariance. Two questions are addressed. First, when do theoreticians and practitioners talk about invariance, and what is it that we are talking about? Second, is invariance only a property of latent variable models such as IRT and is there invariance in classical test theory? If so, what is it for the: observed score, and latent variable formulations.

1 Introduction

This is an overview of my longstanding program of research on the paradox that is measurement invariance (Li and Zumbo 2009; Rupp and Zumbo 2003, 2004, 2006; Sawatzky et al. 2012; Wu et al. 2007; Zimmerman and Zumbo 2001; Zumbo and Rupp 2004; Zumbo 1999, 2007a, b, 2008, 2009). On the one hand, under a mathematical lens, it is a trivial identity but on the other hand, under a historical and conceptual lens, it is probably the most important property of latent variable measurement models and item response theory (IRT), in particular. Furthermore, according to much of the contemporary psychometric literature, invariance is what sets IRT apart from classical test theory (CTT) models. This state of affairs has left me with two perplexing questions that will be the focus of this chapter.

1. When do we talk about invariance? And what is it that we are talking about? This is most often discussed in the context of IRT, sometimes called "modern test theory". I will not be discussing Rasch models, per se.

B.D. Zumbo (✉)
Department of ECPS (also Department of Statistics, and Institute of Applied Mathematics), University of British Columbia, Vancouver, BC, Canada
e-mail: bruno.zumbo@ubc.ca

P. Giudici et al. (eds.), *Statistical Models for Data Analysis*, Studies in Classification, Data Analysis, and Knowledge Organization, DOI 10.1007/978-3-319-00032-9_45,

2. Is "invariance" only a property of latent variable models (such as IRT) and is there "invariance" in classical test theory (CTT)? If so, what is it for the: observed score, and latent variable formulations.

I find it useful in considering the matter of invariance in measurement to distinguish two formulations (parameterizations) of CTT: observed score CTT (e.g., Novick 1966; Lord and Novick 1968), and latent variable CTT (e.g., McDonald 1999). Clearly, these two formulations are interrelated but it is useful, for my purposes herein, to distinguish them.

2 When Do We Talk About Invariance?

Matters of invariance come up very often in contemporary applied and theoretical work, including axiomatic measurement. The minimum that we need is: A (parent) population, and some vector-valued random variable, V, which is an indicator (selection function) of sub-populations or range of conditions of interest selected from the parent on the basis of V. We then talk about invariance with respect to elements of the selection function V—e.g., the elements of V are indicators for age, gender, ethnicity, or other such demographics. The concern is for invariance in sub-populations versus their union. For example, all examinees in grade 6 which has as sub-populations grade 6 boys and girls; the sub-populations of grade 6 boys and girls come together to form the target population of grade 6 students.

2.1 Most Often Discussed in the Context of Item Response Theory (IRT)

The versatility of IRT models has made them the preferred tool of choice in many psychometric settings, but beyond the flexibility of IRT models it is the often misunderstood feature of *parameter invariance* that is frequently cited in introductory or advanced texts as one of their most important characteristics. It is the property of parameter invariance which is the major foundation for their widespread use in equating and adaptive testing and assessment.

As a brief review, IRT can be written in the following way. In conventional descriptions of IRT, examinees are indexed by $i = 1, \ldots, I$; items are indexed by $j = 1, \ldots, J$; θ signifies the unidimensional latent indicator, and $P_{ij}(\theta)$ is the probability of examinee i responding correctly to item j as a function of the continuous latent variable θ . From IRT the unidimensional three-parameter logistic (3-PL) model for dichotomously scored items can be written as follows:

$$P_{ij}(\theta) = \gamma_j + (1 - \gamma_j)\frac{\exp(\alpha_j(\theta_i - \beta_j))}{1 + \exp(\alpha_j(\theta_i - \beta_j))},$$

$$0 \leq \gamma_j < 1, \alpha_j > 0, -\infty < \beta_j, \theta_i < \infty$$

wherein α_j is the "item discrimination" parameter related to the slope of an item characteristic curve (ICC), β_j is the "item difficulty" parameter related to the location of the ICC, and γ_j is the "pseudo-guessing" parameter, which is the lower asymptote of the ICC. In what is often referred to as the two-parameter logistic, 2PL, IRT model, γ_j is zero (or, in some cases, a constant other than zero) and in the one-parameter logistic, 1PL, IRT model: $\alpha_j = 1$ (or some other constant) and γ_j is zero (or, in some cases, a constant of than zero).

The concept of item parameter invariance then stipulates that with a sufficiently large pool of examinees item parameters are independent of the ability distribution of the examinees. Likewise, the concept of person parameter (ability or theta) invariance stipulates that with a sufficiently large set of items respondents' ability score and overall distribution of the ability score are independent of the set of test items.

In describing invariance in IRT models, Lord (1980) states that:

> the probability of a correct answer to item i from examinees at a given ability level θ_0 depends only on θ_0 not on the number of people at θ_0 nor on the number of people at other ability levels $\theta_1, \theta_2, \ldots$ Since the regression is invariant, its lower asymptote, its point of inflexion, and the slope at this point all stay the same regardless of the distribution of ability in the group tested. [...] According to the model, they remain the same regardless of the group tested. (p. 34)

In this citation it is the phrase "according to the model", which is key to an understanding of invariance. The phrase can be translated to "if the model holds" and indeed renders invariance a relatively trivial issue (as the author implies himself), because one can say that if a given model holds perfectly for examinees and items in the respective populations, then the sets of item and examinee parameters are invariant. "In other words, invariance only holds when the fit of the model to the data is exact in the population." (Hambleton et al. 1991, p. 23) In this sense, the model is the "glue" that binds the examinees and items together. Put differently, parameter invariance is a term denoting an absolute state so any discussion about whether there are "degrees of invariance" or whether there is "some invariance" are technically inappropriate (Hambleton et al. 1991). Moreover, the question of whether there is invariance in a single population or under a single condition is illogical as invariance requires at least two (sub-) populations or conditions for parameter comparisons.

Put differently, parameter invariance is not guaranteed by the mere fact that an IRT model—or any other latent variable model for that matter—is fit to data (see Engelhard 1994; van der Linden and Hambleton 1997). This illustrates the paradox that is parameter invariance: On the one hand, under a mathematical lens, it is a trivial identity but on the other hand, under a historical and conceptual lens, it is probably the most important property of IRT models that sets them apart from classical test theory (CTT) models.

Indeed, the property of parameter invariance unifies related investigations of: scaling, differential item functioning (DIF), item parameter drift, and latent class mixture models. In an important sense these are all instantiations of a lack of invariance.

Mathematically, parameter invariance is a simple *identity* of parameters that are on the *same* scale; yet, the latent scale in IRT models is arbitrary so that unlinked sets of parameters are invariant only up to a set of *linear* transformations *specific* to a given IRT model. When *estimating* these parameters in *unidimensional* IRT models with calibration samples, this indeterminacy is typically resolved by requiring that the latent indicator θ be normally distributed with mean 0 and standard deviation 1 (i.e., $\theta \sim N$ (0,1)). In *orthogonal multidimensional* IRT models, the latent scale indeterminacy implies that parameters are identical up to an orthogonal rotation, a translation transformation, and a single dilution or contraction.

2.2 A Familiar Description of Invariance from an IRT Framework

In the IRT framework we refer to item parameters and examinee parameters, and invariance means identical values of parameters in different populations or sub-populations indexed by V, as described above. We can get a sense of the IRT usage of invariance by considering the common 2PL model. In this case, for parameters from two populations to be *invariant*,

$$\alpha' = \alpha, \qquad \beta' = \beta, \qquad \theta' = \theta,$$

but due to the *indeterminacy of the latent scale*, we obtain:

$$\alpha_j^* = \delta^{-1}\alpha_j^{'} \quad , \quad \beta_j^* = \varepsilon + \delta\beta_j^{'} \quad , \quad \theta_i^* = \varepsilon + \delta\theta_i^{'}, \text{ and}$$

$$P_j(\theta_i^*) = \alpha_j^*(\theta_i^* - \beta_j^*) = \alpha_j(\theta_i - \beta_j) = P_j(\theta_i)$$

We can see from the 2PL example that the matter of invariance in IRT is a rather complicated statement involving a solution to the latent scale indeterminacy.

However, there may be situations wherein we do not have any sense of a population or sub-population and in those contexts we are, in essence, not concerned with invariance. This would be a type of calibrative measurement or assessment context. In more common contexts, however, the aim is to use a statistical measurement model to draw inferences from calibration samples to the respective populations from which these were drawn. The additional focus is on the range of possible conditions under which invariance is expected to hold. It depends, then on the type (or strength) of inferences one wants to draw.

2.3 Desired Types of Inferences

Zumbo (2001, 2007) presented the following framework modeled on Draper's (1995) approach to classifying causal claims in the social sciences and, in turn, on Lindley's (1972) and de Finetti's (1974–1975) predictive approach to inference.

The various forms of measurement inference.

		Exchangeability of Sampled and Unsampled Items in the Target Construct / Domain (i.e., sampled tasks or items)	
		EXCHANGEABLE	NOT EXCHANGEABLE
Exchangeability of Sampled and Unsampled Units in Target Population (i.e., sampled individuals)	EXCHANGEABLE	General Measurement Inference	Specific Sampling Inference
	NOT EXCHANGEABLE	Specific Domain Inference	Initial Calibrative Inference

In terms of inferential strength,

(Initial Calibrative, Specific Sampling) < Specific Domain < General Measurement Inference

Fig. 1 Zumbo's Draper–Lindley–de Finetti framework

Unlike Draper, Zumbo focused on the inferences about items and persons made in assessment and testing. The foundation of the approach is the exchangeability of:

- Sampled and un-sampled respondents (i.e., examinees or test-takers); this could be based on the selection function for sub-populations.
- Realized and unrealized items.
- Exchangeable sub-populations of respondents and items.

By exchangeability you can think of it in the purely mechanical sense. I have found this useful to help me think of the various possibilities, whether they happen regularly or not. This also helps me detail the range of conditions under which invariance is expected to hold. Figure 1 is a description of the resulting fourfold table and in the range of invariant claims.

So the answer to the question of whether there is invariance in CTT is "yes", however, not of the flavor and sense of invariance that we get with latent variable models. For example, we proved that measurements that are parallel in a given population are also parallel in any subpopulation. This was a point also emphasized by Lord and Novick (1968). We provide other new results but they are of the flavor seen in the sentence above about parallel tests. Let us then write CTT from a latent variable framework and see what we get in terms of invariance therein.

3 Is There "Invariance" in CTT?

Zimmerman and Zumbo (2001) described and extended a model of tests and measurements that identifies test scores with Hilbert space vectors and true and error components of scores with linear operators. One of the niceties of that level of abstraction is that we were able to clearly show that some of the properties and quantities from CTT hold for the entire population of individuals, as well as any subpopulation of individuals. For our purposes today I will not go into the details of the algebra, but it is important to note that the feel for invariance in the observed score characterization of CTT is strikingly different from that in IRT. Furthermore, many of the examples of invariance we prove in the Zimmerman and Zumbo (2001) paper are of the variety seen in all classical mathematical statistics.

It is well known in CTT that:

$$X_i = \tau_i + \varepsilon_i$$

$$\mathrm{var}(X_i) = \mathrm{var}(\tau_i) + \mathrm{var}(\varepsilon_i)$$

$$\left.\begin{array}{l}\mathrm{cov}(\tau_i, \varepsilon_i) = 0\\ E(\varepsilon_i) = 0\end{array}\right\} \text{Properties of } \tau_i, \varepsilon_i.$$

It should be noted that these last two statements about the covariance and expected value are not assumptions, per se, but rather properties implied from the definition of true and error variables.

I will sketch a bit of a model of CTT to motivate my remarks. See Steyer (2001) for a full description of the details. Let us focus on the essentially tau-equivalent test model because it is the one that is most commonly referred to in observed score CTT. The model is:

X_i and X_j are a pair of tests from the set

$X_1, \ldots., X_m$ with the assumptions :

(1) $\tau_{\mathrm{i}} = \tau_{\mathrm{j}} + \lambda_{ij}, \quad \lambda_{ij} \in \Re$, an additive constant

(2) $\mathrm{cov}(\varepsilon_i, \varepsilon_j) = 0, \quad i \neq j.$

Assumption (1) above implies that there is a latent variable, that is a function of the true score variables such that:

$$\eta = \tau_i + \lambda_i, \quad \lambda \in \Re$$

$$X_i = \eta + \varepsilon_i.$$

Note that it is necessary to set the scale of η; and we can do this by, for example, setting $E(\eta) = 0$. Now with some analytical work it can be shown (see, for example,

Steyer 2001) that the discrepancy between the expected values of two essentially tau-equivalent tests are the same for each and every sub-population—i.e., resulting in equal coefficients, λ_{ij}, in each and every sub-population.

$$E^{(V)}(X_i) - E^{(V)}(X_j) = E^{(V)}(X_i - X_j) = \lambda_{ij}$$
$$\text{for each sub-population indexed by } ''\text{V}''.$$

The discrepancy function is important to note here because it is necessitated by the essential tau-equivalence model. On the other hand, parallel tests have equal expectation (but not variances and covariances) of the test score variables in sub-populations.

4 Closing Remarks

It is, of course, widely appreciated that there are quantities in observed score CTT that are population specific (and lacking invariance): variance of the true scores, and test score reliability because it is bound to the variance of the true scores.

It is, however, less widely appreciated that models such as the essential tau-equivalence model (when cast in the formulation of latent variable theory) has the same invariance properties of item and person parameters as IRT. This suggests that the invariance of item and person parameters is a consequence of considering item-level latent models such as factor analysis and its kin, item response theory.

It is important to end on a clear take-home message. Yes CTT has invariance properties but these depend on how the CTT model is formulated (latent versus observed score). Some latent variable models, whether IRT or CTT, allow for item and person parameter invariance. However, simply going about fitting an IRT model to data does not necessarily give you measurement invariance. Believing that going about fitting a model to data guarantees you measurement invariance is simply magical thinking!

There are several things that one can do and statements that one can make with a latent variable model such as an IRT model that they cannot do or say with an observed score CTT model. In addition, we should be able to do the same things and make the same statements as IRT with certain latent variable CTT models.

Invariance is both, in some senses, a trivial and obvious property and at the same time the cornerstone of theoretical and applied measurement theory. I think I have just begun to get a small sense of how it can be both.

In short, from a data modeling perspective, invariance requires that the model be correct (true) in all corners of the data.

References

de Finetti, B. (1974–1975). *Theory of probability* (Vol. 1–2). New York: Wiley.

Draper, D. (1995). Inference and hierarchical modeling in the social sciences. *Journal of Educational and Behavioral Statistics, 20*, 115–147.

Engelhard, G. (1994). Historical views of the concept of invariance in measurement theory. In M. Wilson (Ed.), *Objective measurement: Theory into practice* (Vol. 2, pp. 73–99). Norwood: Ablex.

Hambleton, R. K., Swaminathan, H., & Rogers, H. J. (1991). *Fundamentals of item response theory*. Newbury Park: Sage.

Li, Z., & Zumbo, B. D. (2009). Impact of differential item functioning on subsequent statistical conclusions based on observed test score data. *Psicologica, 30*, 343–370.

Lindley, D. V. (1972). *Bayesian statistics, a review*. Philadelphia: Society for Industrial and Applied Mathematics.

Lord, F. M. (1980). *Applications of item response theory to practical testing problems*. Mahwah: Lawrence Erlbaum Associates, Inc.

Lord, F. M., & Novick, M. R. (1968). *Statistical theories of mental test scores*. Reading: Addison-Welsley.

McDonald, R. P. (1999). *Test theory: A unified treatment*. Mahwah: Lawrence Erlbaum Associates.

Novick, M. R. (1966). The axioms and principal results of classical test theory. *Journal of Mathematical Psychology, 3*, 1–18.

Rupp, A. A., & Zumbo, B. D. (2003). Which model is best? Robustness properties to justify model choice among unidimensional IRT models under item parameter drift. *Alberta Journal of Educational Research, 49*, 264–276.

Rupp, A. A., & Zumbo, B. D. (2004). A note on how to quantify and report whether invariance holds for IRT models: When Pearson correlations are not enough. *Educational and Psychological Measurement, 64*, 588–599 [Errata (2004): 991].

Rupp, A. A., & Zumbo, B. D. (2006). Understanding parameter invariance in unidimensional IRT models. *Educational and Psychological Measurement, 66*, 63–84.

Sawatzky, R., Ratner, P. A., Kopec, J. A., & Zumbo, B. D. (2012). Latent variable mixture models: A promising approach for the validation of patient reported outcomes. *Quality of Life Research, 21*, 637–650.

Steyer, R. (2001). Classical test theory. In T. Cook & C. Ragin (Eds.), *International encyclopedia of the social and behavioural sciences. Logic of inquiry and research design* (pp. 481–520). Oxford: Pergamon.

van der Linden, W., & Hambleton, R. K. (Eds.). (1997). *Handbook of modern item response theory*. New York: Springer.

Wu, A. D., Li, Z., & Zumbo, B. D. (2007). Decoding the meaning of factorial invariance and updating the practice of multi-group confirmatory factor analysis: A demonstration with TIMSS data. *Practical Assessment, Research and Evaluation, 12*(3), 1–26. http://pareonline.net/genpare.asp?wh=0&abt=12.

Zimmerman, D. W., & Zumbo, B. D. (2001). The geometry of probability, statistics, and test theory. *International Journal of Testing, 1*, 283–303.

Zumbo, B. D. (1999). *A handbook on the theory and methods of differential item functioning (DIF): Logistic regression modeling as a unitary framework for binary and Likert-type (ordinal) item scores*. Ottawa: Directorate of Human Resources Research and Evaluation, Department of National Defense. http://educ.ubc.ca/faculty/zumbo/DIF/index.html.

Zumbo, B. D. (2001). Methodology and measurement matters in establishing a bona fide occupational requirement for physically demanding occupations. In N. Gledhill, J. Bonneau, & A. Salmon (Eds.). *Proceedings of the consensus forum on establishing BONA FIDE requirements for physically demanding occupations* (pp. 37–52). Toronto.

Zumbo, B. D. (2007a). Validity: Foundational issues and statistical methodology. In C. R. Rao & S. Sinharay (Eds.), *Handbook of statistics, vol. 26: Psychometrics* (pp. 45–79). The Netherlands: Elsevier Science B.V.

Zumbo, B. D. (2007b). Three generations of differential item functioning (DIF) analyses: Considering where it has been, where it is now, and where it is going. *Language Assessment Quarterly, 4*, 223–233.

Zumbo, B. D. (2008). *Statistical methods for investigating item bias in self-report measures.* Florence: Universita degli Studi di Firenze E-prints Archive. http://eprints.unifi.it/archive/00001639/.

Zumbo, B. D. (2009). Validity as contextualized and pragmatic explanation, and its implications for validation practice. In R. W. Lissitz (Ed.), *The Concept of validity: Revisions, new directions and applications* (pp. 65–82). Charlotte: Information Age Publishing, Inc.

Zumbo, B. D., & Rupp, A. A. (2004). Responsible modeling of measurement data for appropriate inferences: Important advances in reliability and validity theory. In D. Kaplan (Ed.), *The SAGE handbook of quantitative methodology for the social sciences* (pp. 73–92). Thousand Oaks: Sage.

Index

P. Giudici et al. (eds.), *Statistical Models for Data Analysis*, Studies in Classification, Data Analysis, and Knowledge Organization, DOI 10.1007/978-3-319-00032-9,

D

CPSIA information can be obtained at www.ICGtesting.com
Printed in the USA
LVOW07s2009170813

348398LV00004B/194/P

9 783319 000312